Oxford Series in Ecology and Evolution

Edited by Paul H. Harvey, Robert M. May, H. Charl

The Comparative Method in Evolutionary Biology
Paul H. Harvey and Mark D. Pagel
The Cause of Molecular Evolution
John H. Gillespie
Dunnock Behaviour and Social Evolution
N. B. Davies
Natural Selection: Domains, Levels, and Challenges
George C. Williams
Behaviour and Social Evolution of Wasps: The Communal Aggregation Hypothesis
Yosiaki Itô
Life History Invariants: Some Explorations of Symmetry in Evolutionary Ecology
Eric L. Charnov
Quantitative Ecology and the Brown Trout
J. M. Elliott
Sexual Selection and the Barn Swallow
Anders Pape Møller
Ecology and Evolution in Anoxic Worlds
Tom Fenchel and Bland J. Finlay
Anolis Lizards of the Caribbean: Ecology, Evolution, and Plate Tectonics
Jonathan Roughgarden
From Individual Behaviour to Population Ecology
William J. Sutherland
Evolution of Social Insect Colonies: Sex Allocation and Kin Selection
Ross H. Crozier and Pekka Pamilo
Biological Invasions: Theory and Practice
Nanako Shigesada and Kohkichi Kawasaki
Cooperation Among Animals: An Evolutionary Perspective
Lee Alan Dugatkin
Natural Hybridization and Evolution
Michael L. Arnold
The Evolution of Sibling Rivalry
Douglas W. Mock and Geoffrey A. Parker
Asymmetry, Developmental Stability, and Evolution
Anders Pape Møller and John P. Swaddle
Metapopulation Ecology
Ilkka Hanski
Dynamic State Variable Models in Ecology: Methods and Applications
Colin W. Clark and Marc Mangel
The Origin, Expansion, and Demise of Plant Species
Donald A. Levin
The Spatial and Temporal Dynamics of Host-Parasitoid Interactions
Michael P. Hassell
The Ecology of Adaptive Radiation
Dolph Schluter

Parasites and the Behavior of Animals
Janice Moore
Evolutionary Ecology of Birds
Peter Bennett and Ian Owens
The Role of Chromosomal Change in Plant Evolution
Donald A. Levin
Living in Groups
Jens Krause and Graeme D. Ruxton
Stochastic Population Dynamics in Ecology and Conservation
Russell Lande, Steiner Engen and Bernt-Erik Sæther
The Structure and Dynamics of Geographic Ranges
Kevin J. Gaston
Animal Signals
John Maynard Smith and David Harper
Evolutionary Ecology: The Trinidadian Guppy
Anne E. Magurran
Infectious Diseases in Primates: Behavior, Ecology, and Evolution
Charles L. Nunn and Sonia Altizer
Computational Molecular Evolution
Ziheng Yang
The Evolution and Emergence of RNA Viruses
Edward C. Holmes
Aboveground–Belowground Linkages: Biotic Interactions, Ecosystem Processes, and Global Change
Richard D. Bardgett and David A. Wardle
Principles of Social Evolution
Andrew F. G. Bourke
Maximum Entropy and Ecology: A Theory of Abundance, Distribution, and Energetics
John Harte
Ecological Speciation
Patrik Nosil

Ecological Speciation

PATRIK NOSIL

Department of Ecology and Evolutionary Biology, University of Colorado, Boulder, USA

OXFORD
UNIVERSITY PRESS

Great Clarendon Street, Oxford OX2 6DP
United Kingdom

Oxford University Press is a department of the University of Oxford.
It furthers the University's objective of excellence in research, scholarship,
and education by publishing worldwide. Oxford is a registered trade mark of
Oxford University Press in the UK and in certain other countries

First published 2012
Reprinted 2012

British Library Cataloguing in Publication Data

Data available

Library of Congress Cataloging in Publication Data

Library of Congress Control Number: 2011945449

ISBN 978-0-19-958711-7

Printed and bound by CPI Group (UK) Ltd, Croydon, CR0 4YY

This book is dedicated to Aspen

Preface

I undertook the writing of this book during a yearlong stay at the Institute for Advanced Study, *Wissenschaftskolleg*, in Berlin, Germany. At that point in time, I had been involved in speciation research for about a decade. However, upon commencing the writing of the book, it came clear how little I knew. Thus, writing this book was a true learning experience, and one that was made possible only by the freedom from other academic responsibilities that was provided by the *Wissenschaftskolleg*. For this, I am greatly indebted to the *Wissenschaftskolleg* and to Axel Meyer for inviting me to participate in a working group there. I thank Axel and the other two members of that group, Jim Mallet and Jeff Feder, as well as fellow Robert Trivers for ongoing scientific discussions concerning speciation. I thank Hélène Collin for many discussions during the writing of this book, and for enduring many days and nights in Berlin where all I did was work on the book.

Although I commenced the task of writing that year in Berlin, it took three years to complete the project. Numerous people are to be thanked. In particular, I am most indebted to my mentors over the years. In particular, I thank those involved in my entry into research in evolutionary biology, including Tom Reimchen, Bernie Crespi, Cristina Sandoval, Arne Mooers, Felix Breden, and Dolph Schluter. Tom has provided an endless source of inspiration and motivation. At times in our scientific careers we are lucky enough to meet exceptional people, those with truly thoughtful and unique perspectives. These are the people that when they open their mouths, we know we should listen carefully to what they say, even if ultimately we might disagree. Tom has been that person for me. Others, such as Dolph, have played similar roles. Dolph's work strongly influenced me, and his support and insight over the years cannot be overstated, and is much appreciated. More recently, I had the fortune to engage in a fun and fruitful collaboration with Jeff Feder. Discussions with him influenced me in positive and creative ways. I thank my mentors for their ongoing support, as well as their constructive criticism.

For comments on drafts of specific chapters of the book, I thank Scott Egan, Matt Wilkins, Joey Hubbard, Rudy Riesch, Kei Matsubayashi, Issei Ohshima, Justin Meyer, Sergey Gavrilets, David Lowry, Daniel Ortiz-Barrientos, Rowan Barrett, Scott Pavey, Yael Kisel, Fred Guillaume, and Jim Mallet. My doctoral students, Tim Farkas and Aaron Comeault read a draft from cover to cover, providing much ticism along the way. For this, I am indebted. Dolph Schluter and Ole ovided numerous useful comments on a complete draft. The book m discussion with members of a graduate course on speciation, includ-Wilkins, Joey Hubbard, Loren Sackett, Se Jin Song, Sharon Aigler, Povilus, Brent Hawkins, Timothy Farkas, Christine Avena, Amanda obert Jadin, and Bader AlHajeri. Finally, I thank numerous other co-authors leagues with whom I have worked on and discussed speciation with, includ-elène Collin, Maria Servedio, Daniel Funk, Michael Kopp, Sander Van Doorn, ecca Safran, Samuel Flaxman, Axel Meyer, Scott Egan, Zach Gompert, Tom chman, Alex Buerkle, Daniel Ortiz-Barrientos, Sean Rogers, Christine Parent, eve Springer, Jeff Joy, Rutger Vos, Andrew Hendry, John Endler, Michael Turelli, Trevor Price, Jenny Boughman, Daniel Bolnick, Juan Galindo, Daniel Duran, Gabriel Perron and Howard Rundle. I extend my apologies to anyone forgotten, and am very grateful for the many friends within the scientific community that I have had the pleasure of interacting with.

I also thank my parents and sister for their lifelong support of whatever curiosities engaged me at the time. I thank my editors at Oxford University Press, Helen Eaton and Ian Sherman, for their advice and encouragement. Finally, I note that although this book clearly could not have come to fruition without the support and intellect of numerous colleagues, any errors contained within are my responsibility alone.

Contents

4 A form of reproductive isolation 85

5 A genetic mechanism to link selection to reproductive isolation 109

Introduction

Adaptation to different ecological environments, via divergent natural selection, can generate phenotypic and genetic differences between populations. In turn, these changes might create new species. The general aim of this book is to synthesize the theoretical and empirical literature on the formation of new species due to divergent selection. This process of "ecological speciation" has seen a large body of particularly focused research in the last 15–20 years, and thus it is an excellent time for a review and synthesis. A particularly important goal will be the integration of the ecological and genetic literature. Both ecological studies of the sources of natural selection and genetic studies of the genes and genomic regions affected by divergent selection have accumulated, but they have yet to be strongly integrated. Such integration will hopefully shed new insight into the speciation process.

This book necessarily focuses to a large extent on the predictions and tests of ecological speciation, and on reviewing its three main components: (1) a source of divergent natural selection; (2) a form of reproductive isolation; and (3) a genetic mechanism to link divergent selection to reproductive isolation. However, a critical aspect that sets the book apart from past treatments of speciation will be the inclusion and further development of some recent concepts that particularly apply to ecological speciation. For example, the concept that natural selection against immigrants into foreign environments (i.e., "immigrant inviability") can reduce interbreeding between populations, and thus act as a barrier to genetic exchange, unifies the study of divergent adaptation and speciation (Nosil et al. 2005). The concept of "genomic islands of divergence" is a useful metaphor for considering how the diversifying effects of selection spread throughout the genome during the speciation process (Turner et al. 2005). Similarly, "isolation-by-adaptation" describes how selection affects patterns of genetic divergence, even potentially for neutral genes unlinked to those under divergent selection, by reducing gene flow to the extent that genetic drift can occur at all loci (Nosil et al. 2008). These concepts will all be explored in depth in this book. Finally, a central theme of the book, one that is receiving much current attention in the literature, will be the often-continuous nature of divergence during the speciation process (Mallet et al. 2007, Berner et al. 2009, Peccoud et al. 2009).

The above-mentioned concepts, and the explicit focus on the mechanism of speciation via divergent selection, set this book apart from excellent past books on speciation. For example, Mayr's (1942, 1963) treatments of speciation predate the DNA sequence era. Schluter (2000b) devoted a chapter explicitly to ecological speciation but focused on adaptive radiation more broadly. Coyne and Orr (2004) covered overall patterns of speciation and thoroughly reviewed the genetic basis of reproductive isolation, but the breadth of their book necessarily precluded an exhaustive treatment of ecological speciation. The edited volume by Dieckmann et al. (2004) examined "adaptive speciation" due to frequency-dependent disruptive selection. However, adaptive speciation is a special case of ecological speciation because adaptive speciation, unlike ecological speciation, is restricted to cases where frequency-dependent disruptive selection in sympatry or parapatry drives speciation. Thus, unlike this book, Dieckmann et al. (2004) did not treat cases of allopatric divergence or cases in which divergent selection was not frequency-dependent. Price (2007) covered speciation, but focused on birds. Other books on speciation were edited volumes, and thus contained many useful different views and topics, but did not focus systematically on ecological speciation (Otte and Endler 1989, Howard and Berlocher 1998, Butlin et al. 2009). This book aims to build upon and expand the treatments of ecological speciation that rest within each of these previous books, as well as to provide an empirical counterpart to theoretical treatments of ecological speciation in Endler (1977) and Gavrilets (2004).

This book covers extensive literature, but owing to the rapidly growing nature of the field, not all aspects can be exhaustively covered. In numerous sections, I provide comprehensive and relatively exhaustive reviews, often in the form of tables. However, in many other points of the text I draw upon key examples that illustrate the point at hand. The book itself is organized into three main sections. The first section deals with clarifying what ecological speciation is, its predictions, and how to test for it (chapters 1 and 2). This sets up the second part of the book, which reviews the three components of ecological speciation. It is useful to devote a separate chapter to each component because each can be studied, to a certain extent, independently from the others. This approach also helps highlight areas that have received less attention, and is warranted because data within any given system usually exists for only one or two components. Chapter 3 thus reviews the sources of divergent selection: namely, environmental differences, interactions between populations, and ecologically dependent sexual selection. Chapter 4 reviews the forms of reproductive isolation that occur during ecological speciation, including their relative contributions to total reproductive isolation and the temporal order in which they evolve. Chapter 5 considers the genetic mechanisms that link selection to reproductive isolation, as well as the individual genetic bases of adaptation and reproductive isolation during ecological speciation.

The third and final section of the book deals with other outstanding topics in the study of ecological speciation. Chapter 6 considers the geography of ecological speciation, including how geography affects the sources of selection, how gene flow might promote or constrain divergence, and the possibility of multiple geographic modes of divergence during any single instance of speciation. Chapter 7 reviews the genomics of ecological speciation, focusing on how selection generates variation among genomic regions in their levels of genetic differentiation between populations. The role of gene expression in speciation is also considered. Chapter 8 focuses on variability in how far the often-continuous process of speciation proceeds. This chapter considers whether partial reproductive isolation is often a stable outcome (i.e., an equilibrium) and offers hypotheses for variation in how far ecological speciation proceeds. The final chapter of the book focuses on particularly pressing questions that remain.

Abbreviations

AFLP	amplified fragment length polymorphism
CD	character displacement
ECD	ecological character displacement
ERG	ecology, reproductive isolation, genetic distance
eQTL	expression quantitative trait locus
IBA	isolation-by-adaptation
IBD	isolation-by-distance
IM	isolation with migration
mtDNA	mitochondrial DNA
NIL	near-isogenic line
pQTL	phenotypic quantitative trait locus
QTL	quantitative trait locus
RCD	reproduction character displacement
RI	reproductive isolation
SNP	single nucleotide polymorphism

Part I

Ecological speciation and its alternatives

1

What is ecological speciation?

1.1. The often-continuous nature of the speciation process

"the evidence for continuity between races and species is overwhelming" (Dobzhansky 1940, p. 314)

The process of speciation is often quantitative in nature, as evidenced by numerous data indicating that divergence during speciation varies continuously. For example, the strength of reproductive isolation can vary quantitatively, as can the degree of clustering in molecular markers or phenotypic traits, the sharpness of geographic clines in gene frequencies, and the extent of lineage sorting (Endler 1977, Funk 1998, Lu and Bernatchez 1999a, Jiggins and Mallet 2000, Funk et al. 2002, Coyne and Orr 2004, Crespi et al. 2004, Dopman et al. 2005, Nosil et al. 2005, Rundle and Nosil 2005, Funk et al. 2006, Mallet et al. 2007, Cummings et al. 2008, Seehausen et al. 2008b, Nosil et al. 2009b, Merrill et al. 2011). Thus, the speciation process often represents a continuum of divergence, which I will refer to as the "speciation continuum" (Fig. 1.1).

The different means of quantifying divergence mentioned above can be used to measure different points or arbitrary "stages" along the speciation continuum, ranging from continuous variation to differentiated populations, recent species pairs, and post-speciational divergence. The later stages of divergence along the speciation continuum are characterized by the evolution of strong discontinuities in nature, which can occur for many different types of divergence, and which, in sympatry, generally requires strong barriers to gene flow (i.e., reproductive isolation). Table 1.1 documents some examples of quantitative divergence in the speciation continuum and Figure 1.2 depicts a particularly illustrative one from cichlid fishes in Lake Victoria.

While different species concepts can disagree on when speciation starts and when it is complete, they generally share the characteristic of having stages of divergence along a continuum (Wu 2001b, de Queiroz 2005). Thus, arguments about the quantitative nature of speciation apply across species concepts (Mallet et al. 2007, Nosil 2008a, Nosil et al. 2009b). This book will focus on the role of divergent natural selection in driving the evolution of reproductive isolation. Thus, I adopt the biological species concept (see Coyne and Orr 2004 for a review of species concepts). This focus is useful because it allows an explicit consideration of the role of selection in generating specific forms of barriers to genetic exchange, as well as on the genetic mechanisms (i.e., pleiotropy versus linkage disequilibrium) linking

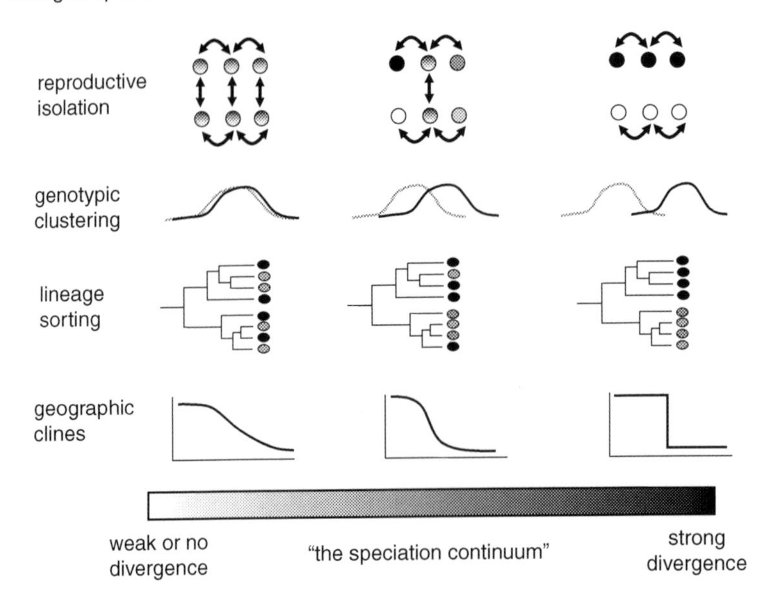

Figure I.1. Schematic illustration of the continuous nature of divergence during speciation. Three arbitrary points along the speciation continuum are depicted. Numerous types of differentiation can vary quantitatively during speciation, with the magnitude of divergence representing a measure of how far speciation has proceeded. Experimental estimates of the strength of reproductive isolation can quantify the speciation continuum between allopatric taxon pairs, whereas the other measures of divergence are best applied to sympatric or parapatric taxa with similar levels of geographic overlap. Two-headed arrows represent mating between individuals. Modified from Nosil et al. (2009b) with permission from Elsevier.

selection to these specific barriers to gene flow (Coyne and Orr 2004, Gavrilets 2004).

I will often consider how divergent selection drives the evolution of even partial levels of reproductive isolation. In such cases we cannot know if speciation will ever go to completion, but we nonetheless learn about the process of speciation (i.e., the evolution of reproductive isolation) in the meantime. Thus, the magnitude of reproductive isolation observed will be used as a measure of how far the process of speciation has proceeded. This approach does not mean that the category of "species" is of no use, but simply that the speciation process is often continuous. This point is well exemplified by the entomologist Bruce Walsh in an analogy he drew between "host plant races" versus "species" of insects, and childhood versus adulthood (Walsh 1864, p. 411). Childhood is one thing and adulthood another; but there are intermediate periods when it is difficult to say whether an individual is a child or an adult. Yet it would be strange logic to argue that, on that account, childhood was the same thing as adulthood. I now turn to introducing the process of ecological speciation in more detail.

Table I.I. Examples of variation in how far speciation has proceeded.

Measure of divergence	Study system(s)	Result	References
RI inferred via experiment	Numerous (hundreds of taxa from disparate groups)	Individual forms of reproductive isolation varied from 0.00 to 1.00, both within and among study systems	(Coyne and Orr 2004, Nosil et al. 2005, Funk et al. 2006)
RI inferred via experiment	*Timema* walking-stick insects	Population pairs vary drastically in levels of reproductive isolation according to exposure to gene flow	(Nosil 2007)
RI inferred via experiment	*Mimulus* monkeyflowers	Among nine reproductive barriers, estimates of the strength of individual barriers ranged from 0.00 to 0.99	(Ramsey et al. 2003)
RI inferred by levels of gene flow	Numerous (1284 studies reviewed)	F_{ST} varied among taxon pairs from 0.00 to 1.00	(Morjan and Rieseberg 2004)
RI inferred by levels of gene flow	*Gasterosteus* sticklebacks	Species pairs vary in the degree of reproductive isolation, with one pair recently collapsed into a single species	(Taylor et al. 2006)
RI inferred by hybridization frequency	*Heliconius* butterflies	Species pairs vary continuously in the frequency of hybridization from relatively common to extremely rare	(Mallet et al. 2007)
Genotypic clustering	17 hybrid zones in numerous different taxa	The distribution of gene frequencies in hybrid zones ranged from unimodal to strongly bimodal	(Jiggins and Mallet 2000)
Lineage sorting in mitochondrial DNA	2319 animal species	Phylogenetic grouping between closely related species ranged from polyphyly to reciprocal monophyly	(Funk and Omland 2003)
Lineage sorting in multiple loci	*Ostrinia nubilalis* corn borer strains	Genealogies for five gene regions are discordant, with only one exhibiting pheromone strain exclusivity	(Dopman et al. 2005)
Cline sharpness	Wide range of systems	Geographic clines range from shallow to extremely steep	(Barton and Hewitt 1985)

Examples were chosen to span different measures of divergence. Gene flow, when inferred, was in sympatry or parapatry. Abbreviations: RI, reproductive isolation.

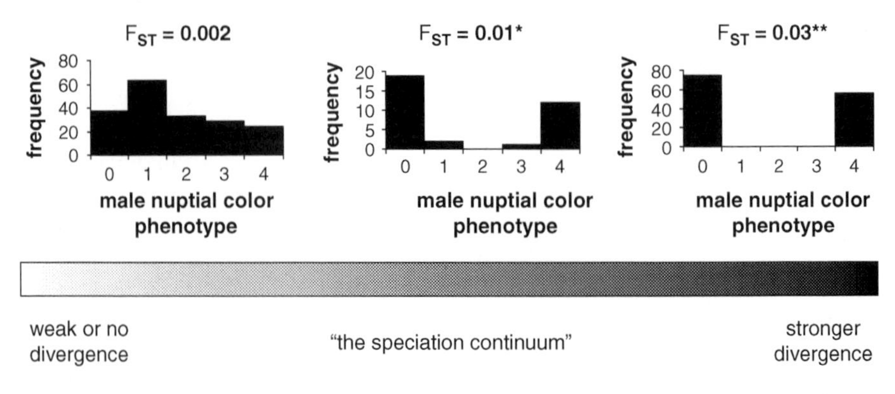

Figure I.2. The quantitative nature of divergence in the species pairs *Pundamilia pundamilia* and *P. nyererei*. In this example, how far speciation has proceeded was inferred using neutral genetic differentiation at microsatellite loci (F_{ST}: $^{*}p < 0.05$, $^{**}p < 0.01$, no asterisk $p > 0.05$), the distribution of male nuptial coloration (frequency histograms above), and experimental data on mating preferences (not shown, but patterns similar to those depicted for male color were observed). Data from Seehausen (2008) and Seehausen et al. (2008b). Modified from Nosil et al. (2009b) with permission from Elsevier.

I.2. Ecological speciation via divergent natural selection

Biologists have long been fascinated with, and sought to explain, the origin and maintenance of biological diversity within and among species. Natural selection is generally recognized as a central mechanism of evolutionary change within species (Darwin 1859, Endler 1986, Kingsolver et al. 2001). Thus, natural selection plays a major role in generating the array of phenotypic and genetic diversity observed in nature. But to what extent is selection, the process driving phenotypic diversification within species, also responsible for the formation of new species? In other words, to what extent do phenotypic and species diversity arise via the same processes?

Recent years have seen renewed efforts to address these questions. For example, populations living in different ecological environments (e.g., desert versus forest habitats) might undergo divergent and adaptive evolutionary change via divergent natural selection (unless otherwise noted, the term "selection" is hereafter used as shorthand for divergent selection). These same evolutionary changes can also result in the populations evolving, perhaps incidentally, into separate species. For example, adaptation to different environments, via divergent selection, might cause the evolution of genetically based differences between populations in the way that individuals tend to look, smell, and behave. In turn, these differences might cause individuals from different populations to avoid mating with one another, or for hybrids to be unfit if mating occurs. Thus, two such populations would cease exchanging genes, thereby diverging into separate species because of the adaptive changes that occurred via natural selection. This is a simple description of the "ecological speciation" hypothesis

(Schluter 1996b, c, Funk 1998, Schluter 1998, 2000b, 2001, Nosil et al. 2002, Rundle and Nosil 2005, Schluter 2009, Juan et al. 2010).

Consistent with past work (see Table 1.2), I define "ecological speciation" as the process by which barriers to gene flow evolve between populations as a result of ecologically based divergent selection between environments. This will often occur because traits under divergent selection, or those genetically correlated with them, incidentally effect reproductive isolation (Muller 1942, Mayr 1963). Under such a scenario, speciation occurs as an incidental "by-product" of adaptive divergence. In other cases, divergent selection might operate on reproductive isolation itself. This occurs, for example, if the evolution of mating signals and preferences is dependent on the ecological environment such that ecologically differentiated populations diverge in signals and preferences, resulting in pre-mating sexual isolation (Boughman 2001, Seehausen et al. 2008b).

Selection itself is considered ecological when it arises as a consequence of the interaction of individuals with their external environment during resource or mate acquisition or from the interaction of individuals with other organisms in their attempt to obtain resources. Selection is divergent when it acts in contrasting directions in the two populations (e.g., large body size confers high survival in one environment and low survival in the other). This includes the special case in which selection favors both tails of the phenotype distribution within a single population, termed "disruptive selection" (Rundle and Nosil 2005). The agents of divergent selection during ecological speciation are thus extrinsic and can include abiotic and biotic factors, such as food resources, climate and habitat, and interspecies interactions, such as disease, competition, and behavioral interference. Figure 1.3 depicts divergent selection, using the example of different colors conferring protection against predators in different populations that vary in the color of the substrate upon which individuals rest.

Table 1.2. Some past definitions or formulations of ecological speciation, and the one used in this book.

Definition	Reference
"the process by which barriers to gene flow evolve between populations as a result of ecologically based divergent selection between environments"	This book
"I will use 'ecological speciation' to refer to the process whereby reproductive isolation evolves from divergent selection"	(Schluter 1996c, p. 807)
"ecological sources of natural selection might cause reproductive isolation to evolve between . . . populations"	(Funk 1998, p. 1754)
"ecological speciation occurs when divergent selection on traits between populations . . . in contrasting environments leads directly or indirectly to the evolution of reproductive isolation"	(Schluter 2001, p. 372)
"the process by which barriers to gene flow evolve between populations as a result of ecologically-based divergent selection"	(Rundle and Nosil 2005, p. 336)
"the evolution of reproductive isolation between populations by divergent natural selection arising from differences between ecological environments"	(Schluter 2009, p. 738)

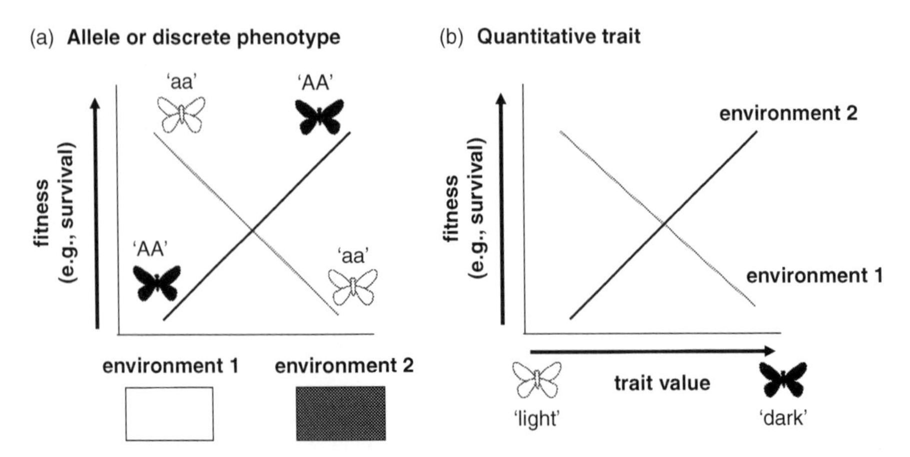

Figure I.3. The pattern generated by divergent natural selection: "crossing fitness functions." In this case, light butterflies are more camouflaged and thus exhibit higher survival in environment one. In contrast, dark butterflies are more camouflaged and exhibit higher survival in environment two. The end result can be phenotypic divergence between populations in different environments. a) Alleles ("a" and "A") or discrete phenotypes. b) Quantitative traits.

It is useful to consider ecological speciation as its own form of species formation because it focuses on an explicit mechanism of speciation: namely divergent natural selection. There are numerous ways other than via divergent selection in which populations might become genetically differentiated and reproductively isolated. For example, genetic divergence could occur via random genetic drift, via the chance fixation of different, incompatible mutations in separate populations that are experiencing similar selective pressures (i.e. "uniform selection"), or via forms of sexual selection that do not necessarily involve ecologically based selection, such as Fisher's runaway (Lande 1981). Ecological speciation is distinguished from these other mechanisms of speciation by involving a key role for divergent selection in population differentiation. Some have proposed that the concept of ecological speciation is of limited utility (Sobel et al. 2010). In contrast, I argue here that the study of ecological speciation, and its alternatives, is clearly a useful endeavor, because it allows one to isolate the roles of specific evolutionary processes (e.g., divergent selection, uniform selection, genetic drift) in the evolution of reproductive isolation, thereby revealing the mechanisms of speciation (Schluter 2009). Ecology can affect speciation in ways other than by causing divergent selection (Sobel et al. 2010 for review), but the generation of divergent selection probably represents a common way, and thus is the focus of this book. In a subsequent section of this chapter, I discuss ways in which ecology can affect speciation that do not fit under the umbrella of ecological speciation, at least in the normal definition and usage of the term.

An alternative definition of ecological speciation would restrict it to situations in which barriers to gene flow are ecological in nature. However, when the goal is to understand mechanisms of speciation (as here), it is of interest when both "ecological" and "non-ecological" forms of reproductive isolation evolve as a result of a specific process (i.e., divergent selection). Thus, the evolution of any form of reproductive isolation via divergent selection constitutes the process of ecological speciation. Indeed, it is critical to note that divergent selection can promote the evolution of any type of reproductive barrier via the by-product mechanism noted above (Mayr 1947, Mayr 1963, Funk 1998, Schluter 2000b, Rundle and Nosil 2005, Funk et al. 2006, Vines and Schluter 2006). For example, divergent selection might drive the evolution of intrinsic genetic incompatibilities between populations (Agrawal et al. 2011). This could occur if the different alleles favored by selection within each population are incompatible with one another when brought together in the genome of a hybrid (Gavrilets 2004, Funk et al. 2006, Dettman et al. 2007). Thus, ecological speciation is not limited only to the evolution of "inherently" ecological forms of reproductive isolation, such as habitat preference or ecological selection against hybrids, although such forms are expected to be common during the process.

Another critical point concerns the geographic mode (allopatry, parapatry, sympatry) of ecological speciation (Butlin et al. 2008, Fitzpatrick et al. 2008, Mallet et al. 2009). Ecological speciation itself can occur under any geographic arrangement of populations, as long as divergent selection is the process driving divergence (Funk 1998, Schluter 2001, Coyne and Orr 2004, Rundle and Nosil 2005). Thus, for example, ecological speciation should not be equated to sympatric speciation. Although instances of sympatric speciation probably often require divergent selection to overcome the homogenizing effects of gene flow, ecological and sympatric speciation are not one and the same. In fact, allopatric ecological speciation may be common because allopatric populations will often find themselves in different environments and can undergo adaptive divergence unimpeded by gene flow. Indeed, examples of allopatric ecological speciation exist (Funk 1998, Nosil et al. 2002, Vines and Schluter 2006, Langerhans et al. 2007). Although ecological speciation can occur under any geographic context, geography is important to consider, because it affects the sources of divergent selection and rates of gene flow (see Chapter 6 for details).

Finally, it is important to note that some, but not all, forms of sexual selection are included within ecological speciation: whenever sexual selection is divergent between ecological environments, it falls within the domain of ecological speciation (Schluter 2001). Examples include scenarios where environments differ regarding which mating signals are most efficiently transmitted, thereby resulting in ecologically based and divergent selection on mating signals (Boughman 2001, Seehausen et al. 2008b, Maan and Seehausen 2011). In contrast, in cases such as Fisherian runaway sexual selection or sexual conflict between the sexes, selection is not necessarily ecologically based and divergent between environments, and thus divergence via these processes may not constitute ecological speciation (Rundell and Price 2009, Sauer and Hausdorf 2009). Given these considerations, some common misconceptions about ecological speciation are presented in Table 1.3.

The process of ecological speciation makes some explicit and simple predictions. For example, ecologically divergent pairs of populations adapted to different ecological environments should exhibit greater levels of reproductive isolation than

Table 1.3. Some common misconceptions about ecological speciation. See text for details.

Misconception
1. Ecological speciation is the same as sympatric speciation
2. Ecological speciation involves only inherently "ecological" forms of reproductive isolation
3. Sexual selection is not involved in ecological speciation
4. Ecology affects everything, so there are no explicit alternatives to ecological speciation
5. Divergent selection is the only way in which ecology contributes to speciation
6. Ecological speciation is speciation involving any role for ecology
7. One has to study distinct species to learn about the process of ecological speciation

ecologically similar pairs of populations adapted to similar environments (Schluter and Nagel 1995, Funk 1998, Räsänen and Hendry 2008). This pattern should exist independently from the amount of time that different population pairs have had to diverge via non-ecological processes such as random genetic drift. An example supporting this prediction is depicted in Figure 1.4, where ecologically divergent pairs of *Timema cristinae* walking-stick insects feeding on different host plant species exhibit greater reproductive isolation than ecologically similar pairs feeding on the

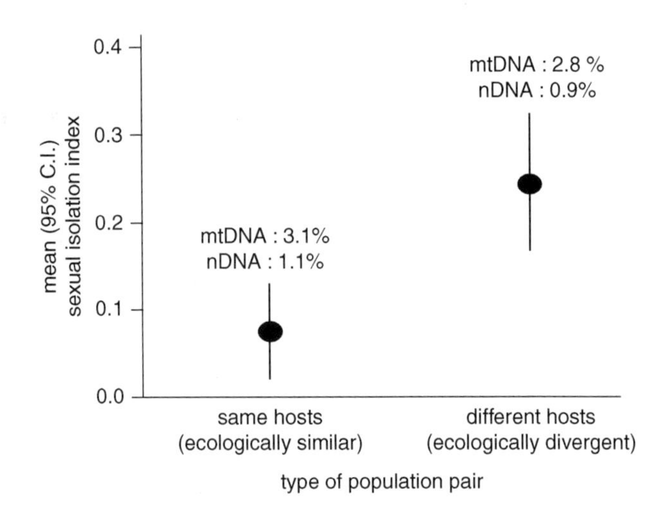

Figure 1.4. Evidence for ecological speciation in host-plant-associated populations of *Timema cristinae* walking-stick insects (individual populations feed on either the host plant *Ceanothus spinosus* or on *Adenostoma fasciculatum*). Pairs of populations feeding on the same host plant species, but in different geographic localities, are ecologically similar and assumed not to be subject to divergent selection. In contrast, pairs of populations feeding on different host plant species are ecologically divergent and subject to divergent selection. Different-host pairs (n = 15 pairs) exhibit significantly greater reproductive isolation due to divergent mating preferences (i.e., sexual isolation) than do same-host pairs (n = 13 pairs). This pattern is independent from neutral genetic divergence, a proxy for time since divergence. Mean divergence is shown for the mitochondrial COI gene (mtDNA) and for the nuclear IT-2 gene (nDNA). Modified from Nosil et al. (2002) with permission from the Nature Publishing Group.

same host plant but living in different localities (Nosil 2007). Another prediction is that traits under divergent selection will often contribute to reproductive isolation (Jiggins et al. 2001, Gavrilets 2004, Maan and Seehausen 2011, Servedio et al. 2011). In the next chapter, I focus on these and other predictions, and how to test them.

At the genetic level, ecological speciation requires a mechanism by which selection on genes conferring divergent adaptation is transmitted to genes causing barriers to genetic exchange. Thus, it is important to consider the relationship between selection and reproductive isolation. Consider two species, which live in different habitats. One species lives in a habitat favoring allele "a" and is fixed for that allele. The other species lives in a habitat favoring allele "A" and is fixed for that allele. Assume no genetic dominance. Allele "a" can be selected against in the habitat it is disfavored in, whether the afflicted allele resides in an immigrant individual of the parental species (homozygote "aa") or in a hybrid individual (heterozygote "Aa"). Traditionally, the former scenario would be considered a case of selection and the latter an example of reproductive isolation (i.e., low hybrid fitness). Here, I consider both scenarios to represent reproductive isolation, because realized gene flow between populations is reduced in both cases (Fig. 1.5). Recognizing that selection against immigrants into habitats to which they are maladapted represents a form of reproductive isolation (Nosil et al. 2005), and that hybrid

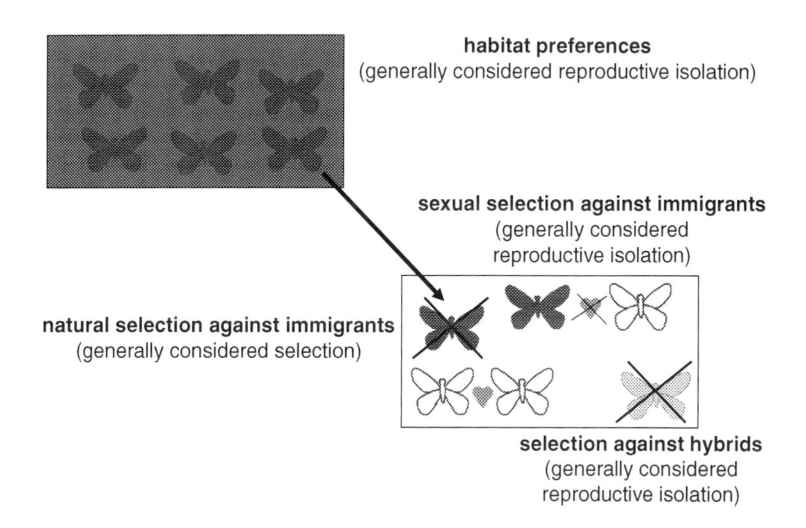

Figure 1.5. The relationship between divergent selection and reproductive isolation. An allele that causes butterflies to be darker in color is selected against in the light environment, whether that allele finds itself in the parental form (dark butterfly) or in a hybrid individual (light grey butterfly). Although the former case is generally considered selection and the latter reproductive isolation, both act as legitimate barriers to realized genetic exchange between populations. Similarly, divergent mating preferences can prevent interbreeding between populations via "sexual isolation." However, the same process might be conceptualized as selection: that is, sexual selection against immigrants. Thus, selection and reproductive isolation are intimately related concepts.

inviability (low fitness of "Aa") is a manifestation of selection, helps clarify the relatedness between selection and reproductive isolation (Mallet 2008a, b, Nosil et al. 2009a).

Ecological speciation is thus perhaps best conceptualized by considering the nature of the genes under selection and those causing reproductive isolation. Under the scenario of pleiotropy, loci under selection and loci causing reproductive isolation are one and the same. For example, the genes that cause hybrids ("Aa") with intermediate phenotypes to be selected against in the environment of each parental species are the same genes causing the form of reproductive isolation often referred to as "extrinsic postmating isolation" (Rundle and Whitlock 2001, Rundle 2002). Another example is when genes under divergent selection, such as those affecting body size, also affect mating preference to cause sexual isolation (Nagel and Schluter 1998, Servedio et al. 2011). In these examples, the distinction between selection and reproductive isolation is erased (or at least highly blurred). In other

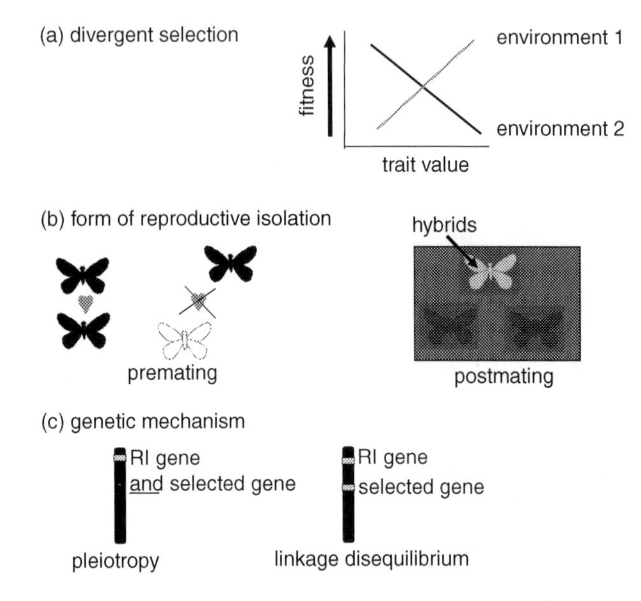

Figure I.6. The three components of ecological speciation (c.f., Rundle and Nosil 2005). a) A source of divergent selection. The commonly recognized sources are differences between environments, interactions between populations, and sexual selection. b) A form of reproductive isolation, which might act before or after mating. The example of pre-mating isolation here concerns divergent mating preferences such that there is assortative mating based upon color. The example of postmating isolation concerns environment-dependent selection against hybrids such that the intermediate phenotype of hybrids renders them more vulnerable to visual predation in each parental environment. c) A genetic mechanism linking selection to reproductive isolation. The genes under selection and conferring reproductive isolation may be one and the same (pleiotropy) or physically different (but statistically associated via linkage disequilibrium, and this can occur without physical linkage on a chromosome). Modified from Nosil and Rundle (2009) with permission from Princeton University Press.

cases, the loci under selection will be physically different from those causing reproductive isolation. In this case, reproductive isolation can still evolve if the different kinds of loci are statistically associated (i.e., in linkage disequilibrium). In this case, selection on fitness loci "spills over" to reproductive isolation loci and causes the correlated evolution of the latter. In both genetic scenarios, divergent selection is intimately involved in causing barriers to genetic exchange (Fig. 1.6).

Ecological speciation thus has three necessary components, which can be to some extent studied independently (Rundle and Nosil 2005): (1) a source of divergent selection, (2) a form of reproductive isolation; and (3) a genetic mechanism linking selection to reproductive isolation (pleiotropy or linkage disequilibrium) (Fig. 1.6). Chapters 3–5 examine each of these components in detail, and subsequent chapters integrate the components.

1.3. A brief history of the ecological speciation hypothesis

I provide here a brief history of the ecological speciation hypothesis, and refer readers to past works for more detailed consideration of the history of speciation research (Coyne and Orr 2004, Mallet 2008b). The concept of ecological speciation dates back at least to *On the Origin of Species* (Darwin 1859). Although it is often stated that Darwin did not explicitly provide a model of how new species arise, Darwin did discuss the causes of speciation in the context of "divergence of character" (i.e., essentially morphological divergence) (Mallet 2008b). Darwin's mechanism for speciation was natural selection, as illustrated by the only figure in *On the Origin of Species* and quotations relating to it:

> "the principle of benefit derived from divergence of character... will generally lead to the most divergent variations... being preserved and accumulated by natural selection" (Darwin 1859, p. 117).

> "... these forms may still be only... varieties; but we have only to suppose the steps of modification to be more numerous or greater in amount, to convert these forms into species... thus species are multiplied" (Darwin 1859, p. 120).

Soon after the publication of *On the Origin of Species*, Walsh (1864, 1867) proposed that when insects shift and adapt to new host plants, they evolve new host preferences and the inability to use non-native hosts, thus forming new species:

> "When... a phytophagic variety has fed for a great many generations upon one particular plant... it is likely to transmit to its descendants... a tendency to select that particular plant... it will cease to be possible that insect... to feed upon any plant other than that to which it has become habituated by the laws of inheritance." (Walsh 1864, pp. 405–6)

An explicit formulation of the ecological speciation hypothesis in terms of how divergent selection might drive the evolution of barriers to genetic exchange occurred during the modern evolutionary synthesis. Dobzhansky wrote about how species were adapted to their ecological niches, and thus hybrids would be selected against:

> "The genotype of a species is an integrated system adapted to the ecological niche in which the species lives. Gene recombination in the offspring of species hybrids may lead to formation of discordant gene patterns" (Dobzhansky 1951, pp. 405–6)

Muller (1942) discussed how adaptation to different environments might incidentally result in speciation (see Johnson 2002 for a review):

> "as automatic by-products of the general differentiation produced by. . . selection" (Muller 1942, p. 100)

Mayr (1942, 1947, 1963) wrote about the role of ecological shifts and local adaptation in the evolution of reproductive isolation:

> "the stronger their need for local adaptation . . . the greater the probability of changes in the components of isolating mechanisms" (Mayr 1963, p. 495).
>
> "Many isolating mechanisms have ecological components. The ecological shifts in incipient species are bound to have an effect on their isolating mechanisms" (Mayr 1963, p. 551).

However, focused studies of the role of divergent selection in speciation during the time of the modern synthesis were few, and the situation remained as such for several decades. A potential reason for this dormancy was the acceptance of the importance of allopatry: given geographic isolation and enough time, speciation was seen as inevitable, whether divergent selection was involved or not. There were some exceptions to this dormancy, both in lab and field studies. For example, early experimental evolution studies on *Drosophila* asked whether divergent selection could drive the evolution of reproductive isolation (Thoday and Gibson 1962, 1970; see Rice and Hostert 1993 for a review). For populations in the wild, botanists recognized that ecologically differentiated plant populations represented intermediate stages in the speciation process (Clausen 1951, Lowry unpublished). Some zoologists also focused on ecology and speciation. For example, Dice (1940) and Blair (1950) asked whether adaptation to different habitats fueled speciation in the mouse genus *Peromyscus*, and outlined evidence for the ecological nature of subspecies:

> "Geographic races . . . appear to be mainly the adaptive products of environmental selection . . . there is a correlation, in some species, between body dimensions, and the vegetational environment occupied. The width of zones of intergradation between geographic races is related to the rate of geographic change in the environment" (Blair 1950, pp. 261–273).

A general area in which ecological speciation received attention at this time was during discussion of sympatric speciation, where disruptive selection was required to counter the homogenizing effects of gene flow (Maynard-Smith 1966, Bush

1969b). For example, Bush's work on apple and hawthorn races of *Rhagoletis* flies was focused on the geographic context of speciation, but nonetheless supported Walsh's earlier statements about how divergent adaptation could generate new species (Bush 1969b). However, these studies of speciation were the exception, rather than the norm, and tended to focus on the geographic context of adaptive divergence, rather than on the build-up of reproductive barriers per se.

The 1990s saw renewed interest in the role of ecology in speciation, fueled by the efforts of Schluter and others (Schluter 1996c, b, Funk 1998, Schluter 1998). At this point in time, definitions of "ecological speciation" similar to the one proposed in this book were put forth and the term itself came into wider use (Table 1.2). Moreover, explicit predictions of the process were formulated, particularly those concerning the evolution of reproductive isolation (Schluter and Nagel 1995, Schluter 1996c, Funk 1998, Schluter 2000b, 2001, Funk et al. 2002). The emergence of these explicit predictions, coupled with the ability to test them using a combination of field, experimental, and molecular data, spurred a renaissance in empirical research on ecological speciation. Thus, most focused studies of the process have emerged in the last 20 years (Rundle and Nosil 2005, Hendry et al. 2007, Schluter 2009). The main question driving these efforts was whether and how adaptation to different ecological environments could promote or accelerate the speciation process.

1.4. Alternatives to ecological speciation

"The . . . evolution of isolating mechanisms as a by-product of the steady genetic divergence is inevitable" (Mayr 1963, p. 581).

As exemplified by the above quotation, genetic divergence is expected to eventually cause speciation, whether or not divergent selection is involved. Thus, ecological speciation is distinguished from other models of speciation in which the evolution of reproductive isolation involves key processes other than deterministic and ecologically based divergent selection between environments. These alternatives tend to involve stochastic events, such as random changes in gene frequencies and stochastic differences among populations in which mutations arise. Such alternatives can be classified into two main categories. The first considers mechanisms of speciation that do not involve selection. The second considers mechanisms that do involve selection, but in which selection is not divergent between ecological environments. All models that do not involve divergent selection are explicit alternatives to ecological speciation and generate different predictions. However, the different models are not mutually exclusive, and more than one model may be operating simultaneously. Table 1.4 presents a classification of these different mechanisms of speciation (following Schluter 2001, 2009). A major strength of this classification scheme is that it can be used to determine which processes are clearly affecting speciation. A weakness is that it does not allow one to easily determine the relative importance of different processes, or how they interact. I return to this weakness later in the book.

Table 1.4. List of alternative mechanisms of speciation and examples of their predictions.

Mechanism of speciation	Description	Example process causing divergence	Example prediction
1. "Ecological speciation"	Divergent selection between ecological environments drives the evolution of reproductive isolation	Divergent selection	Reproductive isolation is correlated with adaptive and ecological divergence
2. "Speciation without selection"	The evolution of reproductive isolation without a key role for selection	Genetic drift in stable populations	Reproductive isolation is correlated with time and not ecological divergence
		Genetic drift in small populations ("founder effect" speciation)	Reproductive isolation is correlated with the occurrence of population bottlenecks, perhaps also time
		Hybridization and polyploidy	Predicts postzygotic isolation due to genetic incompatibilities and rapid speciation
3. "Mutation-order speciation"	Separate populations adapting to similar selection pressures fix different advantageous mutations (alleles) that are incompatible with one another	Selection arising from sexual or genetic conflict	Reproductive isolation is uncorrelated with ecological divergence and correlated with the intensity of conflict

1.4.1. Models of speciation lacking selection

Alternatives to ecological speciation include non-ecological models in which chance events play a central role, including speciation by polyploidization, genetic drift in stable populations, and founder events/population bottlenecks. For a thorough treatment of such models, I refer readers to a past review (Coyne and Orr 2004), which concluded that speciation via genetic drift and/or founder effects was theoretically difficult and rare in nature (but see Gavrilets and Hastings 1996, Templeton 2008, Rundell and Price 2009).

1.4.2. Mutation-order speciation

A potentially powerful alternative to ecological speciation is "mutation-order speciation," defined as the evolution of reproductive isolation by the fixation of different advantageous mutations in separate populations experiencing similar selection pressures, i.e., "uniform selection" (Schluter 2009). In essence, different populations find different genetic solutions to the same selective problem (Mani and Clarke 1990). In turn, the different genetic solutions (i.e., mutations) are incompatible with one another, causing reproductive isolation (Schluter 2001, Price 2007, Schluter 2009, Nosil and Flaxman 2011). During ecological speciation, different alleles are favored between two populations. In contrast, during mutation-order speciation, the same alleles are favored in both populations, but divergence occurs anyway because, by chance, the populations do not acquire the same mutations or fix them in the same order. Divergence is therefore stochastic, but the process involves selection, and thus, is distinct from genetic drift (Mani and Clarke 1990). Selection can be ecologically based under mutation-order speciation, but ecology does not favor divergence as such, and an association between ecological divergence and reproductive isolation is not expected (Fig. 1.7).

How might mutation-order speciation arise? Sexual selection may cause mutation-order speciation if reproductive isolation evolves by the fixation of alternative advantageous mutations in different populations living in similar ecological

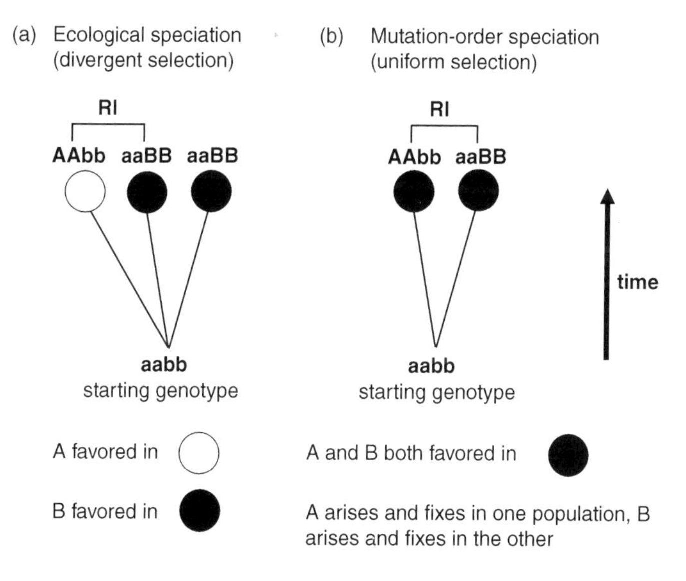

Figure 1.7. Schematic diagram of a) ecological speciation versus b) mutation-order speciation (following Schluter 2009). Distinct environments are indicated by open versus filled circles. Population sizes are large enough that genetic drift can be ignored. Genetic changes are shown at two bi-allelic loci, with the derived alleles A and B being incompatible with one another. "RI" denotes the types of population crosses that result in reproductive isolation. The essential feature depicted is that under ecological speciation only populations in different environments evolve reproductive isolation, whereas under mutation-order speciation populations adapted to the same environment evolve reproductive isolation.

environments. This might be particularly likely via perpetual coevolution between the sexes resulting from sexual conflict in mating. Sexual conflict occurs when characteristics that enhance the reproductive success of one sex reduce the fitness of the other sex (Chapman et al. 2003, Arnqvist and Rowe 2005). For example, females may be selected to choose high-quality mates, whereas males are selected to maximize mating quantity, resulting in constant conflict and coevolution between the sexes. Sexual conflict can cause different populations to undergo divergent coevolutionary trajectories such that strong mating incompatibilities evolve between populations, for example owing to divergence in mating preferences or in gametic interactions that affect fertilization success (Rice 1998, Gavrilets 2000, 2004, Arnqvist and Rowe 2005, Gavrilets and Hayashi 2005, Sauer and Hausdorf 2009). These divergent coevolutionary trajectories represent mutation-order speciation if different, advantageous mutations become fixed in different populations adapting to similar selective pressures. An example of speciation via sexual conflict comes from an experimental evolution study in which allopatric populations of the dung fly *Sepsis cynipsea* were evolved for 35 generations under varying strengths of sexual conflict (Martin and Hosken 2003). Reproductive isolation between populations evolved in treatments where sexual conflict was expected to be moderate or strong, but did not evolve in a treatment where sexual conflict was expected to be weak (Fig. 1.8).

A related scenario involves genetic conflict. The idea is that genetic conflict between elements within the genome causes coevolutionary divergence between populations in a process analogous to the sexual conflict between males and females described above (Burt and Trivers 2006). Thus, genetic conflicts involving selfish

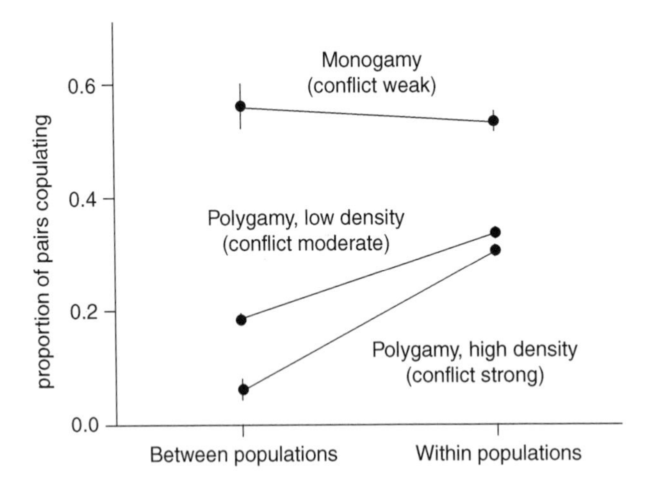

Figure 1.8. Speciation via sexual conflict in the dung fly *Sepsis cynipsea*. Reduced between-population mating, relative to within-population mating, represents reproductive isolation and evolved only in treatments where sexual conflict is expected to be moderate or strong. Modified from Martin and Hosken (2003) with permission from the Nature Publishing Group.

transposons, repetitive DNAs, meiotic drive elements, cytonuclear interactions, etc. could drive the divergence of species. The mutations that continually resolve and then counter these conflicts are unlikely to be the same in different populations—a scenario that can lead to mutation-order speciation. Recent work on hybrid incompatibilities between *Drosophila* species points to such genetic conflict, arising from meiotic drive, as the cause of divergence (Presgraves 2007a, b, Presgraves and Stephan 2007, Phadnis and Orr 2009, Tang and Presgraves 2009). Other potential examples stem from cytonuclear interactions causing hybrid incompatibilities in plants (Fishman et al. 2008, Sambatti et al. 2008). Other forms of conflict, such as that between parents and offspring, present further scenarios for potential mutation-order speciation (Elliot and Crespi 2006, Schrader and Travis 2008, Coleman et al. 2009).

In summary, a useful way to conceptualize the difference between ecological speciation and mutation-order speciation is to consider the sources of selection. The agents of divergent selection during ecological speciation are extrinsic and can include abiotic and biotic factors such as food resources, climate, habitat, and interspecies interactions such as disease, competition, and behavioral interference. The agents of selection during mutation-order speciation might often be "intrinsic," for example, having to do with the internal genetic environment. Interactions occurring between the sexes or among genetic elements might be considered "ecological" by some, but they are not necessarily divergent between external environments, and thus, are distinct from the factors driving ecological speciation. Notably, some reproductive isolation between ecologically similar population pairs has often been documented (Nosil et al. 2002, Coyne and Orr 2004, Price 2007, Schluter 2009), and mutation-order speciation could be responsible for this. However, it is difficult in such cases to rule out ecological speciation via divergence in some unmeasured ecological variable. A more detailed consideration of how to test for ecological speciation, thereby distinguishing it from the alternatives, forms the focus of Chapter 2.

1.5. Other roles for ecology in speciation: population persistence and niche conservatism

There are roles that ecology may play in speciation other than generating divergent or uniform selection (Sobel et al. 2010 for review). These roles concern: (1) the facilitation of the long-term persistence of populations, which is required for the gradual build-up of mating incompatibilities between species; and (2) the long-term geographical separation of populations owing to niche conservatism. Under these scenarios, ecology facilitates speciation, but is not necessarily directly involved in causing the evolution of reproductive isolation.

Under the first "population persistence" view, ecological processes affect speciation mainly through their effects on the long-term viability of populations undergoing the speciation process: new species evolve only when populations can avoid extinction for long enough to evolve reproductive isolation (Mayr 1963, Schluter 1998, Levin 2004). Thus, the occupation of previously unoccupied niches and subsequent local adaptation facilitates speciation because it allows large population sizes and the long-term persistence of populations. Reproductive isolation itself then can gradually evolve via any combination of divergent natural selection, uniform natural selection, sexual selection, and random genetic drift. This view thus predicts that time and population persistence, rather than ecological divergence per se, will be correlated with the degree of reproductive isolation. The importance of this persistence view for speciation depends on the rate at which reproductive isolation evolves. The more slowly reproductive isolation evolves, the more important for speciation are the mechanisms that facilitate long-term persistence of populations.

Under the second view, phylogenetic niche conservatism facilitates speciation by causing populations to become and remain geographically isolated (Wiens 2004). Geographical isolation occurs because populations have limited ecological tolerances such that they cannot occur in regions that would connect the populations via gene flow (Ramsey et al. 2003, Sobel et al. 2010). For example, before speciation is initiated, gene flow might occur between mountains because the ecological conditions suitable to population persistence overlap between the two mountains. At some

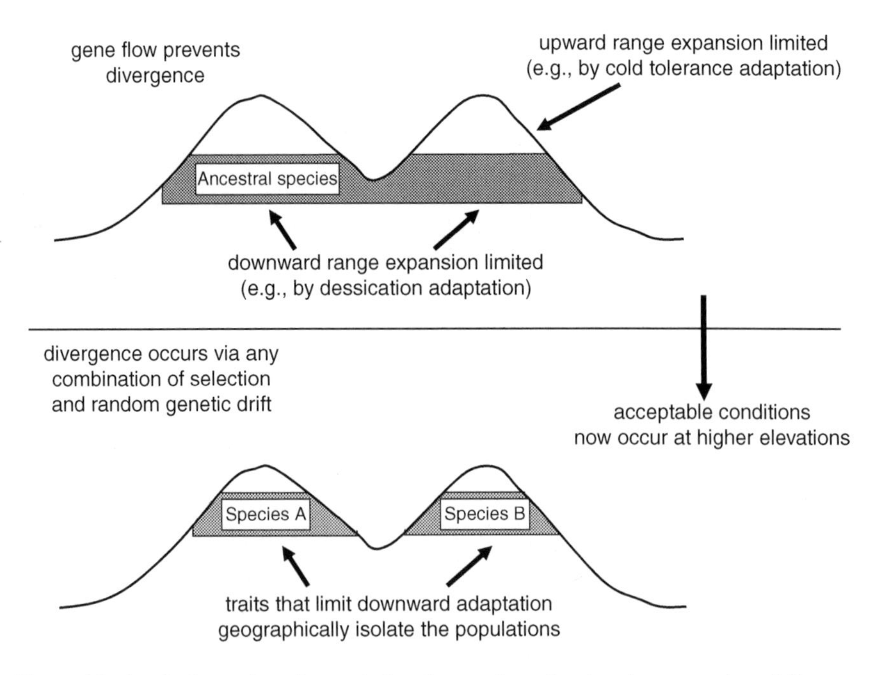

Figure I.9. A role for ecology in speciation that need not involve the generation of divergent selection. "Phylogenetic niche conservatism" can cause geographic isolation that prevents gene flow between populations (in this case on different mountains). In turn, this allows divergence between populations under any combination of natural selection, sexual selection, and genetic drift. Modified from Wiens (2004) and reprinted with permission from John Wiley and Sons.

point, the ecological conditions that allow population persistence might change such that suitable conditions for population persistence now occur at higher elevations. In turn, this results in geographic separation of populations on different mountains. The resulting lack of gene flow between populations on different mountains then allows divergence under any combination of different forms of selection and genetic drift (Fig. 1.9). This view predicts that time in geographical separation, rather than ecological divergence per se, is the main predictor of reproductive isolation. The importance of niche conservatism for speciation depends on the extent to which strong geographical barriers to gene flow are required for speciation. A prediction is that if divergent selection can commonly overcome gene flow to cause speciation, then geographical isolation due to niche conservatism becomes less important for speciation.

The fundamental difference between the roles of ecology in these mechanisms versus its role in ecological speciation is that only in the latter is divergent selection always directly and causally responsible for the evolution of barriers to genetic exchange.

1.6. Summary

Ecological speciation is the process by which barriers to gene flow evolve between populations as a result of ecologically based, divergent selection between environments. Alternatives to ecological speciation are mechanisms of speciation that do not invoke divergent selection. Ecological speciation can involve any type of reproductive barrier and can occur under any geographic arrangement of populations. Now that I have introduced ecological speciation and its alternatives, I turn to predictions and tests of ecological speciation: that is, how can one empirically distinguish ecological speciation from other mechanisms of speciation?

2

Predictions and tests of ecological speciation

In this chapter, I review five classes of approaches for testing for ecological speciation. The explicit predictions associated with each approach are summarized in Table 2.1. These predictions allow ecological speciation to be distinguished from alternative mechanisms of speciation. Having a diversity of approaches is useful because their applicability will vary among groups of organisms; for example according to the number of taxon-pairs available for analysis, availability of *a priori* information on traits subject to divergent selection, experimental tractability, etc.

The first class of "comparative" approaches examines multiple taxon pairs and tests for a positive association between levels of ecological divergence and levels of reproductive isolation, independent from genetic distance at neutral loci (under the

Table 2.1. Approaches for testing ecological speciation.

Test	Prediction	Example of an explicit outcome
1. Comparative approaches: "ERG"	Levels of RI are positively correlated with levels of ecological divergence between population pairs, independent from time	RI is correlated with ecological divergence after the effects of time and phylogenetic relatedness are controlled for
2. Trait-based approaches: "magic traits"	Traits under divergent selection also affect reproductive isolation	Traits under divergent selection contribute to sexual isolation
3. Fitness-based approaches: "selection = RI"	Divergent selection results in ecological selection against immigrants and hybrids	Reduced fitness of immigrants and hybrids relative to parental forms, owing to an ecological mismatch between the phenotype of immigrants and hybrids and the ecological environment
4. Gene flow approaches: "isolation-by-adaptation"	Adaptive divergence reduces gene flow between populations	Among population pairs, a positive correlation between levels of adaptive phenotypic and molecular genetic differentiation
5. Phylogenetic shifts approach	Speciation events require ecological shifts	Branching events in a phylogeny coincide with ecological shifts; recent sister-species differ in ecology

Abbreviation: RI, reproductive isolation.

assumption that such genetic distance is a proxy for time since population divergence). This will be referred to as the "ERG" approach, reflecting the fact that ecological divergence ("E"), reproductive isolation ("R"), and genetic distance ("G") are each measured (following Funk 1998, Funk et al. 2002, Funk et al. 2006). When analysis of multiple taxon pairs in a comparative framework is not possible, "trait-based" approaches can test whether traits involved in divergent adaptation also affect reproductive isolation ("magic traits" c.f., Gavrilets 2004). Sometimes it will not be possible to know *a priori* which traits are subject to divergent selection. For example, it may be difficult to determine which specific physiological traits are involved in adaptation to different altitudes or to feeding on different host plants. In this case, we may use a third class of "fitness-based" approaches to test whether reproductive isolation is a direct consequence of ecologically based selection against immigrant and hybrids. These approaches involve measuring fitness in different habitats, rather than selection on specific phenotypic traits. When experimental data cannot be collected, "gene-flow-based" approaches can test whether gene flow at molecular markers is reduced as adaptive divergence increases. Finally, a fifth "phylogenetic shift" approach can be applied to taxa that are no longer exchanging genes to ask whether branching (i.e., speciation) events in a phylogenetic tree coincide with ecological shifts.

Collectively, one might think of the following questions, each associated with one of the approaches: (1) Is ecological divergence correlated with reproductive isolation? (2) Do traits involved in divergent adaptation affect reproductive isolation? (3) Does ecologically based selection reduce the fitness of immigrants and hybrids? (4) Is ecological divergence correlated with reduced levels of gene flow in nature? (5) Do speciation events coincide with ecological shifts? I turn to reviewing each approach in detail. I conclude by discussing causality, ascertainment bias, and publication bias. The most critical finding that emerges is that although evidence for ecological speciation is abundant, almost all of it is correlative and indirect, or from lab studies. Manipulative field experiments testing whether adaptive divergence reduces gene flow in nature are lacking and represent the "final frontier" in tests of ecological speciation.

2.1. Comparative approaches (ERG)

The most general comparative tests for ecological speciation involve testing for a correlation between ecological divergence and reproductive isolation, independent from time (Funk 1998, Funk et al. 2002, Funk et al. 2006). In such approaches, time is generally controlled for using the molecular genetic distance between taxon pairs, under the assumption that genetic distance is a proxy for time since divergence (i.e., a "molecular clock"). When only a few taxon pairs are available for analysis, qualitative comparisons are possible (Stelkens and Seehausen 2009b). For example, Funk (1998) examined sexual isolation between three population pairs of *Neochlamisus* leaf beetles: the two population pairs feeding on different host plant species

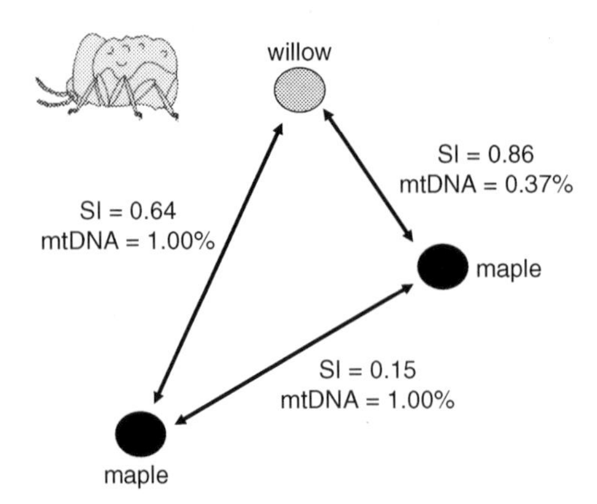

Figure 2.1. Sexual isolation (SI, where 0 = random mating and 1 = complete sexual isolation) between *Neochlamisus bebbinae* leaf beetle populations is greater in comparisons involving populations feeding on different host-plant species (maple versus willow) relative to that observed between a population pair feeding on the same host-plant species (maple versus maple). This pattern is not confounded by greater mtDNA sequence divergence between the different-host pairs. Leaf beetle drawn by Chris Brown. Modified from Funk (1998) with permission from John Wiley & Sons.

(willow versus maple) exhibited stronger sexual isolation than the population pair feeding on the same host (two maple populations) (Fig. 2.1). This pattern appears to be independent from the time since population divergence, because mitochondrial DNA (mtDNA) differentiation between the different-host population pairs was similar, or even weaker, than mtDNA differentiation between the same-host population pair. Thus, ecological divergence in host plant use, rather than time, was the predictor of reproductive isolation.

When a greater number of taxon pairs are available for analysis, a regression approach suggested by Funk et al. (2002) can be used. The approach adds an ecological dimension to comparative studies investigating the relationship between reproductive isolation and divergence time (Coyne and Orr 1989). The approach was implemented by Funk et al. (2006) in a study that quantified ecological divergence for >500 species pairs from eight plant, invertebrate, and vertebrate taxa. Using the ERG approach, they statistically isolated a significant and positive association between ecological divergence and reproductive isolation across the eight taxa, independent from time (Fig. 2.2). The findings are consistent with the hypothesis that ecological adaptation plays a major role in promoting speciation, across disparate taxa (see also Bolnick et al. 2006). A number of further questions could be investigated, including the types of reproductive barriers that tend to be correlated with ecological divergence, the role of geography in mediating the correlation between ecology and reproductive isolation, and the time course or rate of speciation.

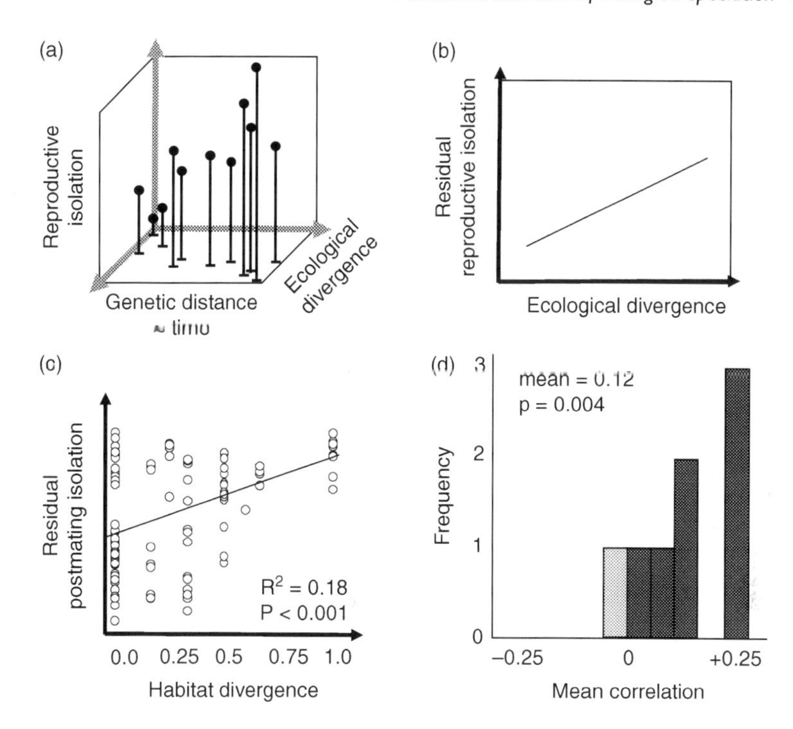

Figure 2.2. Isolating the relationship between ecological divergence and reproductive isolation, independent from time, following Funk et al. (2002, 2006). a) A hypothetical scenario in which reproductive isolation increases with both genetic distance (a proxy for time) and ecological divergence. b) A hypothetical best-fit line illustrating the predicted positive association between ecological divergence and residual reproductive isolation upon statistical removal of the contributions of time. c) An empirical positive association between habitat divergence and residual postmating isolation (effects of time removed via regression of reproductive isolation and genetic distance) for the angiosperm dataset (Moyle et al. 2004). d) The distribution of correlation coefficients from analyses of residual reproductive isolation against ecological divergence across the eight taxa examined by Funk et al. (2006). Each data point contributing to the bars is the mean correlation coefficient from a particular taxon. The mean of the depicted distribution was significantly greater than zero using a one-sample t-test. Modified from Funk et al. (2006) with permission from the National Academy of Sciences USA.

A major caveat associated with the ERG approach is that controlling for time using genetic data works best for allopatric species pairs. For hybridizing taxa, the degree of gene flow confounds the estimate of divergence time because it directly affects genetic distance. For example, gene flow can deflate genetic distance, leading to underestimates of divergence time. A potential solution is to apply coalescent-based techniques to estimate divergence time independent from gene flow (Hey and Nielsen 2004, Hey 2006, Hey and Nielsen 2007, Wakeley 2008a, Hey 2010b). However, the efficacy of these methods, particularly when assumptions of the underlying models are violated, needs consideration (Becquet and Przeworski 2009, Strasburg and Rieseberg 2010, 2011).

2.1.1. Parallel speciation

A specific type of ERG test for ecological speciation is termed "parallel speciation" (following Schluter and Nagel 1995). Parallel evolution of similar traits in independent populations that inhabit ecologically similar environments strongly implicates natural selection as the cause of evolution, as random genetic drift is unlikely to produce such a pattern. Parallel speciation is a special form of parallel evolution in which traits that determine reproductive isolation evolve repeatedly, in closely related populations, as by-products of adaptation to ecological conditions. In essence, reproductive isolation evolves multiple times independently and in correlation with ecological divergence. The outcome of such parallel evolution is that ecologically divergent pairs of populations exhibit greater levels of reproductive isolation than ecologically similar pairs of populations of similar age (Funk 1998). In other words, reproductive isolation between ecologically similar forms is low, despite their distant phylogenetic relatedness. In contrast, reproductive isolation between ecologically divergent forms should be strong, despite their close phylogenetic relatedness (Fig. 2.3). Parallel speciation provides some of the strongest

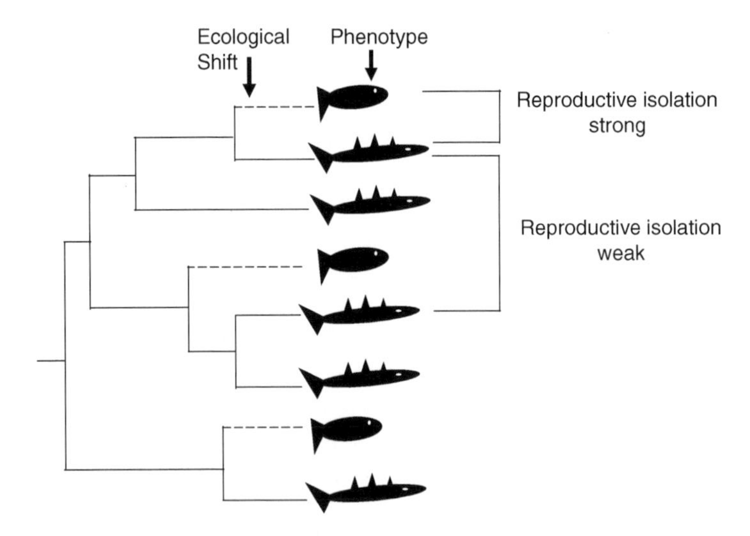

Figure 2.3. A schematic illustration of parallel speciation. Parallel evolution is indicated by the correlated evolution of ecological shifts and phenotype. During parallel speciation there is the added element that phenotypic divergence underlies reproductive isolation. Thus, the predictor of reproductive isolation is ecological and phenotypic divergence, rather than phylogenetic relatedness. Following Schluter and Nagel (1995).

evidence for natural selection in the process of speciation, and a number of putative examples now exist.

2.1.2. Parallel speciation via experimental evolution in the lab

Laboratory selection experiments provide ideal opportunities for testing ecological speciation. Replicate lines can be selected divergently and at the end of the experiment one can test whether lines adapted to different environments exhibit greater reproductive isolation than different lines adapted to the same environment. In this manner, laboratory evolution experiments using *Drosophila* fruit flies have shown that parallel speciation is feasible: when replicate populations are independently adapted to one of two environments, stronger reproductive isolation has sometimes arisen between populations from different environments than between populations evolved in similar environments. Classic examples of such experiments are depicted in Figure 2.4.

A large number of studies have now applied divergent selection in the lab in attempts to evolve sexual reproductive isolation. Rice and Hostert (1993) reviewed 38 such selection experiments. Nosil and Harmon (2009) updated that review, using a total of 59 studies. The studies were subdivided into whether they imposed selection directly on premating isolation or on a non-isolation trait (e.g., sternopleural bristle number), and according to whether selection was imposed with or

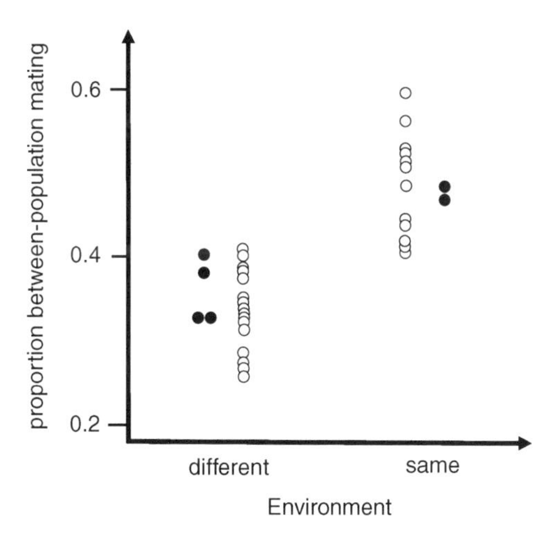

Figure 2.4. Evidence for ecological speciation from laboratory evolution studies. Each dot represents a replicate experimental line. Shown is the proportion of matings occurring between independently evolved lines of *Drosophila* as a function of the similarity of their environments. Between-population mating is less common when populations have been adapted to different environments. Open circles are from Dodd (1989) and closed circles from Kilias et al. (1980).

without gene flow (following Kirkpatrick and Ravigné 2002). Table 2.2 provides details concerning this scheme. The main conclusion emerging from these reviews is that divergent selection applied in the lab often supports the process of ecological speciation. Specifically, roughly half (51%) of the experiments saw the evolution of at least some sexual isolation between divergently selected lines. Experiments that selected directly on reproductive isolation and those which selected on multiple traits were particularly successful at evolving sexual isolation (Table 2.3).

It is important, however, to note that nearly half of the experiments failed to generate any sexual isolation between divergently selected lines. Furthermore, 20% of experiments reported sexual isolation that varied strongly among lines. Finally, even among the experiments that observed the consistent evolution of sexual isolation among lines adapted to different environments, 81% saw the evolution of only partial (i.e., <90%), rather than strong or complete, sexual isolation. Thus, laboratory experiments demonstrate that parallel speciation is a feasible, but not

Table 2.2. Summary of 59 laboratory speciation experiments.

	No. studies	No. divergently selected traits	Result
1) No gene flow direct	7	1	Mixed RI
	5	1	Partial RI
	1	1	Strong RI
2) No gene flow indirect	7	1	No RI
	2	1	Mixed RI
	8	1	Partial RI
	1	4	No RI
	1	2	Partial RI
3) Gene flow direct	3	4	Strong RI
4) Gene flow indirect	21	1	No RI
	1	1	Mixed RI
	3	1	Partial RI

"RI" refers to sexual reproductive isolation in all cases. "No RI" refers to significant RI never observed. "Mixed RI" refers to cases in which partial RI was detected in some cases, but RI varied among years, replicates, and strains, such that RI was sometimes lacking. "Partial RI" refers to consistent patterns of incomplete RI. "Strong RI" refers to consistent patterns of over 90% assortative mating. 1) No gene flow direct. These experiments "destroy-all-hybrids" and impose selection directly on reproductive isolation during the course of the experiment. 2) No gene flow indirect. These experiments impose selection on one or more non-isolation trait(s). 3) Gene flow direct experiments. These impose selection on reproductive isolation, but not all hybrids are destroyed. 4) Gene flow indirect experiments. These impose selection on one or more non-isolation trait(s), but not all hybrids are destroyed. See Nosil and Harmon (2009) for original references.

Table 2.3. Summary of the main trends in the speciation experiments reported in Table 2.2. Numbers refer to the proportion of experiments where the denoted outcome was observed.

	Multiple selected traits	Single selected trait	Outcome
Direct selection	3/3	1/13	Near complete RI
Indirect Selection	1/2	11/42	Partial RI

inevitable, outcome of divergent selection. Further studies could usefully focus on the causes of these variable outcomes.

2.1.3. Parallel speciation in the wild

A number of putative cases of parallel speciation in nature also exist. A prime example stems from freshwater stickleback (*Gasterosteus*) fishes (Rundle et al. 2000) (Fig. 2.5), which come in two main forms: a slender limnetic form, which feeds primarily on plankton in the open water of a lake; and a more robust benthic form, which feeds on invertebrates in the shallows. Sympatric limnetic–benthic pairs occur in a number of lakes in western Canada, and molecular genetic evidence suggests that the pairs have arisen independently. Thus, the present-day phenotypic similarity of limnetics from separate lakes is likely to be the result of parallel evolution and not shared ancestry, and the same applies for the phenotypic similarity

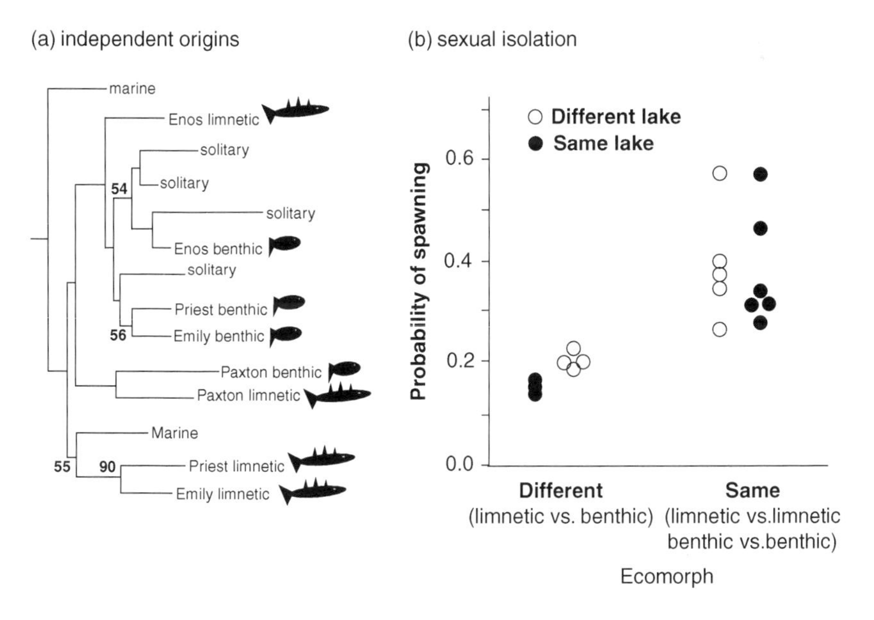

Figure 2.5. Parallel speciation in sympatric limnetic and benthic ecomorphs of freshwater stickleback fishes. a) The phylogenetic tree, inferred from microsatellite genetic distances, illustrates how the different ecomorphs probably arose independently more than once. Modified from Taylor and McPhail (2000) with permission of the Royal Society of London (several outgroup marine populations are omitted for clarity; "solitary" refers to lakes where only a single form of stickleback is found). Numbers at nodes represent cases where bootstrap support exceeded 50. b) The probability of interbreeding between versus within ecomorphs, for ecomorphs from the same versus different lakes. Modified from Rundle et al. (2000) with permission of the American Society for the Advancement of Science.

of benthics from different lakes (Taylor and McPhail 2000). Mating trials demonstrate that reproductive isolation between limnetics and benthics has likewise evolved in parallel: premating isolation is strong between limnetics and benthics, whereas premating isolation is absent or weak within a form, even when they derive from different lakes (e.g., premating isolation is weak between limnetic forms from different lakes).

Similar patterns were observed in a number of other systems, including freshwater versus anadromous forms of sticklebacks, host forms of *Neochlamisus* leaf beetles and *Timema* walking-stick insects, and ecotypes of *Gambusia* fish, *Hyalella* amphipods, and *Littorina* snails (Table 2.4 for details). Some cases of parallel speciation involve sympatric or parapatric divergence (Quesada et al. 2007, Conde-Padin et al. 2008), others allopatric divergence (Funk 1998, Langerhans et al. 2007), and others a mixture of geographic modes of divergence (Rundle et al. 2000, Nosil et al. 2002, 2003, McKinnon et al. 2004). Despite these putative examples, the overall frequency of parallel speciation in nature remains unknown.

2.1.4. Difficulties with tests of parallel speciation

At least three major limitations for demonstrating parallel speciation exist (Fig. 2.6). First, multiple, independent origins of ecological divergence will not always exist, thereby precluding tests of parallel speciation. Second, gene flow between ecologically divergent forms could obscure historical relationships among population pairs such that a true single origin of each ecological form is mistaken for multiple origins (or in less extreme scenarios, ecologically divergent forms appear more closely related than they truly are). More specifically, gene flow between geographically adjacent populations can result in them grouping close together in phylogenetic analyses, independent from how closely related they are historically, making it difficult to empirically distinguish single from multiple origins (Quesada et al. 2007). Thus, as previously mentioned for ERG tests in general, tests for parallel speciation should attempt to distinguish the effects of history versus gene flow on levels of current genetic differentiation. Third, incomplete lineage sorting (retention of ancestral polymorphism) can also result in patterns consistent with multiple origins of ecological forms in different localities, even if there was only a single origin. These problems are unlikely to be trivial, particularly for recently diverged taxa. For example, Funk and Omland (2003) reviewed hundreds of studies and demonstrated that over 20% of animal species are not monophyletic for mtDNA, in part owing to gene flow and incomplete lineage sorting.

2.1.5. Explicit criteria for parallel speciation

Given the problems mentioned above, the following four criteria should be met to confidently conclude that parallel speciation has occurred:

Table 2.4. Putative examples of parallel speciation, via the parallel evolution of sexual isolation (SI), and their fit to the four criteria. Criterion 1. Multiple origins: refers to the data used to reconstruct phylogeny. Criterion 2. Gene flow/lineage sorting: provides information on the degree to which gene flow and incomplete lineage sorting (inferred indirectly as being more likely when origins of taxa are recent) may contribute to non-monophyly of ecologically similar forms. Criterion 3. SI: Y' denotes cases where SI was shown to be greater between ecologically divergent forms than between ecologically similar forms, independent from phylogenetic relatedness. "N" denotes cases where parallel evolution of traits likely to contribute to SI (e.g., mating behavior, body size, etc.) was reported, but SI itself was not measured. Criterion 4. Genetic basis: evidence that SI has a genetic basis (e.g., study organisms were reared under common-garden conditions).

		1.	2.	3.	4.	
Study system	Divergent forms	Multiple origins	Gene flow/lineage sorting	SI tested	Genetic basis[1]	References
1. *Gasterosteus* sticklebacks	Limnetic and benthic forms	msats, mtDNA	GF, RO	Y	NT	(Rundle et al. 2000, Taylor and McPhail 2000, Boughman et al. 2005)
2. *Gasterosteus* sticklebacks	Freshwater and marine forms	msats	GF	Y	NT	(McKinnon et al. 2004)
3. *Timema* stick insects	*Ceanothus* and *Adenostoma* host ecotypes	mtDNA, nDNA, AFLPs	GF, A, NRO	Y	CG, RR	(Nosil et al. 2002, 2003, Nosil 2007, Nosil et al. 2008)
4. *Hyalella* freshwater amphipods	Predator and predator-free forms	allozymes	NGF, RO	Y	NT	(McPeek and Wellborn 1998)
5. *Gambusia* fishes	Predator and predator-free forms	mtDNA, allozymes	A	Y	NT	(Langerhans et al. 2007)

Table 2.4. *Cont.*

		1.	2.	3.	4.	
Study system	Divergent forms	Multiple origins	Gene flow/lineage sorting	SI tested	Genetic basis[1]	References
6. *Neochlamisus* leaf beetles	Willow and maple host forms	mtDNA, AFLPs	GFP, A	Y	NT	(Funk 1998, Egan et al. 2008)
7. *Littorina* intertidal snails	Upper- versus lower-shore ecotypes	mtDNA, msats, allozymes, AFLPs	GF	Y	NT	(Quesada et al. 2007, Rolan-Alvarez 2007, Conde-Padin et al. 2008, Galindo et al. 2009)
8. *Asellus aquaticus* freshwater isopods	Ecotypes in different lake habitats	AFLPs, plus geographic evidence	GF, RO (but contemporary data implicate recent independent origins)	N	NT	(Eroukhmanoff et al. 2009a, 2009b, Karlsson et al. 2010)
9. *Eumeces* skinks	Variable body size forms	mtDNA	GF (but apparently restricted)	N	N/A	(Richmond and Reeder 2002)
10. *Enallagma* damselflies	Fish versus dragonfly predation regimes	morphology	no data	N	N/A	(Stoks et al. 2005)

Abbreviations: A, allopatric populations examined for which gene flow between ecological forms is unlikely; AFLPs, amplified fragment length polymorphisms; CG, common garden; GF, gene flow known for some populations; GFP, gene flow between some populations possible, but unconfirmed (e.g., sympatric taxa were studied but molecular data on gene flow are lacking); msats, microsatellites; mtDNA, mitochondrial DNA sequences; nDNA, nuclear DNA sequences; NGF, gene flow shown to be absent, or very restricted; NRO, not recent origin (probably more than a 100,000 years ago); NT, not tested; RO, recent origin (roughly 10,000 or fewer years ago); RR, reciprocal rearing.

[1] Note that common-garden experiments are conservative in their ability to detect genetically based reproductive isolation. For example, genetically based habitat preference in nature results in organisms being reared in different habitats. This could results in plastic differences in trait expression between populations in different habitats, which then contribute to SI. Such contributions to SI would not be identified under common-garden conditions.

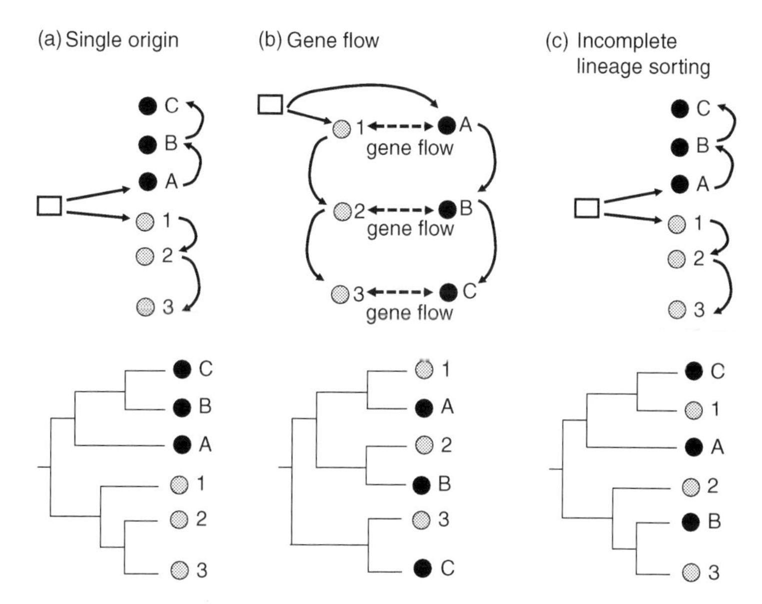

Figure 2.6. Limitations and caveats for testing parallel speciation. Populations in different environments (ecotypes hereafter) are depicted by gray versus black circles. The ancestral population that gave rise to these populations in different environments is depicted as an unfilled rectangle. Single-headed arrows represent the sequence of colonization (i.e., the origins of new populations). Double-headed, dashed arrows represent gene flow between ecologically divergent forms. a) If there is a true single origin of each ecotype, a test for parallel speciation cannot be conducted. b) A single origin of each ecotype, but gene flow between geographically proximate populations leads to phylogenetic patterns that might be erroneously interpreted as multiple origins. c) A single origin of each ecotype, but incomplete lineage sorting leads to phylogenetic patterns that might be erroneously interpreted as multiple origins. Unlike the gene-flow scenario, geographically proximate populations will not necessarily group closely together.

1) evidence that ecologically similar forms do not form a monophyletic group, preferably from multiple independent genes, because different genes can yield different phylogenetic patterns;
2) non-monophyly of ecologically similar forms represents multiple origins, rather than the signature of gene flow or incomplete lineage sorting;
3) reproductive isolation is positively correlated with ecological divergence, independent from phylogenetic relatedness;
4) reproductive isolation is genetically based.

Table 2.4 outlines the degree to which putative cases of parallel speciation fit these criteria and demonstrates that even many of the strongest candidates to date do not meet all of the criteria. In particular, criterion 4 has rarely been satisfied, at least when it comes to tests for sexual isolation under common-garden conditions. Also, the effects of gene flow on phylogenetic reconstruction have not been explicitly quantified in most cases, although this poses less of a problem for studies that examined allopatric populations undergoing little or no gene flow.

2.2. Trait-based approaches ("magic traits")

"Pleiotropism. The gene pair . . . adapting individuals to different niches may themselves cause assortative mating . . . This seems very unlikely." (Maynard-Smith 1966, p.643)

Another approach for detecting ecological speciation involves identifying traits under divergent selection that also contribute to non-random mating, i.e., premating isolation (Maynard-Smith 1966, Schluter 2001). Such traits have been referred to as "magic traits" (Gavrilets 2004), to reflect the fact that a single trait is performing the functions normally attributed to two separate traits. At the genetic level, magic traits involve pleiotropy, because one set of genes is subject to divergent selection (i.e., affects ecological traits) and pleiotropically also affects non-random mating (Bolnick and Fitzpatrick 2007). As the above quotation by Maynard Smith demonstrates, the concept behind magic traits has long been controversial. Indeed, the adjective "magic" can be interpreted to imply that such traits may be unusual and, thus, rare in nature. I focus on whether this assumption is met, and find that because divergent adaptation can involve a variety of mechanisms, there is a corresponding diversity of magic traits. In some cases, the evolution of traits under divergent selection leads to non-random mating as an automatic by-product, indicating that magic traits may be more common than currently thought. In other cases, the connection between selection and non-random mating is less straightforward. I therefore draw heavily from the framework of Servedio et al. (2011) in distinguishing between "automatic" and "classic" magic traits.

2.2.1. "Automatic" magic traits

There are many scenarios where we can think of non-random mating as being "built into" traits under divergent selection (i.e., "automatic" magic traits). Although not usually thought of in this way, such traits are technically magic traits, because one underlying set of genes controls both divergent adaptation and non-random mating. For example, divergent selection on phenological traits automatically leads to assortative mating via temporal isolation, as occurs for flowering time differences in plants (Lamont et al. 2003, Lowry et al. 2008a, 2008b) and diapause emergence differences in insects (Filchak et al. 2000, Dambroski and Feder 2007). Similarly, selection on habitat choice can generate premating isolation when populations use different habitats and mating takes place in the habitat (Rice 1984, Rice and Salt 1990, Gavrilets 2004), as can occur for host plants in phytophagous insects (Bush 1969b) and lake environments in fish (Seehausen and Magalhaes 2010).

Finally, a very general class of magic traits arises whenever the ability of individuals to survive in a specific location generates assortative mating. In such cases, traits conferring local adaptation are magic traits. This is related to the idea that selection against immigrants constitutes a powerful reproductive barrier (Nosil et al. 2005). An extreme example of this is provided by ascomycete fungi (Giraud 2006, Giraud et al. 2008, 2010). These fungi can infect only specific host plants, and mating takes place on these host plants. Thus, if a strain evolves the ability to infect a new host and becomes specialized on it, mating will automatically be assortative with respect to host use. Note, however, that traits involved in local adaptation and habitat choice can be different. For example, in phytophagous insects, host adaptation ("performance") and host choice ("preference") might involve different traits—say morphology versus behavior—that are each independently subject to divergent selection, rendering them both "magic." Indeed, most models of speciation via host shift assume separate genetic control of performance and preference (Diehl and Bush 1984, 1989, Johnson et al. 1996, Berlocher and Feder 2002, Fry 2003), and such separate genetic control appears common in nature (Matsubayashi et al. 2010). Thus, habitat choice constitutes a magic trait only if choice itself is subject to selection. In general, to convincingly demonstrate that any trait is a magic trait, evidence must be obtained that divergent selection does indeed act on the trait itself (and not, for example, on a correlated trait).

2.2.2. "Classic" magic traits

In contrast to the examples above, magic traits are more typically thought of as being caused by divergent selection acting on mating cues, such as color or body size ("classic" magic traits). Intuitively, this connection strikes many as unlikely, leading to the term "magic trait" ringing true. Servedio et al. (2011) reviewed the evidence for classic magic traits. They report numerous putative examples, but only a single conclusive one. Finding conclusive evidence for a magic trait is difficult, because two different traits can be mistaken for a single magic trait if they are strongly correlated. Thus, two criteria must be met for a trait to qualify as a magic trait. First, the magic trait, not a correlated trait (controlled by different genes), must be subject to divergent selection. Second, the magic trait, not a correlated trait, must generate non-random mating. Table 2.5 reviews the strength of evidence supporting each putative example, with the strongest evidence requiring experimental manipulation. The level of support varies widely, and the evidence is weaker for the first criterion than for the second.

In only one case have both criteria been met by manipulative experiments. For mimetic wing color patterns in tropical *Heliconius* butterflies, Jiggins et al. (2001) showed that individuals prefer to mate with live individuals and paper models of the same color pattern (Fig. 2.7). Thus, divergence in color pattern generates sexual isolation. Furthermore, both mark–recapture experiments (Mallet and Barton 1989) and manipulative experiments with paper models (R. Merrill personal communication) indicate that coloration itself is subject to divergent selection, with different

Table 2.5. Examples of putative "magic traits," restricted to those putatively affecting sexual or pollinator isolation. Each example was evaluated according to the two criteria required to demonstrate a magic trait, and categorized as to how strongly each criterion was met.

Study system	Divergent forms	Putative magic trait(s)	Putative cause of selection	Criterion 1: ecological selection[1]	Criterion 2: mate choice[2]	References
1. *Gasterosteus* sticklebacks	Limnetic and benthic freshwater forms	Body size	Foraging niche, competition	Experiment	Experiment	(Schluter 1994, Nagel and Schluter 1998, Boughman et al. 2005; Conte and Schluter personal communication)
2. *Gasterosteus* sticklebacks	Freshwater and marine forms	Body size	Foraging niche	Observational	Manipulative experiment	(Snyder and Dingle 1989, McKinnon et al. 2004)
3. *Gambusia* fishes	Predator and predator-free forms	Body shape	Predation regime (predators present versus absent)	Observational	Experiment	(Langerhans et al. 2004, 2007)
4. *Littorina* intertidal snails	Upper- and lower-shore ecotypes	Body/shell size	Crab predation	Experiment	Experiment	(Rolan-Alvarez et al. 1997, Rolan-Alvarez 2007, Conde-Padin et al. 2008)
5. *Heliconius* mimetic butterflies	Different mimetic forms	Color pattern	Visual predation (mimicry)	Manipulative experiment	Manipulative experiment	(Mallet and Barton 1989, Jiggins et al. 2001, Jiggins 2008; Merrill personal communication)
6. *Dendrobates pumilio* poison-dart frogs	Different aposematic forms	Color and color pattern	Visual predation	Observational	Manipulative experiment	(Summers et al. 1999, Reynolds and Fitzpatrick 2007, Maan and Cummings 2008)

Study system	Divergent forms	Putative magic trait(s)	Putative cause of selection	Criterion 1: ecological selection[1]	Criterion 2: mate choice[2]	References
7. *Mimulus* monkeyflowers	Bumblebee-pollinated *Mimulus lewisii* and hummingbird-pollinated *M. cardinalis*	Flower color	Divergent habitat types	Observational	Manipulative experiment	(Schemske and Bradshaw 1999, Bradshaw and Schemske 2003)
8. *Geospiza* Darwin's finches	Ecologically divergent taxon forms	Beak morphology and body size	Foraging niche, competition	Experiment	Manipulative experiment	(Ratcliffe and Grant 1983, Podos 2001, Huber et al. 2007, Grant and Grant 2008a, Hendry et al. 2009b)
9. *Hypoplectrus* marine hamlet fishes	Variable color morphs and species	Color pattern	Aggressive mimicry	Observational	Observational	(Puebla et al. 2007)
10. *Lycaeides* butterflies	Wet-meadow-adapted *Lycaeides idas* and dry-habitat-adapted *L. melissa*	Wing color pattern	Unclear	Observational	Manipulative experiment	(Fordyce et al. 2002)
11. *Gasterosteus* sticklebacks	Unimodal solitary populations	Diet	Foraging niche, competition	Observational	Observational	(Snowberg and Bolnick 2008)
12. *Loxia curvirostra* crossbill birds	Different "call types"	Foraging rate, performance	Foraging niche	Experiment	Manipulative experiment	(Benkman 2003, Snowberg and Benkman 2009)
13. *Carpodacus mexicanus* house finch	Native Sonoran desert and urban areas	Bill morphology	Foraging niche	Experiment	Observational	(Badyaev et al. 2008)
14. *Hippocampus subelongatus* western Australian seahorse	A variable population	Body size	Mating system linked to male pregnancy	Observational	Observational	(Jones et al. 2003)

Table 2.5. *Cont.*

Study system	Divergent forms	Putative magic trait(s)	Putative cause of selection	Criterion 1: ecological selection[1]	Criterion 2: mate choice[2]	References
15. Mormyridae African weakly electric fish	Different electric discharges	Electric organs discharge	Electrolocation, communication	Observational	Manipulative experiment	(Feulner et al. 2009)
16. *Satsuma* snails	Chiral forms	Direction of shell coiling ("chirality")	Snake predation	Experiment	Experiment	(Hoso et al. 2007, Hoso and Hori 2008)
17. *Hyalella azteca* amphipods	Size ecotypes	Body size	Presence or absence of fish predation	Observational	Observational	(McPeek and Wellborn 1998)
18. *Rhinolophus philippinensis* horseshoe bats	Different size morphs	Echolocation	Ability to attack different types of prey during foraging	Observational	Observational	(Kingston and Rossiter 2004)

[1] For selection, these categories in order of increasing strength of evidence were: (1) observational evidence stemming from trait divergence between habitats, often bolstered by functional considerations; (2) experimental evidence stemming from measurements of selection on the trait, but where manipulations were not applied to rule out selection on correlated traits; and (3) manipulative experiments were used to control for correlated traits, demonstrating the trait itself was subject to selection.

[2] For mate choice, these categories were: (1) observational evidence stemming from assortative mating based on the trait in nature or indirect inferences about preferred trait values during mate choice; (2) experimental evidence stemming from mate choice experiments in the lab, but where manipulations were not applied to rule out mate choice on correlated traits; and (3) manipulative experiments were used to control for correlated traits, demonstrating that the trait itself affected mate choice.

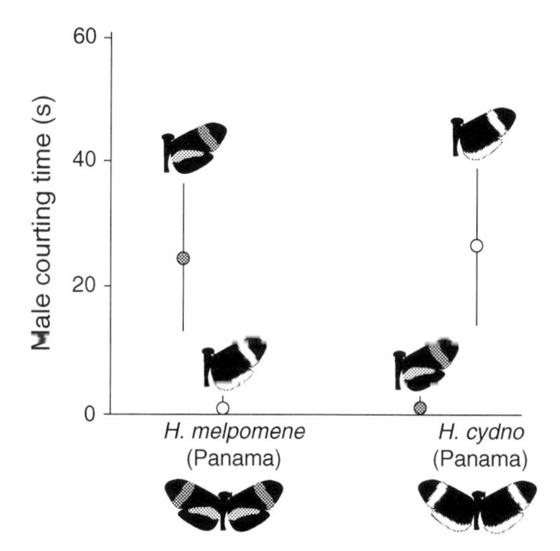

Figure 2.7. An example of a "magic trait": mimetic color patterns in *Heliconius* butterflies are under divergent selection to adapt to different models, and these color patterns also affect mate choice such that individuals prefer to mate with individuals of the same color pattern. Shown here is the mean (±95% confidence intervals) time spent courting live females in ten-minute trials. Similar results were observed using paper color-pattern models, confirming that mate choice was based upon color pattern. Modified from Jiggins et al. (2001) with permission of the Nature Publishing Group.

color patterns favored in different mimicry rings. In summary, both halves of the required evidence are present in this case (Jiggins 2008 for review), whereas in most other potential examples of magic traits (e.g., body size and shape in fish; Fig. 2.8), more work is needed.

A final point is that classic magic traits may also arise when divergent selection acts directly on a mating preference, rather than on a mating cue (Maan and Seehausen 2011; see Fig. 1 of Servedio et al. 2011). For example, if different local environments exert different selection pressures on the sensory system, then these may in turn affect the way individuals perceive potential mates. Viewed this way, putative examples of speciation by "sensory drive" involve magic traits (reviewed in Chapter 3; e.g., see Table 3.1). Thus, many potential examples of magic traits can be found, but in almost all cases more work is needed to provide unequivocal evidence.

2.2.3. Magic traits in the lab

Despite the large number of experimental evolution studies examining the evolution of reproductive isolation in response to divergent selection (Table 2.2), there are essentially no experiments in which the traits underlying reproductive isolation were

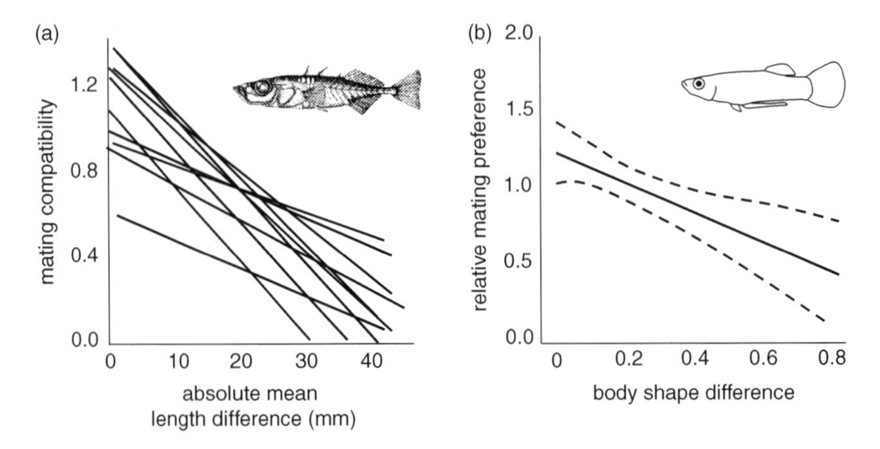

Figure 2.8. Putative "magic traits" in fish. a) Body size as a magic trait contributing to reproductive isolation between marine and freshwater forms of *Gasterosteus* stickleback fishes. Mating compatibility is shown as a function of the mean length difference between individuals. A separate line is drawn for each female population tested. Modified from McKinnon et al. (2004) with permission of the Nature Publishing Group. Stickleback drawing provided by L. Nagel. b) Body shape as a magic trait contributing to reproductive isolation between ecotypes of *Gambusia* fishes living in predator-present versus predator-absent environments. Shown is the relationship between mating preference and body shape differences between sexes. Modified from Langerhans et al. (2007) and with permission of John Wiley & Sons Inc. Drawing provided by B. Langerhans.

identified. An exception concerns an experiment with *Drosophila serrata*, in which Rundle et al. (2005) explored the role of divergent selection between environments in the evolution of both male mating traits (in this case cuticular hydrocarbons) and female mating preferences. Specifically, replicate populations of *Drosophila serrata* were propagated in one of three resource environments: two novel environments and the ancestral laboratory environment. Adaptation to both novel environments involved changes in male cuticular hydrocarbons (some of which predict mating success) and changes in female mating preferences. Further experiments of this kind are needed.

2.2.4. Magic traits: conclusions

In summary, until further definitive examples emerge, it is premature to make firm conclusions about the frequency and importance of magic traits. Nonetheless, evidence is accumulating that magic traits are "magic" in the sense of effectively promoting speciation, but not in the sense of being rare in nature (Servedio et al. 2011). However, it is important to note that there are numerous counterexamples to magic traits; i.e., cases in which traits under divergent selection do not affect mate choice. For example, color pattern, body size, and body shape in host-plant ecotypes of *Timema cristinae* are all subject to strong divergent selection between host species, but none of these traits contributes to sexual isolation (Nosil 2004, 2007). It would be of great interest to determine the proportion of traits subject to divergent selection that incidentally affect sexual isolation. The greater the proportion, the more likely that the process of ecological speciation will be a common phenomenon. Ultimately, however, the importance of magic traits for speciation will depend on the amount of reproductive isolation that results from their presence (Servedio et al. 2011), i.e., their "effect size" (Nosil and Schluter 2011). Magic traits that strongly affect components of premating isolation that evolve early in the speciation process could result in large increases in reproductive isolation, and thus be of critical importance for speciation. In contrast, magic traits that only weakly affect components of premating isolation that evolved late in the process will cause little increase in reproductive isolation, and could be of trivial importance.

2.3. Fitness-based approaches (selection = RI)

Ecological speciation predicts ecologically based selection against immigrants and hybrids. Thus, testing whether such selection reduces the fitness of immigrants and hybrids represents an approach for detecting ecological speciation. In such cases, reproductive isolation is the direct consequence of divergent selection (Rundle and Whitlock 2001, Nosil et al. 2005). I treat these scenarios separately from magic traits because they can be implemented by measuring fitness of different classes of genotypes (parental, hybrids, backcrosses, etc.), even if underlying traits causing reduced fitness are unknown, and because they, unlike magic traits, pertain to both pre- and postmating isolation. It is important to consider selection against immigrant and hybrids, because they represent true tests of ecological speciation, and may represent the usual manner in which ecological speciation proceeds, especially in the early stages of the process when non-ecological forms of reproductive isolation are yet to evolve (Hendry 2004, Gavrilets and Vose 2005, Thibert-Plante and Hendry 2009, Feder and Nosil 2010).

2.3.1. Immigrant inviability

Selection against immigrants ("immigrant inviability") is expected to occur whenever there is divergent adaptation (Nosil et al. 2005, Tobler et al. 2009). I review this process in more detail in Chapter 4, which considers different forms of reproductive isolation. Here I provide two examples for illustrative purposes. The first involves striped versus unstriped color-pattern morphs of *Timema cristinae* walking-stick insects, in which a manipulative field experiment showed that the relative survival of these different color-pattern morphs in the face of visual predation depends on which host-plant species they are resting (Nosil 2004) (Fig. 2.9). Natural populations of these insects adapted to different host plants have diverged in color-pattern morph frequencies such that immigrants onto a non-native host are more likely to be the

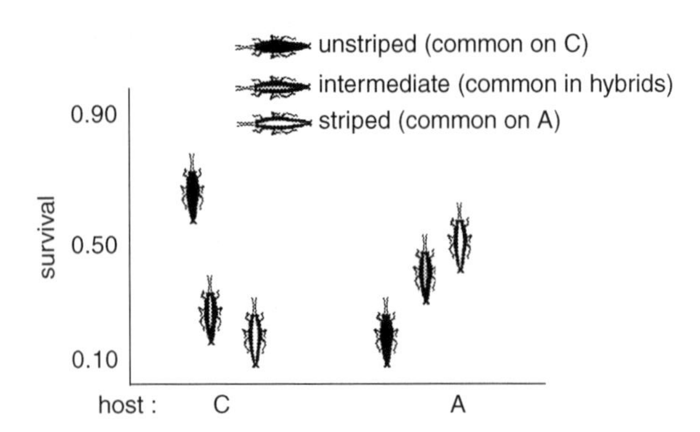

Figure 2.9. An example in which divergent selection on color pattern, inferred using mark–recapture methods, probably contributes to reproductive isolation between populations of *T. cristinae* walking-stick insects. Populations on different host-plant species have diverged in color-pattern morph frequencies, with the unstriped morph being common on *Ceanothus* (C), the striped morph being common on *Adenostoma* (A), and intermediate phenotypes being common in "hybrids." Thus, selection against less-cryptic immigrants and hybrids probably reduces gene flow between populations adapted to different hosts. Selection in the absence of visual predation (not shown) was weak and independent of color-pattern morph, confirming predation as the agent of selection. Modified from Nosil (2004) with permission of the Royal Society of London.

locally less-cryptic morph than are residents. Collectively, these data indicate that high rates of visual predation on less-cryptic immigrants are likely to reduce encounters, and thus interbreeding, between host-associated populations. However, it yet remains to be demonstrated that such selection against less-cryptic individuals actually reduces gene flow between populations in nature. Another example involves river- and beach-spawning populations of sockeye salmon (*Oncorhynchus nerka*) in Lake Washington (Hendry et al. 2000, Hendry 2001, Hendry et al. 2007). Populations in these different habitats differ in traits, probably reflecting local adaptation. Specifically, river females are larger than beach females, river males are shallower-bodied than beach males, and river embryos survive better than beach embryos at incubation temperatures typical of the river. Hendry et al. (2000) report that adult physical dispersal is higher than gene flow inferred using microsatellite loci, such that selection against immigrants probably reduces gene flow.

2.3.2. Ecologically dependent selection against hybrids

Ecologically dependent postmating isolation can arise because of a mismatch between intermediate hybrid phenotypes and the environment, resulting in hybrids exhibiting low fitness in the environment of each parental species. It is important to note that ecological selection against hybrids need not necessarily arise, because hybrids can exhibit high fitness in intermediate environments, as has been observed in big sagebrush (*Artemisia tridentata*) (Wang et al. 1997) and in Darwin's finches (Grant and Grant 2002). Additionally, unequivocally demonstrating ecological selection against hybrids is difficult, as one must disentangle the effects of intrinsic genetic incompatibilities (which can arise via any mechanism of speciation) versus true ecologically dependent hybrid fitness (Rundle and Whitlock 2001, Rundle 2002, Egan and Funk 2009). A more systematic review of this topic is provided in Chapter 4, but I consider a few examples here for illustrative purposes. The first stems from ecologically dependent reductions in the fitness of hybrids between limnetic and benthic forms of sticklebacks (Hatfield and Schluter 1999, Rundle 2002). Hybrids between the limnetic and benthic forms exhibit high fitness in the laboratory. In contrast, the fitness of hybrids in the wild is reduced relative to parental forms. Using various types of hybrid crosses, it was shown that this reduction was a direct result of their intermediate phenotype and was not caused by genetic incompatibilities between the two forms. Other potential examples involve mimetic color patterns of *Heliconius* butterflies (Mallet and Barton 1989, Mallet et al. 1990) and cryptic color patterns of *Timema* walking-stick insects (Nosil 2004). In such cases, intermediate color patterns of hybrids render them susceptible to visual predation, causing ecologically based postmating isolation (Fig. 2.9).

2.4. Gene-flow-based approaches (isolation-by-adaptation)

A gene-flow-based approach for testing ecological speciation involves measuring variation among population pairs in levels of neutral genetic differentiation (e.g., F_{ST}) at loci not linked to those under selection. All else being equal, the following pattern is predicted under ecological speciation: the degree of adaptive phenotypic divergence between populations will be positively correlated with the degree of molecular genetic differentiation (Foll and Gaggiotti 2006, Grahame et al. 2006, Faubet and Gaggiotti 2008) (Fig. 2.10). I will refer to this pattern as "isolation-by-adaptation" (IBA, following Nosil et al. 2008, 2009a). This pattern is expected because divergent selection during ecological speciation reduces gene flow between populations (Barton and Bengtsson 1986, Pialek and Barton 1997, Gavrilets and Cruzan 1998, Gavrilets 2004). In turn, this gene-flow reduction potentially allows neutral divergence across the genome via genetic drift. For a given degree of geographic separation, more adaptively divergent populations are expected to experience greater gene-flow reduction and associated neutral differentiation than less adaptively divergent populations, leading to IBA. The pattern of IBA is analogous to the well-known pattern of isolation-by-distance (IBD, Wright 1943, Slatkin 1993, Rousset 1997), in which genetic differentiation increases with geographic distance, rather than adaptive divergence. Thus, IBA should be tested for while controlling for the geographic distance between population pairs and can be thought of as a molecular signature of ecological speciation (Thibert-Plante and Hendry 2010).

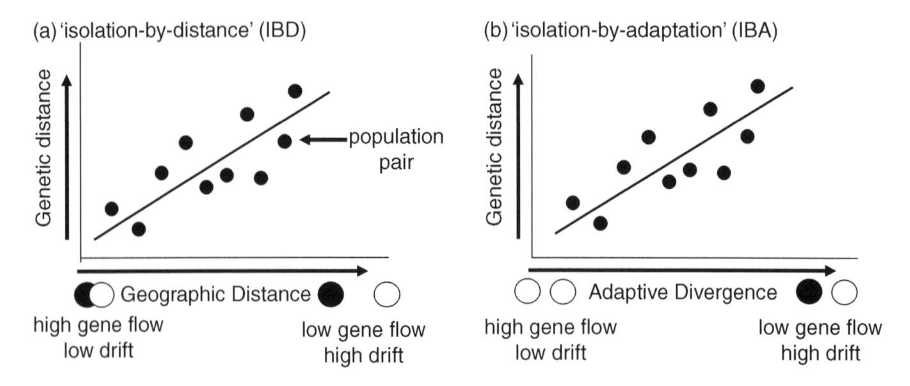

Figure 2.10. Hypothetical examples of isolation-by-distance (IBD) and isolation-by-adaptation (IBA) at putatively neutral loci. a) IBD. Neutral gene flow decreases with increasing geographic distance between population pairs. Thus, genetic distance increases with geographic distance. b) IBA. Neutral gene flow decreases with increasing adaptive divergence ("adaptive distance") between population pairs. Thus, genetic distance increases as levels of adaptive divergence increase, representing a molecular signature of ecological speciation (Nosil et al. 2008). When testing for IBA, geographic distance should generally be experimentally or statistically controlled for.

Reductions in gene flow that generate IBA could arise via various mechanisms, including selection against immigrants and hybrids (Mallet and Barton 1989, Funk 1998, Via et al. 2000, Hendry 2004, Nosil et al. 2005, Tobler et al. 2009), which can create a "general barrier" to the spread of neutral alleles between populations (Barton and Bengtsson 1986, Pialek and Barton 1997, Gavrilets and Cruzan 1998, Gavrilets and Vose 2005, Thibert-Plante and Hendry 2009, Feder and Nosil 2010). The effective immigration rate of neutral alleles is slowed even further under assortative mating (Gavrilets 2004, p. 148). In this chapter, only neutral loci unlinked to those under selection are considered, because the pattern of IBA at such loci tests for general barriers to gene flow across the genome.

2.4.1. Isolation-by-adaptation: evidence

Nosil et al. (2009a) reviewed 22 studies relevant to the evaluation of IBA, and of these 15 (68%) showed evidence for IBA, independent from IBD. This survey was not a formal meta-analysis, and many new examples have since emerged (e.g., Vonlanthen et al. 2009). Thus the results should not be overinterpreted. However, at the very least IBA appears not uncommon in nature. Consider some case studies, such as classic ones by Smith et al. (1997) and Thorpe and Richard (2001), which both report that morphological divergence was correlated with neutral molecular distance, independent from geographic distance, in a rainforest passerine bird and an island lizard, respectively. Similarly, Ogden and Thorpe (2002) report a primary role for ecological divergence, rather than geographic distance, in the

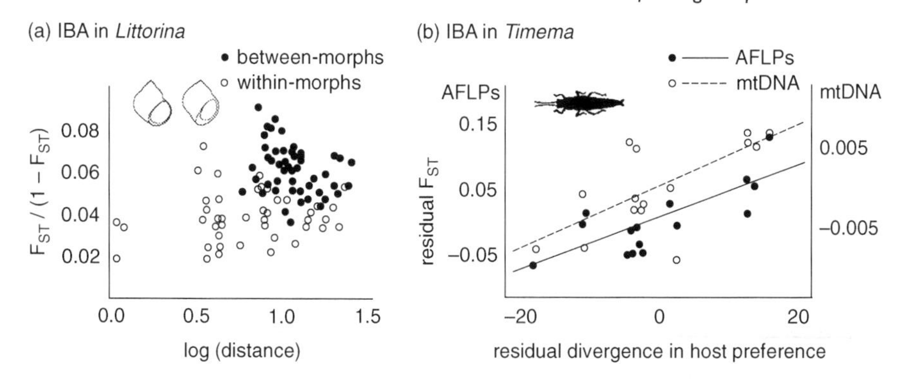

Figure 2.11. Empirical examples of IBA at putatively neutral molecular markers. a) Greater differentiation at putatively neutral AFLP markers between versus within shore ecotypes of *L. saxatilis* intertidal snails, for a given geographic distance. Modified from Grahame et al. (2006) with permission of John Wiley & Sons Inc. b) Evidence for IBA in *T. cristinae* at putatively neutral AFLP loci (filled circles, solid line) and mtDNA (open circles, dashed line), based on analyses controlling for geographic distance. Modified from Nosil et al. (2008) with permission of John Wiley & Sons Inc.

neutral genetic divergence of a Caribbean lizard (*Anolis roquet*). For putatively neutral amplified fragment length polymorphism (AFLP) loci, Grahame et al. (2006) report higher F_{ST} between *Littorina* snail ecotypes relative to within-ecotype comparisons, independent of geographic distance (Fig. 2.11). Finally, studies on two herbivorous insect systems each report IBA among populations associated with either of two host plants (Nosil et al. 2008, Funk et al. 2011). The observation that IBA is not uncommon in nature suggests that selection is strong and greatly reduces gene flow, because even small amounts of gene flow will overwhelm the ability of drift alone to cause differentiation at neutral loci unlinked to those under selection (Wright 1931, 1940, Barton and Bengtsson 1986, Thibert-Plante and Hendry 2009, Feder and Nosil 2010). Alternatively, the loci examined are actually linked to those under selection.

2.4.2. Isolation-by-adaptation at putatively neutral genes: caveats

Some studies tested for IBA but failed to detect it (e.g., Crispo et al. 2006). A number of explanations could explain a lack of IBA. First, adaptive divergence and ecological speciation might be occurring, but selection does not reduce gene flow strongly enough to leave a signature in the neutral genome (Saint-Laurent et al. 2003, Hendry and Taylor 2004, Smith et al. 2005, Crispo et al. 2006, Yatabe et al. 2007, Nosil 2009, Thibert-Plante and Hendry 2009, Feder and Nosil 2010). Second, low gene flow can also preclude IBA (Stelkens and Seehausen 2009b). For example, at a spatial scale greater than that at which gene flow occurs (i.e., allopatric divergence), neutral divergence between any type of population pair can occur without barriers to gene flow caused by divergent selection. Indeed, simulations have shown that detection of IBA with neutral loci is most likely when migration rates are intermediate (Thibert-

Plante and Hendry 2010). Third, the particular phenotypic and ecological traits evaluated might be poor proxies for the major sources of divergent selection acting on study populations, resulting in a lack of detection of IBA.

In conclusion, IBA probably represents a relatively one-sided test of ecological speciation. If a robust pattern of IBA is detected in the face of gene flow, one can conclude fairly confidently that ecological speciation is occurring, because variation in gene flow represents the predominant force affecting levels of genetic differentiation, even under low migration rates (Wright 1931, 1943, Slatkin 1993, Beaumont and Nichols 1996, Rousset 1997, Balloux and Lugon-Moulin 2002, Hedrick 2005). In contrast, failure to detect IBA does not exclude the possibility that ecological speciation is occurring and leaving little or no signature in the neutral genome (Feder and Nosil 2010, Thibert-Plante and Hendry 2010).

2.4.3. Mosaic hybrid zones, phylogeographic studies, divergence-with-gene-flow models

I mention here three other gene-flow-based approaches that can provide evidence consistent with ecological speciation: (1) mosaic hybrid zones; (2) phylogeographic tests for genetic breaks coincident with ecological discontinuities; and (3) divergence-with-gene flow studies generally. All three approaches can documents patterns analogous to IBA.

The first approach concerns mosaic hybrid zones, which are defined as hybrid zones that are spatially patchy, rather than clinal (Rand and Harrison 1989). When a close correspondence is observed between particular genotypes and discernible environmental patches, the patchiness of a hybrid zone is most readily attributed to some combination of active habitat preference and ecological selection against immigrant alleles (Barton and Hewitt 1985). A review of mosaic hybrid zones by Nosil et al. (2005) found that 20 of 27 (74%) zones identified had neutral genetic structure that was habitat-associated (i.e., IBA), as predicted by ecological speciation.

The second approach involves studies that test for associations between ecological divergence and phylogeographic breaks. The concept is similar to IBA, but is generally not implemented in the explicit context of examining correlations among multiple population pairs. An example comes from tropical reef fishes (wrasses) in the genus *Halichoeres*. In a survey of mtDNA sequences of five species, Rocha et al. (2005) report strong phylogeographic partitions between adjacent and ecologically distinct habitats, but high genetic connectivity between similar habitats separated by thousands of kilometers. The concordance of phylogeographic partitions with habitat types, rather than conventional biogeographical barriers, is consistent with ecological speciation. A similar analogy can be drawn for the width of clinal hybrid zones. Under ecological speciation, the width of hybrid zones should be related to the rate of geographic change in the environment (Dice 1940, Blair 1950, Endler 1977, Barton and Hewitt 1985, 1989). Quantitative tests for such a pattern in clinal zones are warranted.

The third type of data involves gene flow generally. The logic is that if divergence can be shown to have occurred in the face of gene flow, then selection must be involved in order to counter gene flow (Rice and Hostert 1993, Kirkpatrick and Ravigné 2002, Niemiller et al. 2008, Nosil 2008b). The problem with this approach is that it might be unknown whether selection is ecologically based and divergent between environments. Nonetheless, demonstrating moderate to high gene flow during divergence might rule out mutation-order speciation, the main alternative to ecological speciation. The logic here is that gene flow will move different mutations (alleles) between populations and the one mutation with the highest selective advantage will fix in all populations, precluding mutation-order speciation (Barton 2001, Church and Taylor 2002, Kondrashov 2003, Gavrilets 2004, Price 2007, de Aguiar et al. 2009, Schluter 2009, Nosil and Flaxman 2011). Nosil and Flaxman (2011) used simulation models to test this prediction. They studied two patches that experience uniform selection for the same optimum. They found that strong genetic divergence (e.g., fixation of different alleles in different populations) via the mutation-order process can occur only when there is very little or no gene flow (Fig. 2.12). However, their models also demonstrate that it is important to know

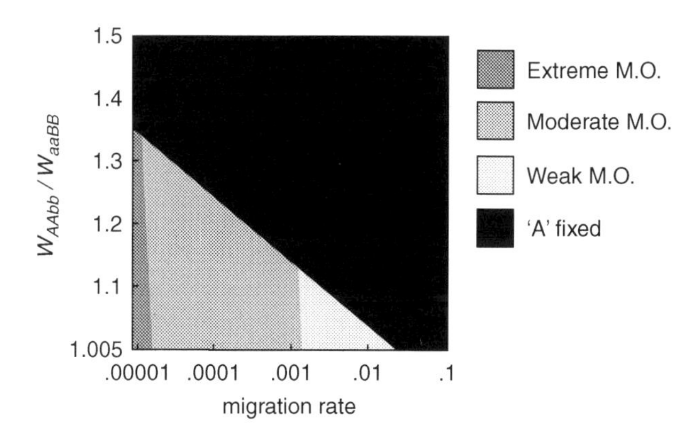

Figure 2.12. A two-locus simulation model of mutation-order (M.O.) speciation showing that gene flow (i.e., migration rate) strongly constrains M.O. divergence. Each of two patches is initially fixed for the ancestral alleles *a* (at locus 1) and *b* (at locus 2). During the course of the simulations, two new mutations arise, denoted *A* and *B*. Both new mutations confer a selective advantage over the ancestral genotype, with the condition that *A* always has a greater selective advantage than *B*. "Hybrids" carrying both *A* and *B* alleles exhibit reduced fitness. The *y*-axis represents the ratio of the fitness advantage each of two new mutations (*A* and *B*) has over ancestral alleles (*a* and *b*). The differently shaded regions correspond to different outcomes. Extreme M.O. = fixed or essentially fixed differences between populations. Moderate M.O. = alternative, incompatible alleles reached and maintained frequencies >0.9, but <0.99, in their respective patches. Weak M.O = *A* and *B* both maintained but divergence was weaker than for Moderate M.O. "*A*" fixed = allele *A* fixed in both patches (and thus allele *B* was eliminated). Results shown here are for simultaneous emergence of the two mutations, weak selection against hybrids, and no period of allopatric divergence preceding gene flow. Qualitatively similar patterns were observed for other parameter combinations. Modified from Nosil and Flaxman (2011) with permission of the Royal Society of London.

whether gene flow occurred during divergence, because allopatric divergence before the onset of gene flow can allow incompatible alleles to diverge and then be maintained upon secondary contact (Barton 2001, Nosil and Flaxman 2011).

2.5. Phylogenetic shifts method

A final method for testing ecological speciation is to estimate the proportion of speciation events in phylogenetic trees that are accompanied by an ecological change (Crespi et al. 2004). Surprisingly, the first such analysis applied to multiple study systems emerged only recently, where Winkler and Mitter (2008) report that roughly half the speciation events between 145 sister-species pairs of phytophagous insects were accompanied by a shift in host-plant use. This represents the fraction of speciation events that *could* have been driven by ecological shifts, but provides no way of knowing if ecological divergence actually accompanied speciation, or arose before or after it. This problem of the timing of an ecological shift in relation to the speciation process is illustrated by a detailed study of a group of sawflies. In this study, roughly 50% of speciation events were estimated to involve niche shifts, but a time-calibrated analysis revealed that immediately after speciation only 22% of sister-species pairs had non-overlapping niches (Nyman et al. 2010). The discrepancy between these estimates is most likely explained by post-speciational host shifts, which can inflate the apparent frequency of ecological speciation in time-uncorrected comparisons. This issue warrants careful attention in future studies adopting phylogenetic approaches. Nonetheless, the results from these studies are thus consistent with a major, but not ubiquitous, role for ecological divergence in insect speciation. These phylogenetic methods are indirect compared with experimental methods, but can be applied to most groups where ecological and phylogenetic information exists.

2.6. Inferring causality when testing for ecological speciation

A general issue when testing ecological speciation is causality: does the degree of adaptive divergence affect levels of gene flow, or vice-versa? This pressing question has been the focus of a number of recent articles (Hendry et al. 2001, Hendry and Taylor 2004, Nosil and Crespi 2004, Räsänen and Hendry 2008, Nosil 2009). As reviewed in this chapter, there are numerous examples now where inverse correlations have been detected between the degree of adaptive phenotypic divergence between populations and levels of genetic exchange between them (e.g., Mayr 1963, Riechert 1993, Sandoval 1994b, King and Lawson 1995, Lu and Bernatchez 1999b, Crespi 2000, Nosil and Crespi 2004, Bolnick and Nosil 2007, Bolnick et al. 2008, Nosil 2008a). Indeed, such correlations could represent ecological speciation. However, reverse causality, in which gene flow constrains adaptive divergence, could also explain such correlations (Fig. 2.13). Data are required to disentangle the extent to which gene flow constrains adaptive divergence, adaptive

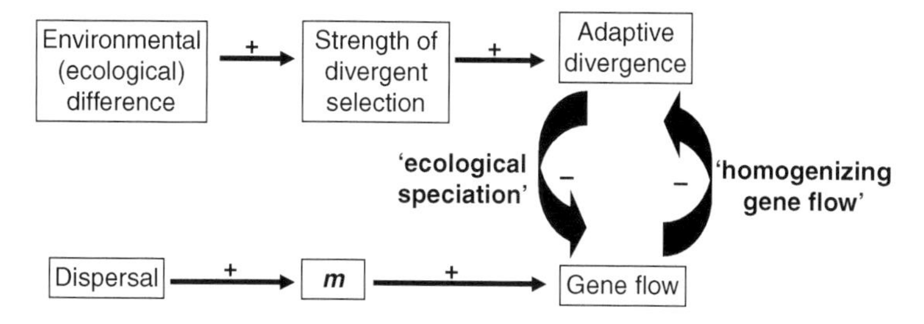

Figure 2.13. A schematic representation of some of the main factors influencing the relationship between adaptive divergence and gene flow. "+" and "–" symbols indicate positive and negative relationships, respectively. *m* represents the proportion of immigrants into a population. Modified from Räsänen and Hendry (2008) with permission of John Wiley & Sons Inc.

divergence constrains gene flow, or both occur (Räsänen and Hendry 2008). A number of approaches have been proposed, but few are yet to be implemented.

The first involves finding traits under selection and measuring whether they affect reproductive isolation. If a trait can be shown to be subject to divergent selection and to also affect reproductive isolation, it is likely that adaptive divergence in that trait reduces gene flow. Thus, tests of ecological speciation involving magic traits or selection against immigrants and hybrids may be less prone to erroneous causal interpretation than tests for IBA.

Another approach is to experimentally manipulate levels of gene flow and levels of adaptive divergence, and then examine how these factors affect each other. For example, one might cause a cessation of gene flow between natural populations. Only two such experiments exist, neither was highly replicated, but both report that gene flow constrains divergence (Riechert 1993, Nosil 2009). To test the reverse scenario, one could simulate secondary contact between pairs of populations that differ in their degree of adaptive divergence. The prediction is that levels of gene flow, which could be measured using molecular markers, will be inversely correlated with the degree of adaptive divergence upon secondary contact. To my knowledge, no such experiment has been conducted. Another approach is to generate experimental treatments that manipulate the strength of selection (Schluter 1994, Bolnick 2004, Nosil and Crespi 2006b) and then measure whether gene flow varies among treatments. The key point is to manipulate the factors of interest experimentally to infer causality.

A different solution involves studying only pairs of populations that are known to undergo no gene flow owing to geographic barriers to genetic exchange. In such scenarios, divergent selection can drive the evolution of reproductive isolation, but the absence of gene flow precludes an effect of gene flow on adaptive divergence. However, this approach requires experimental estimates of reproductive isolation, because neutral genetic divergence between allopatric populations provides little or no information on the degree of reproductive isolation. A final approach involves measuring multiple factors (selection, physical dispersal, gene flow, degree of phenotypic divergence) and then using multivariate approaches such as path analysis to infer causal associations between factors (Räsänen and Hendry 2008).

2.6.1. Divergence after speciation is complete

To maximize the chance that a causal role for divergent selection in speciation is identified, tests for ecological speciation are best implemented using divergent conspecific populations or recently formed species pairs (Drès and Mallet 2002, Funk et al. 2002, Coyne and Orr 2004). This approach avoids the study of post-speciational divergence; but that is not to say that examining species pairs that are somewhat older has not informed our understanding of the factors driving the evolution of reproductive isolation. Consider the seminal paper by Coyne and Orr (1989) that plotted levels of reproductive isolation between species pairs of *Drosophila* against genetic distance (a proxy for time since divergence). Many of the pairs examined were anciently diverged, yet this study generated influential insight, confirming that reproductive isolation increases with time and showing that premating isolation evolves faster than postmating isolation, but only in sympatry (see Coyne and Orr 2004 for a review). Additionally, when studying conspecific populations, there is no guarantee that divergent populations will actually continue to differentiate to form new species. In the end, the most powerful approach is to study many different points along the speciation continuum in an effort to reconstruct how the process of speciation unfolds, from beginning to end. Such an approach forms the focus of Chapter 8.

2.7. Tests and predictions of ecological speciation: conclusions and future directions

The process of ecological speciation generates explicit predictions, and there are numerous ways to test them. Each major class of test outlined in Table 2.1 has been implemented, and support for ecological speciation has been detected. All methods assume that relevant ecological variables or traits were measured. If this is not the case, ecological speciation will not be detected, even if it is occurring.

Four major lines of further work are required. First, each test has been rigorously conducted only a handful of times. It is important to ensure that further tests meet the criteria and assumptions of the methods used. Second, when associations between adaptive divergence and gene flow are robustly identified, a major avenue forward is to address difficult questions concerning causality. Third, only with a larger set of rigorous tests will robust generalities emerge. A particularly important point when searching for generalities is that there is likely to be a strong bias in the published literature, with studies failing to detect ecological speciation never being published (Hendry 2009). Another issue is that there may be an ascertainment bias in published studies, because the taxa involved were chosen *a priori* owing to their obvious ecological differences and suitability for testing ecological speciation (Hendry 2009). In this context, methods that can use random selections of taxa and studies within framework of the ERG approach hold particular promise in the search for generalities. However, random selections of taxa may not allow detailed case studies. Thus, it will ultimately be a combination of detailed case studies and broad comparative studies that yield the most comprehensive understanding. Fourth, there is essentially no experimental evidence that adaptive divergence reduces gene flow in the wild. Almost all existing tests of ecological speciation are correlative (e.g., IBA, parallel speciation). The few manipulative experiments that have been conducted are limited to showing reproductive isolation in the lab or, if conducted in the wild, infer reduced gene flow indirectly from fitness estimates of immigrants and hybrids. Manipulative field experiments testing the extent to which adaptive divergence reduces gene flow are the "final frontier" in tests for ecological speciation.

Part II

Components of ecological speciation

3

A source of divergent selection

Ecological speciation arises because of divergent selection. Thus, a critical question for understanding ecological speciation is: Where does divergent selection stem from? This chapter reviews the three commonly recognized sources of divergent selection: (1) differences between environments; (2) interactions between populations; and (3) ecologically based sexual selection. A general review of divergent selection can be found elsewhere (e.g., Schluter 2000b, pp. 84–122). I thus focus here on how each source of selection might contribute to the evolution of barriers to genetic exchange. Two forms of reproductive isolation—namely, ecologically based selection against immigrants and hybrids—automatically arise when divergent selection acts, and thus are probably major components of ecological speciation. In contrast, it remains less clear how the different sources of divergent selection promote other forms of reproductive isolation; i.e., those that are not direct consequences of selection. I focus on cases in which divergent selection has been explicitly tested, via direct measurements of selection or fitness in two environments. I sometimes also refer to indirect inferences stemming from ecological differences between populations. I conclude the chapter by considering interactions between the different sources of selection.

3.1. Differences between environments

Perhaps the most obvious source of divergent selection stems from differences between environments. Differences among populations in altitude, temperature, habitat, etc., can result in different traits being favored by selection in different populations. As classified here, differences between environments can involve biotic agents such as predators and parasites, so long as environmental differences, rather than interactions between populations per se, are the cause of selection. Consider an example in which there are two habitats that differ in their structure, color, or other environmental characteristics, and divergent selection between the habitats acts on traits related to crypsis and camouflage. This scenario constitutes differences between environments, rather than interactions between populations, because divergent selection on traits arises owing to differences between habitats (e.g., in their structure, color, etc.), rather than via actual direct interactions between predator and prey populations. Although the concept of divergent selection between environments is simple, its actual detection is complicated by the fact that not all traits will be subject to divergent selection. Thus, if divergent selection is not detected, this might represent a true lack of any selection, insufficient statistical power to detect selection on measured traits, or selection acting on traits that were not measured (Fry 1996, 2003).

3.1.1. Examples

A good example of differences between environments concerns insect populations living on different host-plant species (Funk et al. 2002, Matsubayashi et al. 2010). The hosts might vary in their chemistry and morphological structure, resulting in different traits being favored on different hosts. For example, divergent selection might act on digestive and physiological traits related to processing different plant chemicals (hereafter, "selection on physiology"). Reciprocal transplant experiments have been conducted in many groups of insects (Schluter 2000b, Funk et al. 2002, Nosil et al. 2005). In some cases, divergent selection on physiology was detected (Rausher 1984, Via 1984b, a, 1991, Mackenzie 1996, Funk 1998, Sandoval and Nosil 2005, Nosil and Sandoval 2008) (Fig. 3.1), and in other cases it was not (Wiklund 1975, Kibota and Courtney 1991, Gratton and Welter 1998). When selection was not detected, future work could focus on distinguishing among explanations for why this was so. When selection was detected, future work could identify why poor growth and survival on non-native hosts occurred. For example, did low fitness arise because of low levels of ingestion or the inability to digest ingested material?

A lack of selection on physiology does not preclude the existence of selection on other traits. For example, Bernays and Graham (1988) argued that predation might be a major source of divergent selection between host species. Studies of *Timema* walking-stick insects support this claim: all three taxon pairs studied experimentally to date are subject to divergent selection between hosts on cryptic morphology (Sandoval 1994a, Sandoval and Nosil 2005, Nosil and Crespi 2006b, Nosil and Harmon 2009). For example, in a mark–recapture experiment using *Timema cristinae*, divergent selection between hosts on morphological traits was detected, but only in the presence of visual predation (Nosil and Crespi 2006b). Further studies that estimate selection on various types of traits are needed. In particular, traits other than physiology and cryptic morphology warrant examination. For example, selection might act on parasite resistance (Feder 1995, Eizaguirre et al. 2009) or on traits related to biomechanics and maneuvering on different types of plant substrates (Moran 1986, Bernays 1991, Soto et al. 2008). In all cases of divergent host-adaptation, the relevance for ecological speciation increases if links can be made to reproductive isolation (e.g., Grace et al. 2010).

Divergent selection arising from differences between environments has been documented in numerous organisms other than herbivorous insects, including vertebrates and plants. For example, tests for divergent local adaptation have a long history in the botanical literature (Hiesey et al. 1971), and a recent meta-analysis in plants found that divergent local adaptation was reported in 45% of the 1032

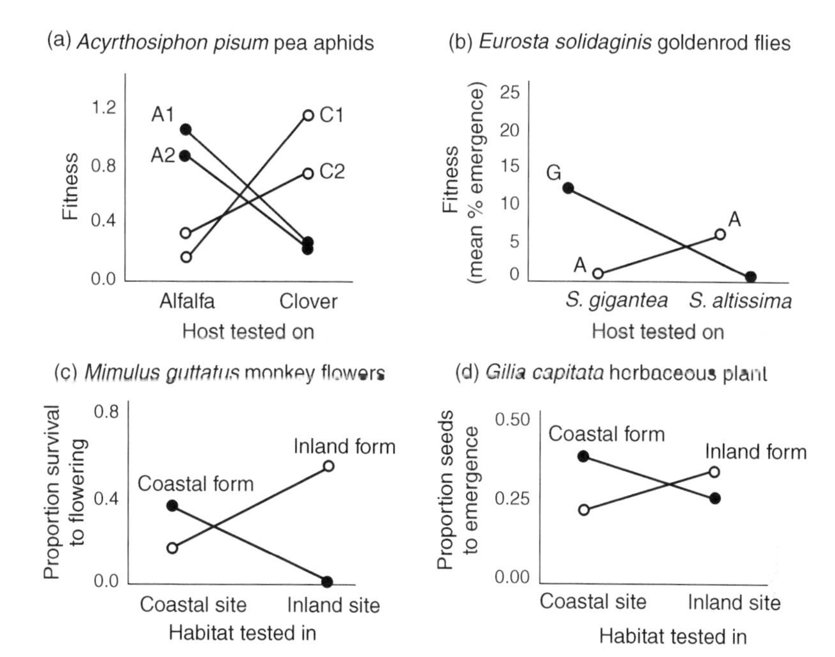

Figure 3.1. Examples of divergent selection due to differences between environments, inferred using reciprocal transplants. In all cases, the y-axis represents fitness. a) Alfalfa and clover host races of *Acyrthosiphon pisum* pea aphids. Shown is mean fitness of aphid clones obtained from two alfalfa fields (A1, A2) and two clover fields (C1, C2), and transplanted between these crops. Fitness is population growth rate. Modified from Via (1991) with permission of John Wiley & Sons Inc. b) Fitness (mean) of *Solidago altissima* (A) and *S. gigantea* (G) host forms of *Eurosta solidaginis* goldenrod flies transplanted to each of these hosts. Modified from Craig et al. (1997) with permission of John & Wiley Sons Inc. c) *Mimulus guttatus*. Modified from Lowry et al. (2008b) and reprinted with permission of Wiley-Blackwell. d) *Gilia capitata*. Modified from Nagy and Rice (1997) with permission of John Wiley & Sons Inc.

population pairs examined (Leimu and Fischer 2008). Another review found that the overall frequency of local adaptation was 71%, and that the magnitude of the native population advantage in relative fitness was 45% (Hereford 2009). The overall results indicate that divergent selection arising from environmental differences is relatively strong and common. A good case study stems from the yellow monkeyflower (*Mimulus guttatus*) in California. In this system, Lowry et al. (2008b) report divergent selection stemming from environmental differences between coastal versus inland environments, related to variation in tolerance to seasonal drought, year-round soil moisture, and salt spray (Fig. 3.1).

3.1.2. Inferring divergent selection from adaptive landscapes

Divergent selection is usually inferred based upon the fitness of individuals, using reciprocal transplant experiments or via measurement of selection in different environments. Another approach concerns reconstructing adaptive landscapes, which examine the relationship between mean population (rather than individual) trait values and mean population fitness. Such landscapes can be reconstructed from ecological mechanisms (e.g., food availability) upward instead of extrapolating from fitness functions (Schluter 2000b). For example, the distribution of seed sizes on the Galápagos Islands was used to reconstruct the adaptive landscape for bill characteristics of granivorous Darwin's finches. The results show multiple distinct peaks of predicted high population fitness. Mean bill characteristics of actual populations corresponded with these peaks (Fig. 3.2) (Schluter and Grant 1984b, Grant 1986, Grant and Grant 2006). Such "rugged" adaptive landscapes with multiple distinct peaks and deep valleys are usually thought of as evidence that divergent selection drove divergence by pulling populations apart to occupy different peaks. Recent work challenges this view by arguing that evolutionary divergence instead occurs via nearly neutral drift along ridges of high fitness in "holey" adaptive landscapes (Fig. 3.3) (Gavrilets 2003, 2004).

3.1.3. An alternative to divergent selection: evolution on "holey" adaptive landscapes

The basic premise underlying the concept of holey landscapes is that phenotypic and genetic divergence in nature is highly dimensional. Gene combinations conferring

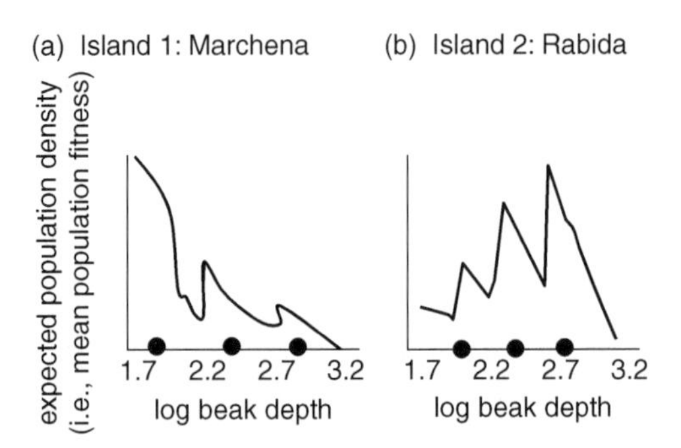

Figure 3.2. Evidence for divergent selection from reconstructions of adaptive landscapes. Depicted lines represent the expected population density of a solitary granivorous finch species on two Galápagos islands. Distinct peaks in this inferred adaptive landscape indicate divergent selection. Black circles on the x-axis show mean log beak depths of actual finch populations. These means correspond to peaks in the landscape, supporting the claim that they evolved via divergent selection. Modified from Schluter and Grant (1984a) with permission of the University of Chicago Press.

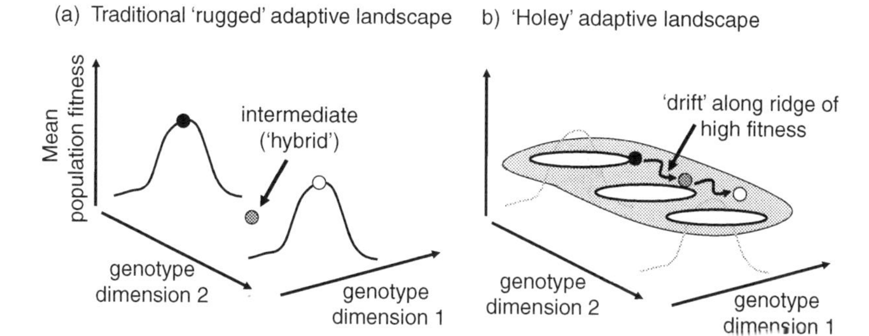

Figure 3.3. Schematic depiction of rugged versus holey adaptive landscapes. a) A traditional rugged adaptive landscape with distinct peaks and valleys. b) A holey adaptive landscape (c.f., Gavrilets' formulation) with ridges of high fitness (light gray oval) and holes representing regions never attained because they represent low fitness or unattainably high fitness. Notice that the fitness of intermediates is low in rugged landscapes, but not always so in holey landscapes.

high fitness in some contexts yield low fitness in others, and thus many different gene combinations have similar overall fitness values (the same argument applies to phenotypic trait combinations, but here I arbitrarily focus on genes). In the context of the metaphor of adaptive landscapes, different combinations with similar fitness form "ridges" of moderate to high fitness, which extend through genotype space across the landscape. Scattered among these ridges are "holes," which represent either: (1) high fitness combinations ("tips" of peaks), which are never achieved because of limits to adaptation—for example, deleterious mutations which push the population downhill or a lack of beneficial mutations; or (2) low fitness combinations ("slopes and valleys") from which natural selection quickly pulls populations up out of, or populations go extinct. Thus, in the metaphor of holey landscapes, peaks and valleys exist, but they are not attained, and the metaphor emphasizes the ridges of relatively high (but not maximal) fitness over other features.

In contrast to rugged landscapes, in holey landscapes population means are not pulled apart by divergent selection; rather, evolutionary divergence occurs via drift along ridges. Evolution is not strictly neutral because the diverging genes are not neutral. However, divergent selection does not pull populations apart to occupy different peaks, and thus differences between populations themselves are neutral, or nearly so.

How might the rugged (divergent selection) versus holey (drift along ridges) views be distinguished? As depicted in Figure 3.3, one prediction concerns the fitness of intermediates between divergent genotypes. Schluter (2000b, p. 121) notes that the concept of divergent selection as a major source of phenotypic differentiation "*goes beyond the proposition that different species are adapted to exploiting different environments. Rather, under divergent selection trait means of populations and species are pulled apart and intermediate phenotypes have lower fitness.*" Rugged landscapes predict low fitness of intermediate forms, because they fall in a fitness valley. In contrast, in holey landscapes intermediates need not exhibit low fitness, because they can fall in other regions of high fitness along ridges.

Most tests for ecological speciation were designed to distinguish divergent selection from no selection (pure neutral genetic drift), and thus did not test the fitness of intermediate forms. Debates about adaptive landscapes cannot be resolved until more studies test intermediate forms and reconstruct landscape features themselves. The few existing studies yield mixed results. For example, Whibley et al. (2006) report a holey landscape in *Antirrhinum* flowers (but see Rausher 2007), whereas McBride and Singer (2010) report a landscape with distinct peaks and valleys. In such tests, the issue of dimensionality is critical, because combinations that confer high fitness in some dimensions may yield low fitness in others, resulting in ridges of similar fitness only when many dimensions are considered.

Finally, the rugged and holey views represent different ends of a continuum, with more intermediate scenarios possible. Holey landscapes might still have some topography, with some regions of ridges being slightly elevated or depressed, because some genetic combinations are slightly, but not radically, better than others. Populations will not occupy global peaks, but will rest on somewhat higher regions of ridges. This view seems realistic, because although the highly dimensional nature of divergence is intuitive, some genetic and ecological combinations will probably be slightly more fit than others, at least in some places and during some time periods.

3.1.4. Linking differences between environments to reproductive isolation

Divergent selection between environments will necessarily lead to two forms of reproductive isolation: ecologically dependent reductions in immigrant and hybrid fitness (Rundle and Whitlock 2001, Rundle 2002, Nosil 2004, Nosil et al. 2005). However, even for these forms, inferences about reductions in gene flow are indirect. There are few data linking specific forms of divergent selection between environments to the evolution of forms of reproductive isolation that are not direct consequences of divergent selection, such as divergent mating preferences or intrinsic postmating isolation. Although some examples of such a link exist, as in the context of the "magic traits," further studies are needed. Finally, many studies testing for divergent selection have not actually identified the phenotypic traits subject to selection. Future studies might focus on doing so, because knowledge of the traits subject to selection will inform how selection contributes to reproductive isolation.

3.2. Interactions among populations

Divergent selection can also arise owing to interactions between co-occurring populations in a single locality. For example, competitive interactions between individuals can generate "disruptive selection," a special form of divergent selection in which both extremes of a phenotype distribution are favored within a single population. Population interactions are distinguished from other sources of divergent selection by two factors. First, they tend to occur only when populations overlap with each other in sympatry, although exceptions could entail parapatry or allopatric populations interacting indirectly via a separate, mobile species. Second, disruptive selection arising from ecological interactions is generally frequency-dependent because an individual's fitness depends on the frequency of other phenotypes in the population (Taper and Case 1992, Schluter 2000b). I focus here on the two scenarios that are the best studied in the context of speciation: selection arising from interspecific competition and selection against maladaptive interbreeding (i.e., hybridization) between populations. However, I touch upon other processes, such as intraspecific competition, apparent competition arising from shared predators, mutualisms, etc.

3.2.1. Competition between closely related species

Disruptive selection can arise from competition between closely related species. The key feature of selection arising from resource competition is that the fitness of phenotypes depends on the total density of individuals and on the frequency of different phenotypes, with competition between individuals increasing with their phenotypic similarity (Roughgarden 1972, Slatkin 1979, 1980, Taper and Case 1992, Doebeli 1996, Schluter 2000b). This can result in common and intermediate phenotypes undergoing stronger competition, and thus exhibiting lower fitness, relative to the tails of the phenotype distribution. For competition to be an important source of disruptive selection, it must generate differences beyond those expected by resource distributions alone. A clear example is where competition generates a highly bimodal phenotype distribution despite a unimodal underlying resource distribution: in effect, competition generates two adaptive peaks from a resource distribution that has a single peak. The effects of interspecific competition on adaptive divergence in general are reviewed elsewhere (Gurevitch et al. 1992, Denno et al. 1995, Schluter 2000b). The main conclusion was that interspecific competition appears to be relatively common in nature. I focus here on summarizing the main evidence, and highlighting how competition might affect ecological speciation. Following Schluter (2000b), I consider three major lines of evidence: observational, predictive, and experimental.

Observational evidence refers to inferences made using patterns of divergence in field populations. The pattern most often used as evidence that interspecific competition promotes divergence is greater divergence between species when they co-occur together (i.e., sympatry = interspecific competition present) versus when they do not (i.e., allopatry = interspecific competition absent). This pattern is called ecological character displacement (ECD) (Pfennig and Pfennig 2005, Rice and Pfennig 2010). Schluter (2000a, b) outlines six criteria that should be met to infer that the pattern of ECD arose from interspecific competition: (1) phenotypic differences between taxa should have a genetic basis; (2) chance should be ruled out as an explanation of the pattern; (3) population and species differences must represent evolutionary shifts and not just species sorting; (4) shifts in resource use should match changes in morphology or other phenotypic traits; (5) environmental differences between allopatric and sympatric sites should be controlled for; and (6) independent evidence should be obtained that similar phenotypes compete for resources. Of 64 cases reviewed for these criteria, 23 satisfied at least four, and five cases fulfilled all six (Schluter 2000b, a). The least commonly satisfied criterion was number 6.

The second line of evidence for the importance of competition stems from a "predictive approach" (Schluter 2000b). This involves the successful prediction of mean phenotypes in models that incorporate interspecific competition. This approach requires an estimate of selection stemming from resources alone. One then tests whether the incorporation of competition into the model significantly improves predictive abilities over that expected from resources alone. This approach has only been implemented a few times (Schluter and Grant 1984a, Schluter 2000b, pp. 147–150 for review).

The third approach involves experimental studies of competition, which are relatively numerous, but generally restricted to species exhibiting character displacement (CD) (Schluter 2000b). One approach tests for the "ghost of competition past": because similar phenotypes compete most strongly, competition between species should decline as character divergence proceeds, yielding descendants whose present-day interaction is a "ghost" of its former strength. Two studies have documented such an effect (Pritchard and Schluter 2001, Gray and Robinson 2002) (Fig. 3.4). Other experimental studies have reported that competition

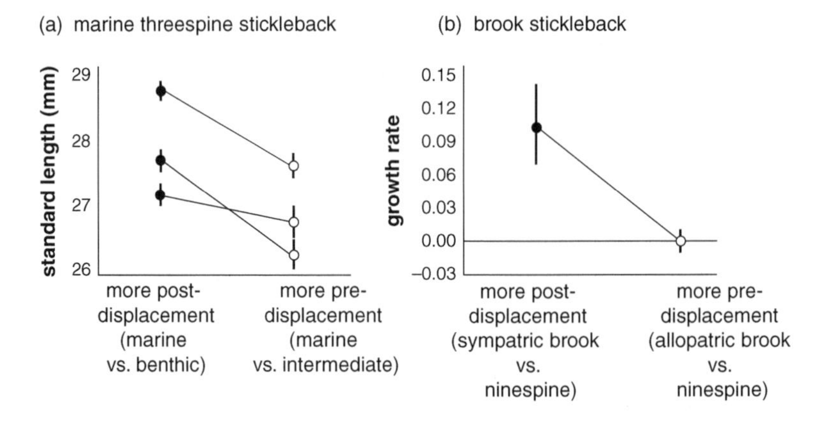

Figure 3.4. Examples of the ghost of competition past. *Y*-axes represent measures of fitness and *x*-axes represent scenarios where the focal population was competed against either a population that had already undergone character displacement (more post-displacement) or a population that had not yet undergone strong character displacement (more pre-displacement). a) Fitness of marine sticklebacks when competed against benthic versus intermediate forms. Modified from Pritchard and Schluter (2001) with permission of the authors. b) Fitness of brook sticklebacks competed against ninespine sticklebacks when the brook sticklebacks derive from populations sympatric versus allopatric with ninespine sticklebacks. Modified from Gray and Robinson (2002) with permission of John Wiley & Sons Inc.

increases, or fitness decreases, with increasing phenotypic similarity (Schluter 1994, Pfennig et al. 2007). Finally, there is some direct evidence for frequency-dependent selection arising from competition (Schluter 2003, Tyerman et al. 2008). In all instances, effects on reproductive isolation were not examined.

3.2.2. Intraspecific competition within a single population

Disruptive selection can also arise when common and phenotypically intermediate individuals within a species compete more strongly for resources than those at the tails of the distribution, resulting in lower fitness of intermediate phenotypes. In this manner, individuals who undergo niche expansion onto new and less-exploited resources might exhibit the highest fitness (Van Valen 1965, Roughgarden 1972, Slatkin 1980, Wilson and Turelli 1986, Taper and Case 1992, Doebeli 1996, Bolnick 2001). Theory predicts that such a process, acting within a species, can drive adaptive divergence and speciation (Rosenzweig 1978, Dieckmann and Doebeli 1999, Kondrashov and Kondrashov 1999, Gavrilets 2004, Bolnick and Fitzpatrick 2007). In fact, disruptive selection underlies the process of "adaptive speciation," which is a special case of ecological speciation. From an empirical perspective, intraspecific competition has received less attention than competition between species, some notable exceptions aside (Fig. 3.5) (Smith 1993, Bolnick 2004, Bolnick and Lau 2008, Calsbeek and Smith 2008, Calsbeek 2009, Hendry et al. 2009b).

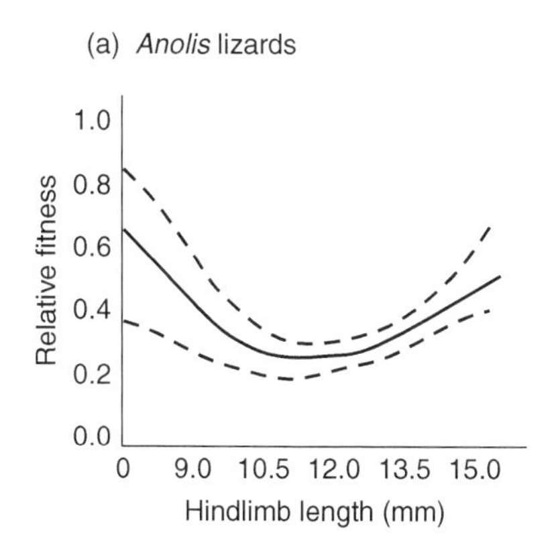

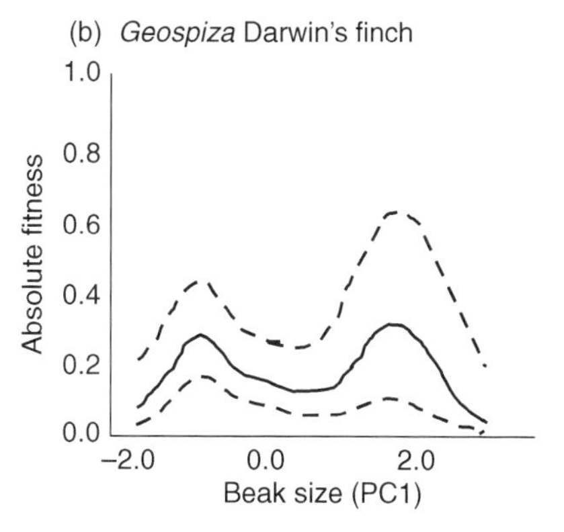

Figure 3.5. Examples of disruptive selection within natural populations. In all cases, the fitness functions were estimated using the cubic spline approach of Schluter (1988) and dashed lines represent 95% confidence intervals from bootstrap replicates. a) *Anolis sagrei* lizards. Relative fitness is probability of survival to maturity. Modified from Calsbeek and Smith (2008) with permission of John Wiley & Sons Inc. b) *Geospiza fortis* medium ground finch. Fitness stems from interannual recaptures. Modified from Hendry et al. (2009b) with permission of the Royal Society of London.

A critical point is that the existence of disruptive selection is not necessarily evidence that such selection is driven by intraspecific competition (Abrams et al. 2008). For example, high fitness of both tails of the phenotype distribution could arise simply from a bimodal resource distribution (i.e., "differences between environments"). In this case, disruptive selection might be considered divergent selection, with each mode in the resource distribution representing a different "environment." As for the case of interspecific competition, to know that intraspecific competition drove divergence, it must be shown that it generates divergence beyond that expected via the resource distribution alone. One way to test whether intraspecific competition generates disruptive selection is to test whether the strength of disruptive selection increases with rising population density. This prediction was supported in *Gasterosteus aculeatus* stickleback fishes, where Bolnick (2004) created paired high- versus low-density treatments using enclosures in four natural lakes, and found that disruptive selection was consistently stronger in the high-density treatment (Fig. 3.6). Multi-generation laboratory experiments using microorganisms have also shown that frequency-dependent competition can drive sympatric, ecological diversification of single strains of asexual taxa (see Buckling and Rainey 2002, Friesen et al. 2004, Meyer and Kassen 2007, Tyerman et al. 2008, Kassen 2009 for reviews), but direct implications for ecological speciation are limited because reproductive isolation does not apply.

Another prediction of divergence driven by disruptive selection is that the rate of adaptive divergence will increase as the strength of intraspecific competition increases. This could occur, for example, if intraspecific competition favors niche

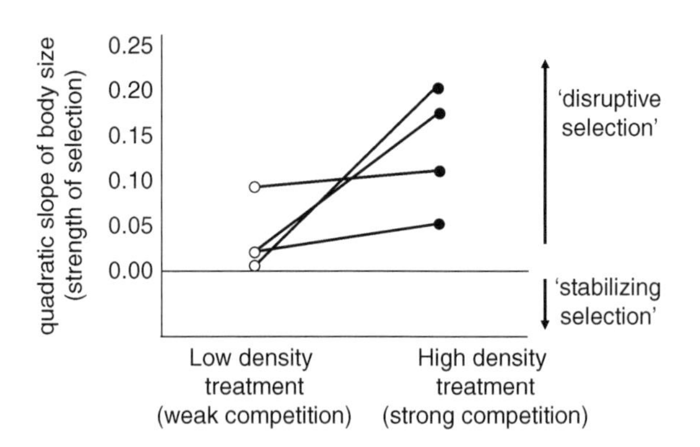

Figure 3.6. Evidence from a field experiment that intraspecific competition generates disruptive selection within four lake populations of *Gasterosteus aculeatus* stickleback fishes. The *y*-axis represents values for the quadratic slope of the fitness function, with increasingly positive values indicative of stronger disruptive selection. Here, relative body size was used as the proxy for fitness, but similar trends were observed for another fitness proxy: relative gonad mass. Lines connect the slopes from high and low density enclosures paired within a given lake ($p < 0.05$ in a paired t-test). Modified from Bolnick (2004) with permission of John Wiley & Sons Inc.

width expansion onto new resources for which competition is less severe (Roughgarden 1972). An experimental evolution study in *Drosophila melanogaster* supported this prediction, by demonstrating that adaptation to a novel resource was faster in high- versus low-intraspecific-competition treatments (Bolnick 2001). Despite these elegant experiments, it remains unclear when disruptive selection can actually split a single unimodal distribution into two, thus driving speciation. For example, in the same stickleback system discussed above, a population known to be subject to strong disruptive selection remains a phenotypically unimodal, single species (Snowberg and Bolnick 2008). Future work needs to focus on what is required for disruptive selection to actually drive speciation (Rueffler et al. 2006)

3.2.3. Reinforcement

"The grossest blunder in sexual preference, which we can conceive of an animal making, would be to mate with a species different from its own and with which the hybrids are either infertile or, through the mixture of instincts and other attributes appropriate to different courses of life, at so serious a disadvantage as to leave no descendants." (Fisher 1930, p. 130).

Another type of interaction between populations that can generate divergent selection is selection against heterospecific (or between-population) mating. If heterospecific mating reduces the fitness of the individuals involved or their hybrid offspring, selection will favor the genes of individuals that mated within their own population (and thus different sets of genes might be favored within different populations). This will strengthen prezygotic isolation in a process known, in the broad sense, as reinforcement (see Dobzhansky 1937, 1951, Servedio and Noor 2003 for a review). Reinforcement predicts reproductive character displacement (RCD): increased prezygotic isolation in geographic regions where hybridization occurs (sympatry or parapatry) relative to regions where it does not (allopatry). Many cases of RCD have been documented (e.g., Noor 1995, Saetre et al. 1997, Rundle and Schluter 1998, Pfennig and Ryan 2006, Ortíz-Barrientos et al. 2009), yet few studies have distinguished among alternative hypotheses to reinforcement, or the role of different types of selection against hybridization (Butlin 1995, Servedio and Noor 2003).

Although reinforcement features prominently in many models of speciation, it is difficult to categorize because it can be involved in any mechanism of speciation, ecological or not (Schluter 2001). Thus, whether reinforcement is a component of ecological speciation depends upon the specific circumstances. If the cost to heterospecific mating originated from divergent selection and ecological causes, then it is reasonable to consider reinforcement as a component of ecological speciation (Kirkpatrick 2001). How often reinforcement has an ecological basis in nature is not well understood, although this should be common when populations are divergently adapted, such that there is ecological selection against hybrids. Indeed, ecological causes have been implicated in the reinforcement of *Poecilia reticulata* guppies (Schwartz et al. 2010), *Gasterosteus aculeatus* stickleback fishes (Rundle and Schluter 1998), and *Timema cristinae* walking-stick insects (Nosil et al. 2003). In the latter two examples, it has been shown that hybrids between ecological forms appear fully fertile and viable in the lab, but exhibit intermediate phenotypes that exhibit low fitness in the environment of each parent species. At face value, this suggests reinforcement due to ecological selection against hybrids. However, theory predicts that other costs to hybridization, such as direct costs to hybridizing females, can also contribute to mate preference evolution (Kirkpatrick and Ryan 1991, Kirkpatrick 1996, Servedio 2001, Servedio and Noor 2003), and indeed such costs occur in the stickleback and walking-stick systems (Nosil et al. 2003, Albert and Schluter 2004, Nosil et al. 2007). Thus, understanding the contribution of reinforcement to ecological speciation will require careful consideration of all costs to heterospecific mating and the mechanisms (ecological or not) by which they evolved.

3.2.4. Distinguishing competition from reinforcement

Separating the effects of reinforcement and competition on ecological speciation may be difficult because both occur in sympatry due to the interaction of populations and can produce the same evolutionary outcome: stronger prezygotic isolation between sympatric than allopatric populations (Servedio and Noor 2003). The extent of this problem will not be known until we determine how frequently prezygotic isolation is strengthened as a by-product of ECD. The problem could be serious, as exemplified by studies of *Gasterosteus aculeatus* stickleback fishes, in which multiple processes simultaneously contribute to character displacement: reinforcement was implicated in RCD of mating preferences (Rundle and Schluter 1998, Albert and Schluter 2004), competition in ECD of trophic traits (Schluter and McPhail 1992, Schluter 1994, 2003), and predation in CD of armor (Vamosi and Schluter 2004) (Fig. 3.7). Although different processes can yield similar patterns, thereby confounding each other's effects, the different processes might also complement one another. For example, ECD due to competition could result in reduced hybrid fitness (e.g., Rice and Pfennig 2008), which then drives reinforcement. Studies of reinforcement are beginning to consider these issues. The results of one study suggest that ECD was not involved in reinforcement (Nosil et al. 2003), and two others attempted to control for its contribution (Rundle and Schluter 1998, Albert and Schluter 2004). The control facilitated by laboratory experiments may be especially useful for testing how different interactions between populations drive speciation (Higgie et al. 2000).

3.2.5. Other interactions: mutualisms, tri-trophic interactions, and coevolution

Interactions not covered above could also generate divergent selection. One major class is indirect interactions such as "apparent competition," where two species interact indirectly with each other because they are both preyed upon by the same

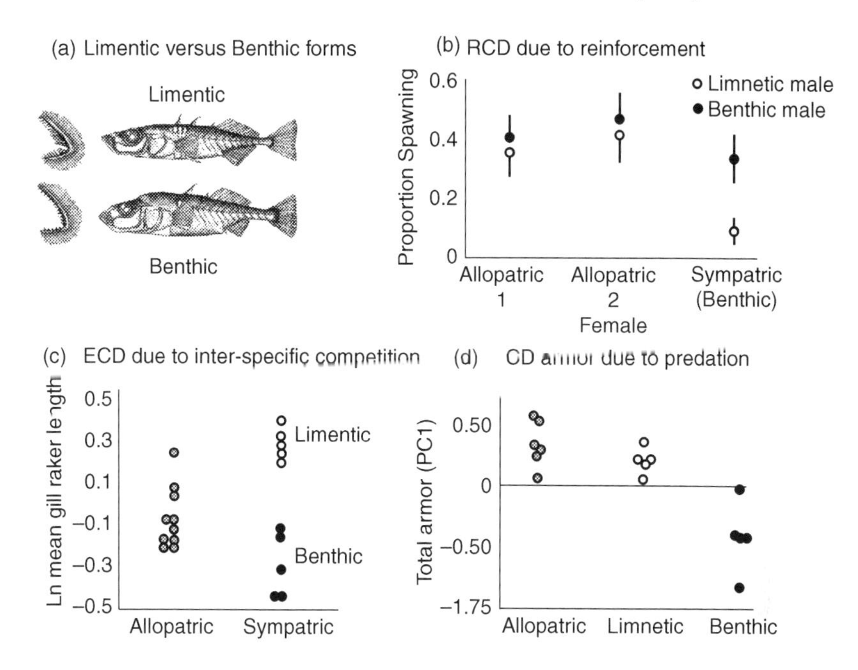

Figure 3.7. Multiple types of interactions between species might contribute to phenotypic divergence and speciation within a single system, exemplified here by sympatric limnetic and benthic forms of *Gasterosteus aculeatus* stickleback fishes. a) Illustrations of the two forms, courtesy of L. Nagel. b) Reproduction character displacement (RCD) due to reinforcement. Modified from Rundle and Schluter (1998). c) Ecological character displacement (ECD) in trophic traits due to interspecific competition. Modified from Schluter and McPhail (1992). d) Character displacement (CD) in defensive armor due to predation. Modified from Vamosi and Schluter (2004). All reprinted with permission of John Wiley & Sons Inc.

predator (Holt 1977). If one species increases in population size, this can then incidentally increase the predation rate on other prey species in the same community. Apparent competition has been well examined theoretically (Holt and Lawton 1994, Abrams 2000), but empirical studies linking it to divergent selection and speciation are essentially absent. Empirical studies might yield interesting results, as apparent competition is predicted to often yield convergence, rather than divergence, and thus might sometimes act to oppose other sources of divergent selection. Likewise, other types of interactions, such as mutualisms, exploitative interactions, and harmful interactions such as interspecific aggression or cannibalism deserve further attention in speciation research. Some interest is emerging: for example, Janson et al. (2008) provide a conceptual framework for how insect-microorganism mutualisms might affect divergent selection. Further work on all these interactions is sorely needed.

I conclude by considering two case studies in which interactions other than competition or reinforcement have been examined in the context of speciation.

The first concerns interactions across trophic levels. The basic idea is that as new species form, they create new niches for others to exploit, leading to a cascade of speciation events, potentially across trophic levels (Emerson and Kolm 2005, Erwin 2005). This idea is perhaps best exemplified by tri-trophic interactions between plants, herbivorous insects, and insect parasites (see Abrahamson and Blair 2008 for a review). As insects adapt divergently to different host plant species, they diverge in morphology, physiology, behavior, etc. In turn, parasites and parasitoids specializing on different herbivorous insect forms are subject to divergent selection between insect host forms. The adaptive divergence of the parasitoid, in response to differences between insect host forms, can then drive speciation of the parasites. An example of this process comes from the parasitic wasp *Diachasma alloeum*, which has diverged into partially reproductively isolated ecotypes as a result of specializing on apple versus hawthorn host forms of *Rhagoletis pomonella* flies (Forbes et al. 2009).

The second example concerns coevolutionary arms races. In a classic paper, Ehrlich and Raven (1964) proposed that coevolution between plants and their insect herbivores drives evolutionary radiation in both systems. In this scenario, plants that evolve defenses against herbivores are freed from attack, and diversify, as do any insect lineages that can subsequently colonize and adapt to these new "empty niches." This cycle of continual "escape and radiation," and coevolution, was proposed to drive diversification. Despite the inherent appeal of this model, empirical support remains elusive and mixed (Futuyma and Agrawal 2009a, b, Fordyce 2010). An example in which divergent selection and coevolution appear to promote speciation stems from arms races between crossbill birds and pine trees. In this system, the birds and pines exhibit a series of adaptations and counteradaptations aimed at increasing foraging efficiency versus avoiding seed predation, respectively (Benkman et al. 2001). Divergent selection as a result of coevolution between South Hills crossbills (*Loxia curvirostra* complex) and Rocky Mountain lodgepole pine (*Pinus contorta latifolia*) promoted the evolution of reproductive isolation between crossbill populations (Smith and Benkman 2007).

3.2.6. Linking interactions between populations to reproductive isolation

Interactions between populations have rarely been linked directly to the evolution of reproductive isolation. The exceptions are aforementioned cases of ecologically driven reinforcement, where links to sexual isolation are inherent, and cases where traits involved in competitive interactions also affect sexual isolation, as documented in cichlid fish (Seehausen and Schluter 2004, Dijkstra et al. 2006, Dijkstra et al. 2007, Young et al. 2009).

Another possibility is ecologically dependent postmating isolation between populations that are found in the presence versus absence of a competing species. The logic is that populations sympatric with heterospecific competitors experience a different competitive environment than allopatric populations that are not facing a competitor, and thus that populations from these different competitive environments diverge in resource traits. In turn, ecologically dependent postmating isolation arises if offspring produced by matings between individuals from different competitive environments are competitively inferior. A test of this hypothesis stems from a study examining populations of the Mexican spadefoot toad (*Spea multiplicata*), which exist in the presence versus absence of a competing species, the plains spadefoot toad (*S. bombifrons*) (Pfennig and Rice 2007). Crosses were made between *S. multiplicata* from different populations adapted to the same competitive environments (e.g., both sexes from allopatry), and between individuals from different competitive environments (one sex from allopatry and one from sympatry). Offspring derived from between-competitive environment crosses generally exhibited lower fitness than those derived from crosses within competitive environments, with this effect accentuated in a high-competition treatment (Fig. 3.8). In sum, postmating isolation appears to have arisen as a by-product of interactions between species, and there is molecular evidence that this actually reduces gene flow between competitive environments (Rice and Pfennig 2010).

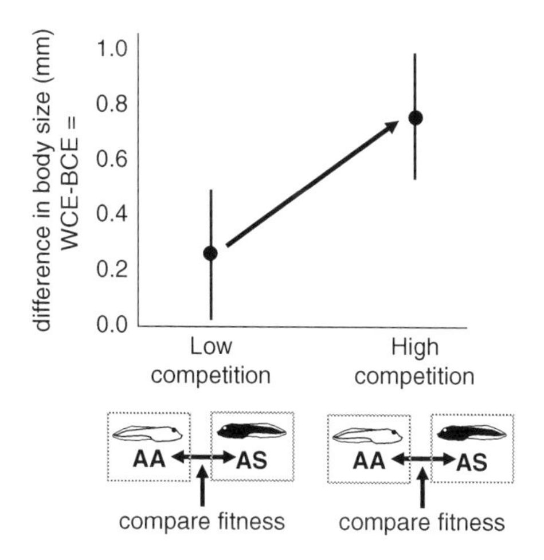

Figure 3.8. Populations of *S. multiplicata* are found in the presence (S, sympatry) versus absence (A, allopatry) of a competing species (*S. bombifrons*), creating two distinct competitive environments. Two types of crosses were considered: (1) those between different *S. multiplicata* populations in the same competitive environment (WCE, within competitive environment: AA, offspring with both parents from allopatry); versus (2) those between *S. multiplicata* populations in different competitive environments (BCE, between competitive environment: AS, offspring with one parent derived from each environment, in the example depicted, the mother was from allopatry). The fitness of offspring from these crosses was then compared in high- and low-competition treatments (individuals raised with competitors or alone, respectively). Positive values on the *y*-axis indicate larger size of WCE relative to BCE offspring. Reduced fitness (i.e., smaller relative size) of BCE offspring was evident in all crosses, but was accentuated in the high-competition treatment. Modified from Pfennig and Rice (2007) with permission of John Wiley & Sons Inc.

3.3. The functional morphology and biomechanics of divergent selection

Although divergence in form and function is implicit in the preceding discussions of the sources of divergent selection, it was not explicitly considered. Functional and biomechanical divergence is expected to underlie sources of divergent selection during ecological speciation. For example, differences between environments might result in divergent selection on traits related to foraging, swimming, running, chewing, etc. Thus, biomechanical considerations can help predict functional morphology in different environments, such as how body shape will diverge in relation to the type of swimming performance required (e.g., sustained or burst swimming) (see Langerhans and Reznick 2009 for a review) and how beak characteristics might diverge in relation to the need for crushing versus chewing, or due to a trade-off between bite force and velocity (Herrel et al. 2009). When selection acts on functional morphology, interrelationships among traits can affect patterns of trait divergence. For example, in *Gasterosteus* stickleback fishes, structural interrelationships between different anti-predatory traits, such as bony lateral plates and spines, can help increase the anti-predator function of each, resulting in specific combinations of traits being particularly favored (Reimchen 1983). The role of functional interrelationships between traits in promoting versus constraining speciation requires attention. If divergence in functional traits affects reproductive isolation, ecological speciation is promoted (see Podos and Hendry 2006 for a review).

3.4. Environmentally dependent sexual selection

The third ecological source of divergent selection involves sexual selection. Because sexual selection acts on traits involved in mate recognition, it may be a powerful force in speciation (Panhuis et al. 2001). Speciation models involving sexual selection can be classified into two types, depending on whether or not differences in mate signals and preferences evolve because of divergent selection between environments (Schluter 2000b, Boughman 2001, Schluter 2001, Boughman 2002). Models involving divergent selection between environments include habitat-specific selection on secondary sexual traits (Lande 1982) and mating or communication systems (Ryan and Rand 1993, Boughman 2002). For example, if habitats differ in their signal transmission properties, different trait values for mating signals might accrue high reproductive success in different habitats, resulting in divergence in signals and preferences (Morton 1975, Endler 1992, Boughman 2002). In contrast, examples of sexual selection that do not necessarily involve divergent selection between environments are models in which sexual selection arises primarily from the interaction of the sexes, such as Fisher's runaway (Lande 1981) and sexual conflict (Rice 1998, Chapman et al. 2003, Arnqvist and Rowe 2005, Sauer and Hausdorf 2009). Sexual selection can thus be involved in both ecological and non-ecological speciation (Schluter 2000a, 2001, Nosil et al. 2007, Maan and Seehausen 2011).

3.4.1. Divergent sexual selection through sensory drive

Divergent sexual selection might contribute to ecological speciation in numerous ways (see Schluter and Price 1993a, van Doorn et al. 2009, Maan and Seehausen 2011). I focus here on the best-known case of "sensory drive" (following Endler 1992, 1993), drawing heavily from a past review (Boughman 2002). In a technical sense, sensory drive is *"the integrated evolution of communication signals, perceptual systems, and communication behavior because of the physics of signal production and transmission, and the neurobiology of perception"* (Boughman 2002, p. 571). More generally, it is the hypothesis that signals evolve to maximize their detectability: easy-to-detect signals are favored by natural selection. In turn, different signals can be most detectable in different environments, because environments differ in their transmission properties. For example, habitat transmission can vary with structural features of the habitat such that a signal that is conspicuous (easy to detect) in one habitat is inconspicuous in another, or a signal that transmits well in one habitat is heavily degraded in another. Thus, divergence in signals and preference between populations in different habitats might occur, thereby generating sexual isolation. I consider hereafter the case of male traits being signaled to female receivers, but the arguments apply to most signal-receiver scenarios.

Speciation by sensory drive considers two main phenomena: (1) signal and preference evolution within populations; and (2) divergence between populations in different signaling environments. Concerning evolution within populations, sensory drive actually involves three processes. First, habitat transmission arises from the physical interaction of signals with the environment. Signals that contrast with the environment (e.g., dark colors in light habitats) are more detectable. As signals travel, they become degraded. Signals that preserve their characteristics through transmission are most likely to attract mates. I will refer to this process, whereby signals evolve to maximize contrast and minimize degradation, as "effective signal transmission" (Fig. 3.9). Second, the habitat can affect not only signal characteristics, but also receiver perception. Because of local adaptation in perception, females can evolve to be more sensitive to some sound frequencies, wavelengths of light, or smells over others. This process can also take into account other signals in the environment, such as that of predators or prey, and has been called "perceptual tuning." Third, the detectability of a signal depends not only on the transmission properties of the environment, but also on how well the signal is matched to the receiver's perception. This process has been called "signal matching." These three components result in the coordinated evolution of signals and perception.

Further requirements are necessary for sensory drive to promote speciation. First, "effective signal transmission," "perceptual tuning," and "signal matching" must operate in different directions in populations occupying different habitats, such that habitat-specific divergence occurs, and sexual isolation results. The most explicit test for sensory drive speciation thus involves showing that divergence in signals, perception, and preference is correlated positively with the degree of sexual

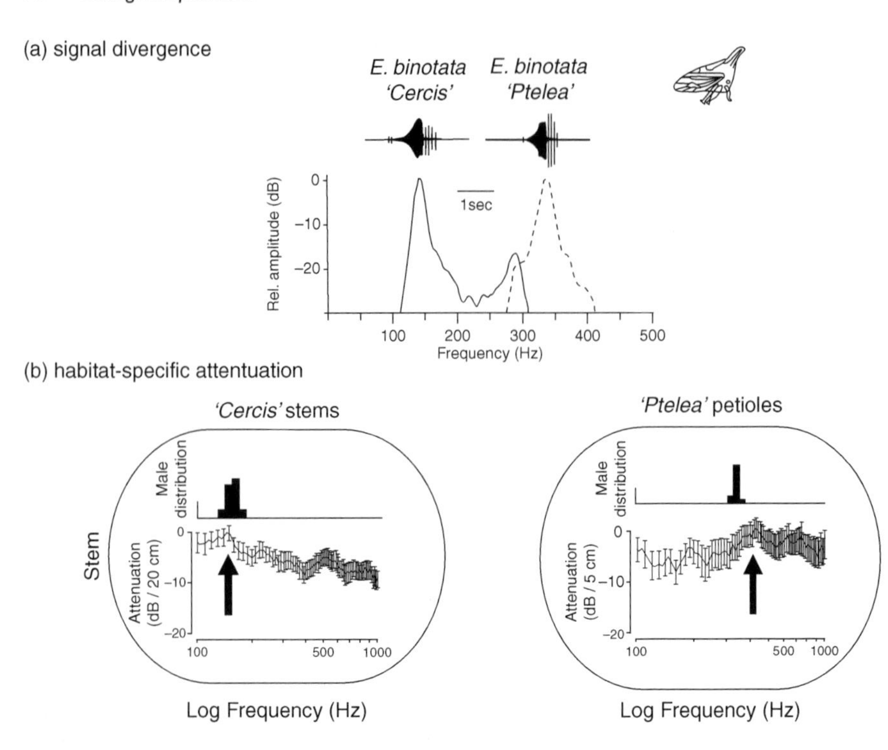

Figure 3.9. Variation in male mating signal frequency between *Cercis* and *Ptelea* host forms of *Enchenopa binotata*. This example focuses on signal divergence. The full process of sensory drive would also involve perceptual divergence, and correlated divergence of signals and perception. a) Signals differ between host forms. Waveforms of each species' signal are shown with the corresponding amplitude spectra showing the frequency difference between forms. b) Matching of male signals to the transmission properties of the environment (shown for the plant part and plant species most commonly used for signaling by each host form). Transmission function curves show differential filtering in *Cercis canadensis* versus *Ptelea trifoliate*, where the value on the *y*-axis represents the efficacy of signal transmission. Histograms above each curve represent the distribution of male signal frequency. Male signals tend to match peak attenuation (denoted by thick upward pointing arrows). Modified from McNett and Cocroft (2008) with permission of Oxford University Press.

isolation (Fig. 3.10). Thus, six criteria emerge for demonstrating sensory drive speciation:

1) habitats differ in transmission properties;
2) signal varies with habitat such that detection of each signal is better in the native habitat (divergent "effective signal transmission");
3) perception varies with habitat (divergent "perceptual tuning");
4) divergence in signals and perception is correlated;
5) preference varies with perception;
6) signal and perceptual divergence results in reproductive isolation.

(a) Variation in male nuptial coloration

absent present extensive

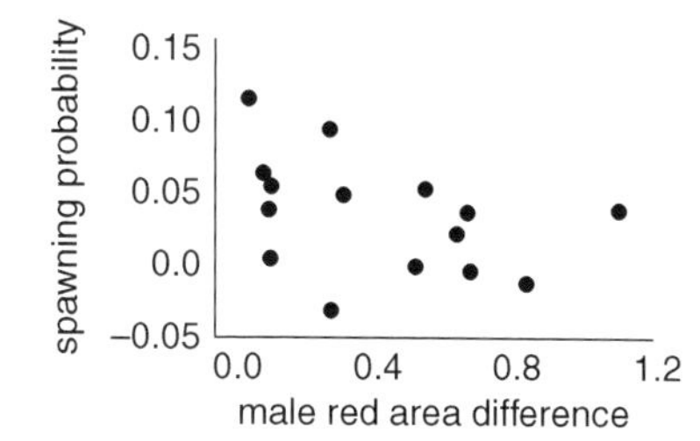

Figure 3.10. Evidence for the evolution of reproductive isolation via sensory drive in *Gasterosteus aculeatus* sticklebacks. a) Variation in red nuptial coloration of male sticklebacks. Modified from Reimchen (1989) with permission of John Wiley & Sons Inc. b) Shown is the relationship between reproductive isolation (y-axis) and divergence among populations in male signal color. A similar relationship was observed between reproductive isolation and female preference for red. Modified from Boughman (2001) with permission of the Nature Publishing Group.

I review examples of sensory drive in Table 3.1, focusing on their fit to the criteria. The list of examples is not exhaustive, but rather was chosen to span a range of organisms and sensory modalities. I find that criterion 6 has been met most rarely. Thus, although numerous candidates for sensory drive speciation exist, only two cases considered in Table 3.1 satisfy all six criteria (i.e., *Gasterosteus* and *Pundamilia* fish). Further evidence that sensory drive affects reproductive isolation is needed.

3.4.2. Examples of divergent sexual selection via sensory drive

A number of empirical examples of sensory drive exist, covering a range of visual, auditory, chemical, and vibrational communication modalities (Table 3.1). One of the best-studied examples concerns nuptial throat color evolution in freshwater populations of *Gasterosteus aculeatus* sticklebacks. In some lakes, males display the ancestral red nuptial color, whereas in other lakes males have lost red coloration and display black. This variation in the expression of coloration is linked to the transmission properties of the environment: the greatest expression of red occurs in habitats with the highest water clarity, whereas the loss of red coloration is generally found in heavily stained waters (Reimchen 1989). Reimchen (1989) thus proposed that this occurs because darker waters mask red coloration (i.e., reduce its contrast against the background), therefore decreasing the efficacy of this signal. This "signal-masking" hypothesis has received support from experiments employing video-imaging (McDonald et al. 1995). Subsequent work on sticklebacks adapted to limnetic versus benthic lake environments revealed that sensory drive is an important mechanism of divergence in male mating signals, female perception of

Table 3.1. Putative examples of sensory drive. See also Boughman (2002). "Ecological context" refers to the putative cause of transmission differences between environments. The six criteria outlined in the text for whether sensory drive contributes to speciation are evaluated here (Y = yes, L = likely, U = unknown, RI = reproductive isolation). Note that criterion six links sensory drive to speciation, but has rarely been definitively met.

Study system	Ecological context	Signal type	1. ΔHT	2. S~H	3. Pe~H	4. S~Pe	5. Pr~Pe (or S)	6. Link to RI	References
1. *Gasterosteus* stickleback fishes	Water color	Color: throat color	Y	Y	Y	Y	Y	Y^{D}	(Reimchen 1989, McDonald et al. 1995, Boughman 2001)
2. *Anolis cristallus* lizard populations	Mesic versus xeric environments	Color: dewlap coloration	Y	Y	U	U	U	U	(Leal and Fleishman 2004)
3. *Anolis cooki* and *A. cristallus* lizard species	Microhabitat light intensity and spectral quality	Color: dewlap coloration, ultraviolet reflectance	Y	Y	Y	Y	U	L^{I}	(Losos 1985, Leal and Fleishman 2002)
4. *Phyllosscopus* warbler species	Forest light intensity/habitat brightness	Color: feather reflectance and number of color patches	Y	Y	U	U	U	L^{I}	(Marchetti 1993)
5. *Carlia* rainbow skinks	Habitat openness	Color: male breeding coloration	U	Y	U	U	U	U	(Dolman and Stuart-Fox 2010)
6. *Pundamilia pundamilia* and *P. nyererei* cichlid fishes	Ambient light in lake environments, along a depth gradient	Color: male coloration (blue to red)	Y	Y	Y	Y	Y	$Y^{D,I}$	(Maan et al. 2006, Seehausen 2008, Seehausen et al. 2008b, Maan and Seehausen 2010)
7. *Neochromis, Mbipia* spp. of cichlid fishes	Ambient light in the lake environment	Color: male coloration (blue to red)	Y	Y	Y	Y	U	L^{I}	(Seehausen et al. 1997, Terai et al. 2006)
8. *Gallotia galloti* lizard	Climate and vegetation	Color: male coloration	U	Y	U	U	U	Y^{I}	(Thorpe and Brown 1989, Thorpe and Richard 2001)

Study system	Ecological context	Signal type	1. ΔHT	2. S~H	3. Pe~H	4. S~Pe	5. Pr~Pe (or S)	6. Link to RI	References
9. Snappers (Lutjanidae)	Water color	Color: unknown (signal not examined)	Y	U	Y	U	U	U	(Lythgoe et al. 1994)
10. *Amblyornis inornatus* bowerbirds	Bower displays, including structure and decoration	Color	U	L	U	U	Y	L^{I}	(Uy and Borgia 2000)
11. *Lucania goodie* bluefin killifish	Water color, clear springs to tea-stained swamps, also canopy cover	Color	Y	Y	Y	Y	U	U	(Fuller et al. 2003, Fuller and Travis 2004)
12. *Telmatherina sarasinorum* polymorphic fish	Water color, beach versus root habitat in the lake	Color: male coloration	Y	Y	U	U	Y	L^{I}	(Gray et al. 2008)
13. Neotropical forest birds (from three genera)	Forest light, shade, gaps in sun, clouds, which also differ by habitat type	Color: plumage coloration	Y	Y	U	U	U	U	(Endler and Thery 1996)
14. *Manacus candei* and *M. vitellinus* manikins	Ambient light, visual background	Color: plumage coloration	Y	Y	U	U	Y	$Y^{I}L^{I}$	(Brumfield et al. 2001, Uy and Stein 2007)
15. *Andropadus virens* little greenbul bird	Ambient noise variation across a rainforest gradient	Song: bird song	Y	Y	U	U	U	Y^{I}	(Smith et al. 1997, Slabbekoorn and Smith 2002b)
16. *Melospiza melodia* song sparrow subspecies	Understory density and heterogeneity	Song: bird song	U	Y	U	U	Y	$Y^{D,I}$	(Patten et al. 2004)
17. *Catharus ustulatus* Swainson's thrush subspecies	Coastal temperate rainforest versus inland coniferous forest habitats	Song: bird song	U	Y	U	U	U	N^{I}	(Ruegg 2008)

Table 3.1. *Cont.*

Study system	Ecological context	Signal type	1. ΔHT	2. S~H	3. Pe~H	4. S~Pe	5. Pr~Pe (or S)	6. Link to RI	References
18. Thamnophilidae antbirds (163 species)	Forest strata; ground, understory, midstory, canopy	Song: bird song	U	Y	U	U	U	U	(Seddon 2005)
19. *Acris crepitans* cricket frogs	Forest habitat	Song: cricket call	Y	Y	L	Y	Y	L^I	(Ryan et al. 1990, Ryan 1991, Ryan and Wilczynski 1991, Ryan 2010)
20. *Enchenopa* host-associated treehoppers	Host-plant species	Vibrational communication	Y	Y	U	U	Y	$Y^D L^I$	(Rodriguez et al. 2004, McNett and Cocroft 2008, Rodriguez et al. 2008)
21. *Nezara viridula* green stinkbug	Host-plant species	Vibrational communication	Y	Y	U	U	U	U	(Miklas et al. 2001, Cokl et al. 2005)
22. *Drosophila mojavensis* cactophilic fruit flies	Host plants	Chemical: epicuticular hydrocarbons	L	Y	U	U	Y	Y^D	(Markow 1991, Stennett and Etges 1997, Etges and Ahrens 2001)
23. *Schizocosa* wolf spiders	Microhabitat, leaf litter complexity and density	Morphology: foreleg ornamentation, but also stridulation	Y	Y	L	Y	Y	L^I	(McClintock and Uetz 1996, Scheffer et al. 1996, Hebets and Uetz 1999)

1. ΔHT, habitats differ in transmission properties (L = likely, but unconfirmed). 2. S~H, signal varies with habitat (L = likely, when signal varies between populations in different habitats, but a quantitative or definitive association between habitat and signal is yet to be established). 3. P~H, perception varies with habitat (L = likely, perception shown to vary between populations with different signals, but perceptual divergence not yet definitively linked to habitat divergence). 4. S~Pe, correlated divergence in signals and perception. 5. Pr~Pe (or S), preference varies with perception (or signal). 6. "Link to RI", data on the association between signal/preference divergence and reproductive isolation (Y^D = direct experimental evidence that signal divergence is positively correlated with the degree of sexual isolation; Y^I, indirect evidence from neutral genetic markers that ecological divergence promotes RI, but RI has not been linked directly to divergence in signal traits/preferences; L^I, indirect experimental evidence signal divergence likely affects aspects that contribute to RI, such as territory size, attractiveness, reproductive success, mate preference, or habitat choice, but the actual degree of RI between populations has not been linked directly to divergence in signal traits/preference; N^I, whether ecological or signal divergence is associated with neutral genetic distance was tested, but a significant association was not found; U, unknown, effect of signal/preference divergence on RI not tested).

color, and female preference for male coloration (Boughman 2001). Specifically, benthics nest in deeper and more densely vegetated areas than limnetics, and thus occupy habitats where the light is more red-shifted. As predicted by the signal-masking hypothesis, male benthics and solitary populations that nest in red-shifted habitats display black nuptial color or reduced red, presumably because the red-shifted light would mask red nuptial color. In addition, female visual sensitivity to red light is lower in populations that nest in dimmer, red-shifted habitats. Thus signals, perception, and preference are correlated such that females with low sensitivity to red light express no preference for red males and females with high sensitivity express strong preference for red males. The extent of divergence in signals and preferences correlates with the extent of reproductive isolation between populations (Fig. 3.10).

Putative examples from different sensory modalities also exist, but these tend to lack evidence that sensory drive generates reproductive isolation. For example, habitat-dependent selection on song characteristics can lead to song divergence between populations (Ryan et al. 1990, Ryan 1991, Ryan and Wilczynski 1991, see Slabbekoorn and Smith 2002a for review, Ryan 2010). Boncoraglio and Saino (2007) conducted a formal meta-analysis of this "acoustic adaptation" hypothesis in birds. Among the 80 studies reviewed, they found a relatively widespread and statistically significant, but weak, effect of habitat structure on bird song. However, direct evidence that this causes reproductive isolation is generally lacking (Table 3.1). In contrast to vertebrates, tests for sensory drive in insects are more rare. An exception concerns host-plant-specific vibrational communication in *Enchenopa* leafhoppers (Rodriguez et al. 2004, McNett and Cocroft 2008, Rodriguez et al. 2008) (Fig. 3.9). In this system, the different host species used by the insects differ in their signal transmission properties, and male signal frequencies on each host match the peak of optimal transmission, thereby maximizing signal detection on each host. This signal divergence generates premating isolation. In two other systems where similar tests have been conducted, one reports matching of insect signals to transmission properties of hosts (Miklas et al. 2001, Cokl et al. 2005), whereas the other reports no such matching (Henry and Wells 2004). McNett and Cocroft (2008) suggest these differences might be explained by the degree of host specificity, with signal divergence being most likely when each insect form feeds on only a single host plant.

A final point is that when mating preferences themselves are subject to divergent selection, as occurs during sensory drive speciation, mate preference itself can act as a "magic trait" (Maan and Seehausen 2011, Servedio et al. 2011). Under this framework, the examples in Table 3.1 constitute further evidence for magic traits.

3.4.3. A note on multimodal mating signals

Many mating signals are multimodal and potentially subject to multivariate selection (Blows et al. 2003, Brooks et al. 2005). Mating signals might involve multiple traits within a sensory modality (e.g., many traits within a visual display) or involve multiple different sensory modalities (e.g., some visual and some auditory traits). For example, subspecies of song sparrow (*Melospiza melodia*) represent a putative example of sensory drive speciation and use both plumage and song in mate choice, but weigh song more heavily (Patten et al. 2004). Likewise, courtship displays of wolf spiders in the genus *Schizocosa* involve simultaneous use of both visual and vibratory signals (Hebets and Uetz 1999). Furthermore, different traits and modalities can be correlated to varying degrees. Little attention has been paid to how the modality or dimensionality of signals affects sensory drive speciation. Likewise, little attention has been paid to how the complexity or the dimensionality of habitats affects sensory drive. Which types of habitats are most likely to generate speciation via sensory drive, those that differ strongly in a single transmission property or those that differ weakly in many aspects of transmission? Given that mating signals can be complex, but are intimately linked to reproductive isolation, future work in this area will likely yield much insight into speciation.

3.4.4. Divergent sexual selection: conclusions and remaining questions

Divergent and ecologically based sexual selection might commonly operate on mating traits and preferences in nature. This finding has major implications for the way we think about both sexual selection and ecological speciation. Concerning sexual selection, if ecology often plays a role, then sexual divergence may not be as arbitrary as argued by non-ecological models of sexual selection. Concerning ecological speciation, pieces of the puzzle are missing, but there is good evidence that sensory drive contributes to the process. Recognition of the ecological context of divergence in sexual traits represents an important advance in our understanding of speciation. That said, only two of the examples in Table 3.1 appear to satisfy all the criteria for sensory drive speciation. A few trends are evident, and shed insight into where further work is needed. First, signals are studied more often than perception. Preferences have often been tied to signal divergence, which indirectly satisfies criterion 4, but further tests for associations between preferences and perception itself are needed. Likewise, most studies noting an association between signal values and habitat type have not taken the further step of demonstrating that signals are actually more detectable within their native habitat. Future work in general should explicitly examine conspicuousness within different habitats, as exemplified by some studies of fish, bird, and lizard coloration (Table 3.1). Finally, the most glaring and critical piece of missing evidence in most systems is that sensory drive generates reproductive isolation.

Relatively unexplored considerations are the factors constraining divergence via sensory drive. For example, conspicuous signals might increase predation risk, as documented in an acoustic signaling moth (Brunel-Pons et al. 2011) and cichlid fish (Maan et al. 2008). Although sensory drive encompasses adaptation to all elements of the signaling environment, factors other than optimal transmission have received almost no attention. It might be that some sensory modalities are less constrained by factors such as predation than others. For example, one could predict that for species attacked primarily by visual predators, chemical cues can undergo sensory drive divergence more easily than visual cues. Another question is the extent to which divergence via sensory drive is constrained by directional preferences for a single male trait value (e.g., bright orange in guppies) (Schwartz and Hendry 2006) or by preexisting biases (Basolo 1995, Basolo and Endler 1995).

Finally, although some work exists (Schluter and Price 1993a, van Doorn et al. 2009), an aspect of sexual selection that needs to be better integrated into thinking about ecological speciation is condition-dependent expression of sexual traits. Environmental conditions can affect the physical condition of individuals, with ecologically maladapted immigrants and hybrids exhibiting lower condition than locally adapted residents (Hamilton and Zuk 1982, Rowe and Houle 1996, Dolgin et al. 2006, Boughman 2007, Servedio 2009, van Doorn et al. 2009). In turn, such low condition of immigrants and hybrids could compromise the expression of their sexual traits, leading to mating discrimination against them. The role of condition-dependence in generating premating isolation deserves further attention.

In general, all these aspects of ecologically based sexual selection are yet to be evaluated in manipulative laboratory experiments, which could be used to distinguish different models of sexual selection (Turelli et al. 2001, Boughman 2002). Additionally, almost all work on sensory drive involves the evolution of mate choice, but the process might also affect premating isolation due to divergent habitat preferences. Tests of this mechanism are lacking.

3.5. Interactions between the different sources of divergent selection

The preceding sections indicate that each major source of divergent selection probably promotes ecological speciation. A relatively unexplored issue is how the different sources of selection interact. I review here a few existing examples, which illustrate how the required further work in this area can feasibly be conducted.

One example concerns the interaction of differences between environments with reinforcement in *Timema cristinae* walking-stick insects. In this system, Nosil et al. (2003) report that sexual isolation was strongest under the combined and relatively additive effects of adaptation to different host plant species and ecologically based reinforcement (Fig. 3.11). A related example stems from a comparative analysis of song structure in 163 species of antbird (Thamnophilidae), where habitat differences and reinforcement were both found to contribute to divergent song evolution (Seddon 2005, see also Schwartz et al. 2010). Theory predicts that the greater the difference between populations upon secondary contact, the more likely that reinforcement will occur (Liou and Price 1994, Kirkpatrick 2001, Kirkpatrick and Ravigné 2002). Thus, it might generally be expected that differences between

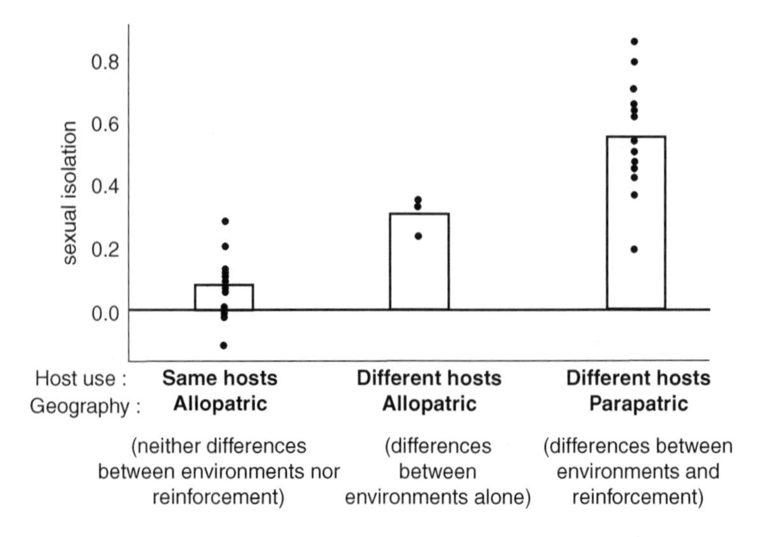

Figure 3.11. The combined effects of differences between environments (i.e., host plants) and ecologically based reinforcement on levels of sexual isolation between pairs of populations of *Timema cristinae* walking-stick insects. On the *y*-axis, zero represents random mating and one represents complete sexual isolation. Bars represent means of population pairs, where each pair is depicted by a black dot. See Nosil et al. (2003) for details.

environments and reinforcement will have combined and complementary effects on promoting speciation.

A more controversial issue is the interaction between predation and competition, where it has been argued that predation may increase, decrease, or have little effect on competition (I use the term "predation" hereafter to refer generally to whole or partial consumption by predators, parasites, disease, etc.). I discuss here some aspects that are particularly relevant for ecological speciation, and refer readers to Chase et al. (2002) for a thorough review of predation and competition. One idea is that predation decreases competition via the reductions in prey density that it causes. Support for this hypothesis stems from a meta-analysis which reports that the effects of competition tend to be weaker in the presence of predation, relative to that observed in its absence (Gurevitch et al. 2000).

Some elegant data on the effects of predation on competition stem from experimental evolution work in microorganisms (Fig. 3.12). For example, Buckling and Rainey (2002) examined the effect of a viral phage (with effects analogous to predation) on diversification of the bacterium *Pseudomonas fluorescens* in spatially structured microcosms. In the absence of phages, bacteria diversified predictably and repeatedly into spatial niche specialists. In the presence of phages, diversity in sympatric bacterial populations was greatly reduced, as a result of phage-induced reductions in bacterial density, which decreased competition for resources. However, in contrast to sympatry, diversity among allopatric populations was increased

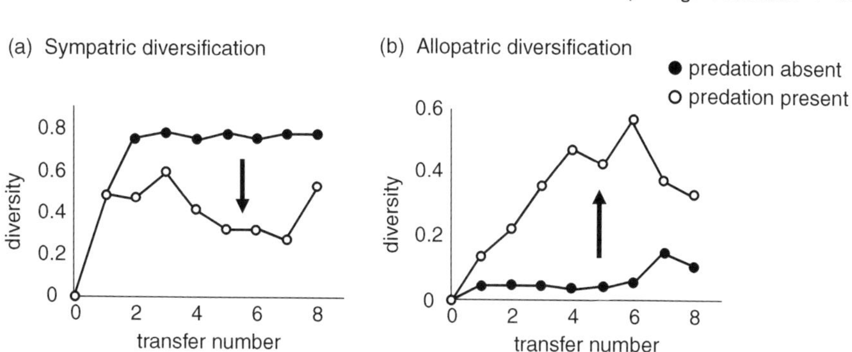

Figure 3.12. Experimental evolution studies in microorganisms examining the effects of the interaction between predation and competition on divergent selection and phenotypic diversification. The diversification of the bacteria *Pseudomonas fluorescens* was studied in the presence versus absence of viral phages. Black arrows indicate whether predation decreased or increased diversity. a) Bacterial diversity within sympatric populations is higher in the absence of phages, presumably because predation reduces bacterial population density and thus weakens competition for resources. b) Bacterial diversity in allopatric populations is higher in the presence of phages, due to phage-induced selection for resistance generating divergent evolutionary trajectories in different populations. Modified from Buckling and Rainey (2002) with permission of the Nature Publishing Group.

in the presence of phages, owing to phage-resistance traits evolving in the genetic background of different niche-specialists in different replicates. Thus, the authors suggest that predation can increase allopatric diversity, while inhibiting sympatric divergence. In another experiment, Meyer et al. (2007) studied how predation by a protist, *Tetrahymena thermophila*, affects competition and diversification in *P. fluorescens*. They report that competition and predation independently generated divergent selection, and each could drive diversification to similar extents. For example, predation increased diversity by promoting the emergence of phenotypes adapted to predation. However, although predation and competition were capable of independently driving prey diversification, the rate of diversification was slowed in the presence of predation, probably due to reductions in prey density. The overall conclusion from microbial studies is that predation can drive divergence in anti-predator traits, while at the same time reducing population density and thus weakening competition (see Kassen 2009 for a review).

Interestingly, the results from microorganisms are in contrast to those from an experiment examining the effects of predation on competition in *Gasterosteus aculeatus* stickleback (Rundle et al. 2003). This study reported that divergent selection on trophic morphology was actually stronger when predation was present, relative to when it was absent (Fig. 3.13). The results further showed that the strength of divergent selection across all treatments was inversely correlated to the strength of competition (see also Abrams et al. 2008). The authors proposed that although predation reduced competition, it promoted habitat segregation, decreased

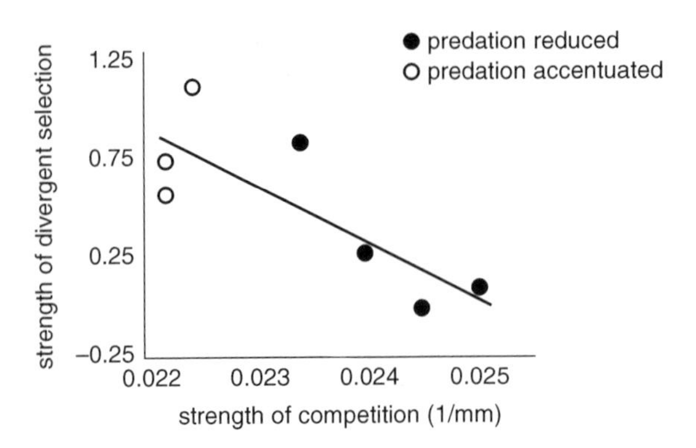

Figure 3.13. For a pond experiment in *Gasterosteus aculeatus* sticklebacks, the relationship between the strength of divergent selection in a pond and the strength of competition in the pond (inferred from the inverse of the average size of fish in the pond such that larger values on the *x*-axis indicate stronger competition). Results are shown for ponds where predation was accentuated (trout predators added and insect predators enhanced) versus where predation was reduced (no trout added and density of insect predators reduced). Predation tended to decrease competition, but result in stronger divergent selection. Modified from Rundle et al. (2003) with permission of the National Academy of Sciences USA.

resource overlap, and increased ecological specialization in a manner that resulted in stronger divergent selection. In other words, "non-consumptive" effects of predation can be important (see Preisser and Bolnick 2008 for a review), with predation sometimes facilitating, rather than constraining, divergence in traits unrelated to predator resistance itself.

A simplified conceptual model of the effects of the interaction between predation and competition on divergent selection and subsequent diversification is presented in Figure 3.14. This model shows how both processes can on their own promote diversification, in foraging and anti-predator related traits, respectively. The model further illustrates how predation can either decrease or increase the effects of competition, depending on the interplay between the effects of predation on ecological specialization versus its effects on prey density.

Further work on interactions between sources of selection is clearly needed in speciation research. In particular, existing studies tend to consider the effects of predation on competition, whereas the reverse effects of competition on predation are less well explored. Finally, interactions between sources of selection other than those covered here are especially poorly understood. For example, in *Anolis segrei* lizards, intraspecific competition and differences in habitat use interact to affect patterns of divergent selection (Calsbeek 2009). A similar interaction, but for interspecific competition, has been documented in *Spea* toads (Martin and Pfennig 2009). A final example is the coevolutionary association between crossbill birds and pines, which is mediated by the presence versus absence of a competitor for pine seeds (*Tamiasciurus hudsonicus* red squirrels). Specifically, crossbills and pines are

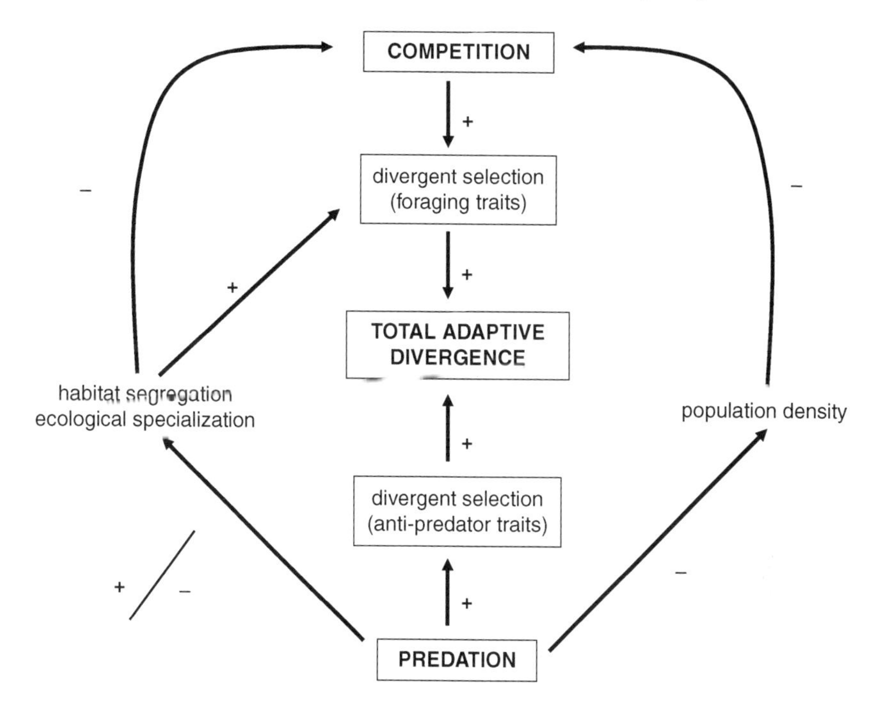

Figure 3.14. The effects of competition, predation, and the interaction between them on divergent selection. The conceptual model is motivated by similarities and differences between studies of experimental evolution in microorganisms versus other organisms (see text for details). Plus signs indicate a positive relationship between two factors and minus signs a negative relationship. Both predation and competition can directly generate divergent selection, for example on foraging and anti-predator traits, respectively. The best-studied interaction between predation and competition concerns reductions in population density caused by predation, which then results in reduced competition. However, predation may also affect levels of habitat segregation and ecological specialization. For example, predation may increase segregation and reduce competition by causing individuals to forage in different habitats. In contrast, predation might decrease segregation if individuals moved into shared refugia. When predation increases segregation, it can lead to reduced competition, but stronger divergent selection arising from increased differences between environments.

in coevolutionary arms races where red squirrels are absent, but not where red squirrels are present (Benkman et al. 2001). Further such studies of interactions between sources of selection are required.

3.6. Sources of divergent selection: conclusions

A number of major advances have been made in understanding the sources of divergent selection during ecological speciation. Nonetheless, debates remain about the extent to which evolutionary divergence on adaptive landscapes proceeds via divergent selection on rugged landscapes versus drift along ridges of high fitness in holey landscapes. These debates cannot be resolved until more studies test the fitness of intermediate forms in multidimensional genotypic space and reconstruct adaptive landscapes. It does seem likely, however, that divergent selection often arises from differences between environments, interactions between populations, and sexual selection. Certain processes remain relatively unexplored, such as interactions between populations other than competition and reinforcement. Important specific advances are recent experiments showing that interspecific competition drives divergence, increased attention to intraspecific competition, and greater appreciation of the role of sexual selection in ecological speciation. However, evidence that each source of selection can cause reproductive isolation is scarce. Generating further such evidence represents perhaps the most important and obvious avenue for future research. Major questions remain about the relative importance of each process, especially in terms of the likelihood that they generate reproductive isolation.

4

A form of reproductive isolation

"For the vast majority of animals, it is still not known which particular isolating mechanisms prevent closely related species from interbreeding" (Mayr 1963, pp. 91–2)

The quotation above is still true today and forms the motivation behind this chapter, which reviews different forms of reproductive isolation with respect to the likelihood that each is involved in ecological speciation (for reviews of reproductive isolation more generally see Coyne and Orr 2004, Nosil et al. 2005, Lowry et al. 2008a). In the context of ecological speciation, one might think of three main classes of reproductive isolation. First, two forms of reproductive isolation, namely immigrant inviability and ecologically dependent postmating isolation, are unique predictions of ecological speciation—their very existence provides evidence for the process, and they must be involved in it. Second, two other reproductive barriers can evolve via any process, but are "inherently" ecological, such that they will probably often evolve via divergent selection. These barriers are habitat and temporal isolation. Third, all other barriers are not inherently ecological and can evolve via many processes. Thus, careful tests of the predictions outlined in Chapter 2 are required to establish whether these barriers evolved owing to divergent selection.

This chapter is organized into three main parts. The first part focuses on cases studies of various types of reproductive isolation during ecological speciation. Different forms of reproductive isolation are presented in the order in which they are expected to act during the life history of an organism. The second part considers the contributions of individual reproductive barriers to levels of total reproductive isolation that exist at any one point in time. This section tackles the issue of how commonly different barriers are observed during ecological speciation. The third part considers the point in time (i.e., early through late) during speciation that different barriers evolve.

4.1. The different forms of reproductive isolation

4.1.1. Divergent habitat preferences and developmental schedules

Prezygotic isolation can arise when populations are separated in space or time (Drès and Mallet 2002, Funk et al. 2002). Habitat isolation occurs when populations exhibit genetically based preferences for separate habitats, reducing the likelihood of heterospecific encounters (Rice and Salt 1990). Divergent habitat preferences are most likely to cause prezygotic isolation when mating occurs in or near the preferred habitat (Johnson et al. 1996, Funk et al. 2002). For example, divergent host-plant preferences cause reproductive isolation between herbivorous insect populations that mate on the plant on which they prefer to feed. Temporal isolation occurs when populations exhibit divergent developmental schedules such that mating occurs at different times in different populations. Both these forms of isolation may be common during ecological speciation because adaptation to different environments will itself generate selection favoring individuals with appropriate habitat preferences and developmental schedules.

Although habitat preferences are commonly documented in laboratory experiments or from observational data in the wild (Table 4.1), there are few cases in which habitat preferences have been shown to actually result in assortative mating (i.e., truly result in reproductive isolation). Some exceptions include cage experiments showing increased assortative mating between host races of *Eurosta solidaginis* goldenrod flies when host plants are present relative to when they are absent (Craig et al. 1993) and mark–recapture studies of host races of *Rhagoletis* flies showing a tendency for flies to reproduce on the same host species that they used in earlier life history stages (Feder et al. 1994). A particularly nice example stems from a transplant experiment that documented native habitat preferences of lake versus stream ecotypes of freshwater *Gasterosteus* stickleback fishes (Bolnick et al. 2009) (Fig. 4.1). The authors used a simple quantitative genetic model to show that

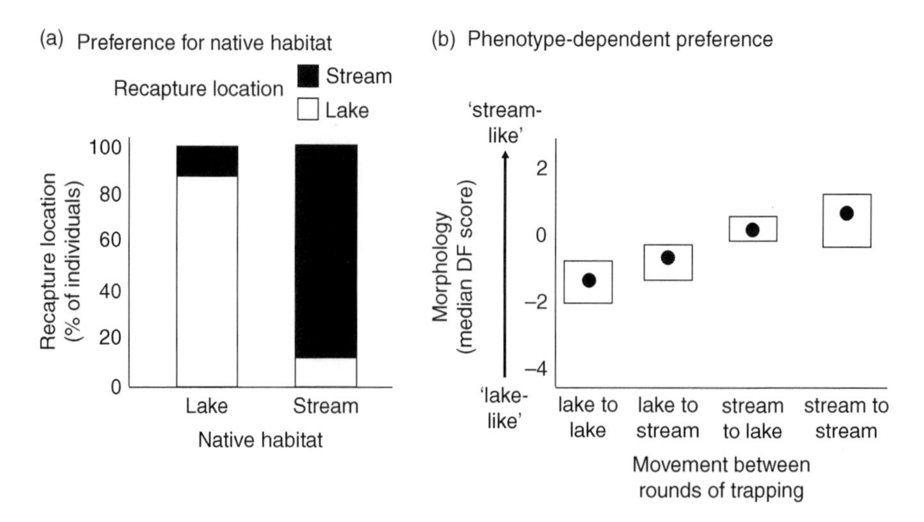

Figure 4.1. Evidence for habitat isolation between lake and stream *Gasterosteus* sticklebacks from a mark–recapture experiment. a) Divergent habitat preference. Shown are the relative frequencies with which lake and stream natives were recaptured in lake or stream habitats, 4 days after release. b) Phenotype-dependent habitat preference. Shown are boxplots of discriminant function (DF) axis scores from relative warps of geometric morphometric landmarks, for recaptured fish that returned to their native habitat (lake to lake, stream to stream) or dispersed between habitats (lake to stream, stream to lake). Modified from Bolnick et al. (2009) with permission of John Wiley & Sons Inc.

Table 4.1. Some examples where divergent habitat preferences and developmental schedules have been observed.

Form of reproductive isolation	Examples
Divergent habitat preferences (i.e., potential habitat isolation)	*Neochlamisus* leaf beetles (Funk 1998), *Acyrthosiphon* pea aphids (Via 1999), *Henosepilachna* ladybird beetles (Katakura et al. 1989, Hirai et al. 2006), *Liriomyza* leaf-mining flies (Tavormina 1982); *Rhagoletis* fruit flies (Feder et al. 1994), *Eurosta* galling flies (Craig et al. 1993), *Gasterosteus* freshwater sticklebacks (Bolnick et al. 2009)
Divergent developmental schedules (i.e., potential temporal isolation)	*Enchenopa* leafhoppers (Wood and Guttman 1982, Wood and Keese 1990), *Rhagoletis* fruit flies (Filchak et al. 2000, Dambroski and Feder 2007), *Banksia* plants (Lamont et al. 2003), *Coregonus* whitefish (Vonlanthen et al. 2009)

adaptive divergence in this system is predicted to be two- to fivefold greater when the documented strength of habitat preference occurs, relative to that when habitat preference is absent. Finally, the study reported that habitat preference was not random with respect to phenotype: stream fish moving into the lake were morphologically more lake-like than those returning to the stream (and the converse for lake fish entering the stream). This is a condition under which habitat preferences might be particularly likely to promote speciation.

Evidence for temporal isolation is similar to that for habitat isolation. Some demonstrations of divergent developmental schedules exist, such as flowering time differences in plants, differences in hatching time in insects, and spawning time differences in fish (Fig. 4.2; Table 4.1). However, direct evidence that these differences cause genetically based reproductive isolation is generally lacking.

For habitat preferences to contribute to speciation they must evolve to differ between populations. On this topic, little attention has been given to the specific mechanisms of evolution. How can selection favor the evolution of behavioral differences within or between populations that are initially monomorphic for a single strategy (e.g., a generalist behavior, or preference for a single habitat)? Habitat preferences can diverge with or without active selection against switching between different, utilized habitats (the term "utilized" refers to habitats that a population uses; other habitats that the population cannot or does not use may exist in the environment as well). Thus, there exist two general hypotheses for habitat preference evolution.

The first "fitness trade-offs" hypothesis is well known and concerns active selection against habitat switching. When switching between utilized habitats is maladaptive (i.e., when local adaptation results in fitness trade-offs between habitats), preferences can diverge via selection against individuals that switch between habitats. Under this scenario, preference for the native habitat is favored because individuals choosing another habitat suffer reduced fitness. Although selection actively favors reduced habitat switching, it may act indirectly on preference loci

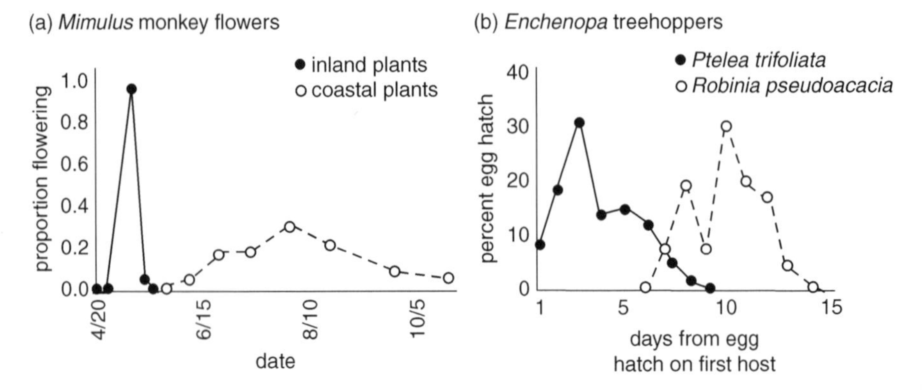

Figure 4.2. Two putative examples of temporal isolation. a) Flowering time differences between coastal and inland ecotypes of *Mimulus guttatus* monkeyflowers. Modified from Lowry et al. (2008b) with permission of John Wiley & Sons Inc. b) Divergent hatching times between host forms of *Enchenopa* treehoppers. Modified from Wood and Guttman (1982) with permission of John Wiley & Sons Inc.

via their genetic association with other loci conferring habitat-specific fitness (e.g., loci affecting color pattern or physiology). Following this hypothesis, selection only acts in populations where there is the opportunity for switching between utilized habitats; that is, when more than one habitat is available in the environment. One prediction is greater preference divergence in geographic regions where multiple habitats are utilized (sympatry or parapatry) than between geographically isolated populations that use a single, yet different, habitat (allopatry). This pattern can be thought of as "character displacement" of habitat preference, with preferences evolving in a reinforcement-like process. Surprisingly, tests for such character displacement are generally lacking, and would clearly be useful for understanding the causes of preference evolution.

In a second "information-processing" hypothesis, selection is for efficiency, rather than against habitat-switching per se. For example, there is no active selection against switching between habitats when only one habitat is utilized in the local environment. Nonetheless, search and efficiency costs can favor increased preference for the single, utilized habitat because individuals without strong preferences accrue lower fitness, but for reasons other than switching to an alternate habitat (Bernays and Wcislo 1994). For example, owing to cognitive constraints associated with information processing, generalized individuals without strong preferences might take longer to locate, or decide whether to feed on, the utilized host (e.g., as observed in *Neochlamisus* leaf beetle host forms; Fig. 4.3) (Egan and Funk 2006), thereby wasting time and energy while increasing predation risk. When preference evolution is driven by such selection, populations in environments in which only a single habitat is utilized can still evolve preference for that habitat. Thus, unlike the "fitness-trade offs" hypothesis, the "information-processing" hypothesis can drive preference evolution within allopatric populations. In fact, when gene flow

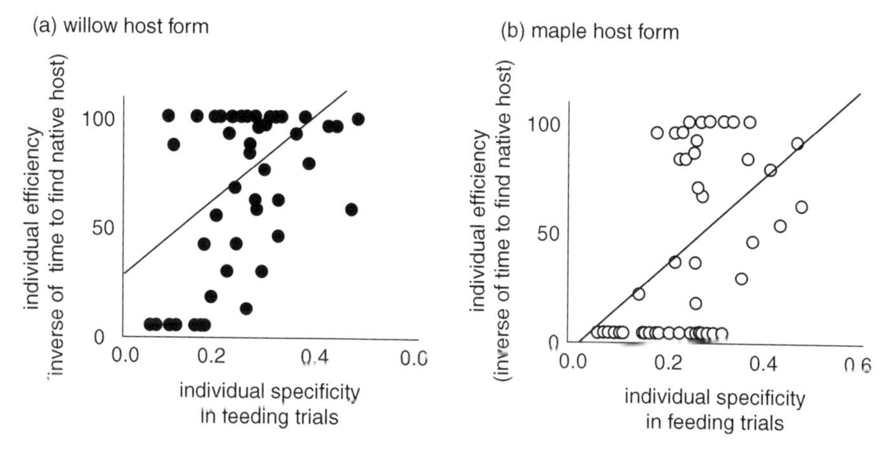

Figure 4.3. Evidence for the "information-processing" hypothesis based upon individual advantages to ecological specialization in two host forms of *Neochlamisus bebbinae* leaf beetles. Shown is the association between individual specificity in feeding trials where choice between hosts was offered and individual feeding efficiency in terms of the time required to locate the native host. Modified from Egan and Funk (2006) with permission of the Royal Society of London.

constrains divergence, the "information-processing" hypothesis predicts the opposite of character displacement: more highly divergent preferences between allopatric populations than between populations in geographic contact, as observed for host ecotypes of *Timema cristinae* walking-stick insects (Nosil et al. 2006b). Finally, a major outstanding question from a behavioral perspective is the degree to which habitat choice involves preference for the native habitat, avoidance of alternative habitats (e.g., Forbes et al. 2005), or both.

4.1.2. Immigrant inviability

Premating isolation can arise when immigrants into populations suffer reduced survival because they are poorly adapted to the foreign, non-native habitat. Although this process was often not traditionally considered as a form of reproductive isolation, such "immigrant inviability" can reduce gene flow between populations by lowering the rate of heterospecific mating encounters (Funk 1998, Via et al. 2000, Nosil 2004, Nosil et al. 2005, Giraud 2006). By reducing gene flow, natural selection against immigrants constitutes a reproductive barrier. In its extreme, immigrant inviability might completely exclude populations from occupying certain geographic and ecological areas, resulting in what has been referred to as "eco-geographic isolation" (Ramsey et al. 2003, Lowry et al. 2008a, Sobel et al. 2010). Additionally, immigrant inviability imposes selection for the evolution of habitat preference, potentially contributing to the evolution of habitat isolation via the aforementioned "fitness trade-offs" hypothesis.

A detailed case study of immigrant inviability stems from populations of the live-bearing fish *Poecilia mexicana* that occupy divergent habitats (Tobler et al. 2008). Some populations live in caves and others on the surface. In addition, some populations live in highly toxic, hydrogen sulfide-containing waters, whereas others do not. Reciprocal transplant experiments demonstrated that migrants between sulfidic and non-sulfidic environments exhibit lower fitness than locally adapted resident forms, and that selection against immigrants can be manifested as quickly as 24 hours (Tobler et al. 2009). In addition, predation experiments demonstrated that giant water bugs predominantly prey on cave fish in light conditions, and surface fish in darkness, leading to low fitness of individuals that migrate between cave and surface environments (Tobler 2009). Finally, selection from piscivorous fishes, which are absent from caves, generates immigrant inviability for cave individuals that immigrate into surface environments (R. Riesch, M. Tobler, M. Plath, unpublished data). This example demonstrates not only that immigrant inviability can occur, but also that it can arise via multiple sources of selection acting within a single system.

Evidence for the importance of immigrant inviability in speciation is not restricted to just a few case studies. For example, quantification of the individual components of reproductive isolation in diverse taxa revealed that immigrant inviability tends to be as strong, or perhaps stronger, than more commonly considered forms of reproductive isolation (Fig. 4.4). I present here two new analyses of published datasets supporting this claim (Tables 4.2 and 4.3). First, Nosil et al.

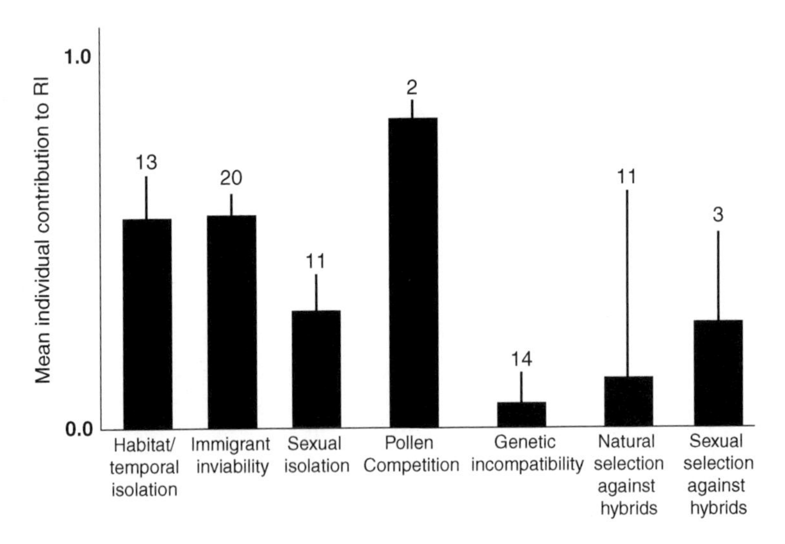

Figure 4.4. Mean individual contributions (± s.e.) to reproductive isolation across reproductive barriers. Each bar represents mean values across those study systems for which data for that barrier were available. Numbers above bars indicate number of systems contributing data. One extreme data point (where hybrid fitness was much higher than parental fitness) was excluded in the calculation for the "natural selection against hybrids" bar. Note the relatively high value for immigrant inviability. Modified from Nosil et al. (2005) with permission of John Wiley & Sons Inc.

Table 4.2. Components of reproductive isolation for the study systems from Nosil et al. (2005). Shown is the strength of reproductive isolation stemming from immigrant inviability, from the mean of all barriers other than immigrant inviability, and from the barrier other than immigrant inviability with maximum strength. To be conservative in comparisons to immigrant inviability, negative values for the case of maximum strength of other barriers were set to zero.

Study system	Common name	Immigrant inviability	Mean of other forms of isolation	Maximum of other forms of isolation	Putative source of selection against immigrants	References
1. *Acyrthosiphon pisum*	Pea aphids	0.97	0.46	0.93	Physiological adaptation	(Via 1999, Via et al. 2000)
2. *Agelenopsis aperta*	Desert spiders	0.63	–0.45	0.06	Food availability, predation risk	(Riechert and Hall 2000, Riechert et al. 2001)
3. *Artemesia tridentate*	Big sagebrush	0.90	–5.25	0.00	Not stated	(Wang et al. 1997)
4. *Bombina* spp.	Toads	0.44	0.43	0.66	Biophysical conditions, predation	(Maccallum et al. 1995, MacCallum et al. 1998, Kruuk et al. 1999)
5. *Eurosta solidaginis*	Galling flies	0.88	0.42	0.96	Physiological adaptation	(Craig et al. 1993, Craig et al. 1997, 2001)
6. *Galerucella nymphaeae*	Water lily leaf beetles	0.77	0.29	0.86	Physiological adaptation, feeding ability	(Pappers et al. 2002a, Pappers et al. 2002b)
7. *Galerucella* spp.	Water lily leaf beetles	0.94	0.07	0.90	Physiological adaptation	(Nokkala and Nokkala 1998)
8. *Gasterosteus aculeatus*	Freshwater sticklebacks	0.36	0.24	0.78	Foraging efficiency, competition, predation	(Schluter 1993, 1994, 1995, Hatfield and Schluter 1999, Rundle 2002, Vamosi 2002)
9. *Gilia capitata*	Herbaceous plants	0.70	0.81	0.81	Biophysical conditions	(Nagy 1997b, a, Nagy and Rice 1997)
10. *Heliconius erato*	Mimetic butterflies	0.52	0.11	0.45	Visual predation	(Mallet and Barton 1989) (Mallet 1989, Mallet et al. 1998)

Table 4.2. *Cont.*

Study system	Common name	Immigrant inviability	Mean of other forms of isolation	Maximum of other forms of isolation	Putative source of selection against immigrants	References
11. *Ipomopsis* spp.	Phlox flowers	0.43	0.01	0.16	Biophysical factors	(Campbell and Waser 2001, Campbell 2003, Campbell et al. 2003)
12. *Iris* spp. 1	Perennial flowers	0.96	–0.21	0.00	Biophysical factors	(Young 1996)
13. *Iris* spp. 2	Perennial flowers	0.09	–0.01	0.00	–	(Emms and Arnold 1997)
14. *Littorina saxatilis*	Intertidal snails	0.69	0.20	0.74	Physical factors, predation	(Johannesson et al. 1993) (Rolan-Alvarez et al. 1997)
15. *Mimulus* spp.	Monkeyflowers	0.59	0.74	0.98	Climatic conditions	(Ramsey et al. 2003)
16. *Mitoura* spp.	Butterflies	0.21	0.35	0.70	Physiological adaptation	(Forister 2004)
17. *Neochlamisus bebbianae*	Leaf beetles	0.99	0.83	0.95	Physiological adaptation	(Funk 1998)
18. *Polemonium viscosum*	Alpine flowers	0.80	0.14	0.14	Climatic conditions, physiological adaptation	(Galen and Kevan 1980, Galen 1985, Galen et al. 1991)
19. *Rhagoletis* spp.	Fruit flies	0.37	0.58	0.56	Physiological adaptation	(Bierbaum and Bush 1990)
20. *Timema cristinae*	Cryptic walking-sticks	0.34	0.24	0.25	Visual predation	(Sandoval 1993, 1994a, b, Nosil et al. 2002, 2003, Nosil 2004, Nosil et al. 2006b)

Table 4.3. Components of reproductive isolation for the study systems from Lowry et al. (2008a). Shown is the strength of reproductive isolation stemming from immigrant inviability, from the mean of all barriers other than immigrant inviability, and from the barrier other than immigrant inviability with maximum strength. Data were taken from Table 1 in Lowry et al. (2008a), averaging across the two values for each taxon pair, and excluding one case where reproductive isolation was not estimated for both taxa in a pair.

Taxon	Type	Immigrant inviability	Mean of other forms of isolation	Maximum of other forms of isolation
1. *Artemisia tridentata*	Basin vs. mountain	0.85	0.77	0.77
2. *Gilia capitata*	Coast vs. inland	0.75	0.88	0.88
3. *Helianthus exilis*	Serpentine vs. riparian	0.97	0.00	0.00
4. *Ipomopsis*	*aggregata* vs. *tenuituba*	0.96	0.30	0.77
5. *Iris* (1)	*douglasiana* vs. *innominata*	0.93	–0.42	0.14
6. *Iris* (3)	*fulva* vs. *hexagona*	0.07	0.27	0.81
7. *Iris* (4)	*fulva* vs. *brevicaulis*	–0.17	–0.32	0.86
8. *Mimulus*	*lewisii* vs. *cardinalis*	0.93	0.41	0.98
9. *Mimulus guttatus*	Coast vs. inland	0.94	0.01	1.00
10. *Phlox*	*cuspidata* vs. *drummondii*	0.19	0.51	0.85

(2005) examined the contribution of immigrant inviability to reproductive isolation in systems where immigrant inviability and at least one other type of reproductive barrier had been measured. Immigrant inviability was estimated from reciprocal transplant experiments. Reproductive isolation of each form was quantified on a common scale ranging from negative infinity to 1, where 0 = no reproductive isolation and 1 = complete reproductive isolation. In 16 of the 20 systems examined, the strength of reproductive isolation caused by immigrant inviability in a system was greater than the strength of reproductive isolation averaged across all other barriers in that system (mean paired difference of immigrant inviability minus mean of other barriers = 0.61; Wilcoxon's signed ranks test, $Z = -2.97$, $p = 0.003$). A parallel and much more conservative test using the maximum (rather than mean) isolation caused by a barrier other than immigrant inviability revealed that immigrant inviability was stronger in 11 of 20 cases. This result is not statistically significant ($Z = -0.37$, $p = 0.79$), but shows that immigrant inviability at least competes in strength with other strong barriers. Notably, immigrant inviability associated with divergent physiological adaptation was particularly strong (Table 4.2).

The study by Nosil et al. (2005) probably suffers from an ascertainment bias, because the systems covered were those for whom reciprocal transplants had been conducted; such systems were probably known *a priori* to be ecologically divergent such that immigrant inviability is expected. A probably less-biased dataset compiled by Lowry et al. (2008a) examined any and all flowering plant systems for which multiple components of reproductive isolation had been examined. Immigrant inviability was examined in about half the cases. Qualitatively similar results to those from the dataset of Nosil et al. (2005) were observed. In seven of the ten systems, the strength of immigrant inviability was greater than the strength of reproductive isolation averaged across all other barriers (mean paired difference = 0.40, $Z = -1.68$, $p = 0.09$). Immigrant inviability was stronger than the other barrier with maximum strength in four of ten cases ($Z = -0.26$, $p = 0.80$).

Collectively, the results indicate that immigrant inviability tends to be a common and strong component of reproductive isolation. Nonetheless, our understanding of how divergent selection generates immigrant inviability is limited, because experimental data addressing the sources and phenotypic targets of selection are few (Schluter 2000b). A more detailed understanding will require experiments that directly manipulate agents of selection and identify the traits involved (e.g., predation, Nosil 2004; water toxicity, Tobler et al. 2008; temperature, Matute et al. 2009).

4.1.3. Divergent mating and pollinator preferences

Prezygotic isolation can arise because individuals from different populations are less attracted to, or do not recognize, one another as potential mates. Such sexual isolation is one of the most commonly recognized forms of prezygotic isolation, but its ecological basis is also one of the most difficult to determine. Differences among populations in signal traits and preferences will generally arise as a by-product of mate choice evolution within populations, a process that necessarily involves sexual selection and may involve natural selection and genetic drift as well (Kirkpatrick and Ryan 1991, Coyne and Orr 2004). Sexual isolation is a part of ecological speciation when ecologically important characters also incidentally influence mate choice, or when environmental differences generate sensory drive. Sexual isolation can also evolve by reinforcing selection within an ecological context.

Sexual isolation has received much empirical attention, and a number of lines of evidence implicate divergent selection in its evolution. For example, pairs of populations independently adapted to different environments have been shown to exhibit stronger sexual isolation than those independently adapted to similar environments (Funk 1998, Rundle et al. 2000, Nosil et al. 2002, McKinnon et al. 2004, Langerhans et al. 2007, Tobler et al. 2009). In addition, traits under divergent selection have been shown to influence mate choice in a number of systems (i.e., "magic traits") (Nagel and Schluter 1998, Jiggins et al. 2001, Jiggins 2008, Maan and Seehausen 2011, Servedio et al. 2011). There is also evidence for sensory drive and ecologically based reinforcement (Rundle and Schluter 1998, Boughman 2002, Nosil et al. 2003, Albert and Schluter 2004, Seehausen et al. 2008b, Schwartz et al. 2010). In plants, populations in different environments can be exposed to selection to adapt to different pollinators, or habitat-specific selection might incidentally act on traits that affect pollinator preferences. The subsequent divergence in pollinator-related traits will generate pollinator isolation. Such pollinator isolation has been strongly implicated in monkeyflowers (Schemske and Bradshaw 1999, Bradshaw and Schemske 2003, Ramsey et al. 2003) and may be common in plants (Lowry et al. 2008a).

Nonetheless, there are many cases where adaptation to different environments does not appear to generate sexual isolation. For example, adaptation to different host plants by some phytophagous insects has apparently not resulted in sexual isolation, as reported in host forms of *Rhagoletis pomonella* fruit flies (Reissig and Smith 1978) and *Acyrthosiphon pisum* pea aphids (Via 1999), and the beetle species pair *Henosepilachna niponica* and *H. yasutomii* (Katakura et al. 1989). The most general explanation for a lack of sexual isolation is that divergent adaptation does not involve traits that incidentally also affect sexual isolation. This will be mediated, in part, by the unresolved issue of how often traits used in within-population mate choice also generate sexual isolation between populations (Claridge and Morgan 1993, Boake et al. 1997, Nosil and Crespi 2004, Higgie and Blows 2007, Arbuthnott 2009, Arbuthnott and Crespi 2009, Head et al. 2009). Other explanations for a lack of sexual isolation involve the opportunity for reinforcing selection. If there is no sympatry, or if strong habitat preferences evolve early in the speciation process, hybridization can be reduced to the extent that reinforcing selection on mating preferences is weak or absent (Nosil and Yukilevich 2008).

4.1.4. Postmating, prezygotic incompatibility

Postmating, prezygotic isolation exists when there is a reduction in the fertilization success of between-population matings, or a reduction in female fitness following between-population copulation. Examples include poor transfer or storage of sperm (Price et al. 2001), failure of fertilization when gametes come into contact (Vacquier et al. 1997), and conspecific sperm or pollen preference (Rieseberg et al. 1995, Howard et al. 1998). Such barriers can evolve via numerous processes, and divergent selection need not be involved. However, divergent selection could drive the evolution of postmating–prezygotic isolation: for example, if habitat-related differentiation affects genitalia, or the physical condition of gametes (which in turn might affect the physical matching of male and female genitalia or of sperm and eggs). Although the extreme diversity of animal genitalic structures suggests that postmating–prezygotic isolation could be common (Shapiro and Porter 1989, Sota and Kubota 1998, Knowles and Markow 2001, Hosken and Stockley 2004, Arbuthnott et al. 2010), there are very few data linking divergent selection to the evolution of this form of reproductive isolation. A clear example stems from ecotypes of *T. cristinae*, in which female fecundity is greatly reduced following between-population mating, relative to fecundity following within-population mating, but only when mating is between populations using different host-plant species (Nosil and Crespi 2006a). Although the mechanism reducing female fitness in this example is unknown, the observation of postmating–prezygotic isolation only between populations adapted to different hosts shows that divergent selection can drive the evolution of this form of reproductive isolation. More attention to the association between divergent selection and these types of barriers is warranted.

4.1.5. Intrinsic hybrid incompatibilities

'the effects in question [hybrid dysfunction] . . . may legitimately be regarded as automatic by-products of the general differentiation produced by a combination of drift, and of selection for other characters than those observed here' (Muller 1942, p. 100)

Postzygotic isolation can arise from intrinsic genetic incompatibilities between loci. Such incompatibilities are often thought of as not evolving due to divergent selection, although as indicated by the quotation above, there is no reason to suspect that they do not. The assumption that extrinsic isolation is a consequence of divergent adaptation, whereas intrinsic isolation is not, can lead to confusion because it defines extrinsic and intrinsic isolation without reference to the process driving genetic divergence. Indeed, theory indicates that divergent selection can readily drive the evolution of intrinsic hybrid incompatibilities, especially when there is little or no gene flow (Gavrilets 2004). The basic premise is that different alleles are favored by divergent selection in each of two environments. Alleles within each environment are tested by selection to work well together. In contrast, alleles from different environments are not tested together, and might be incompatible when brought together in the genome of a hybrid. Recent theory indicates that intrinsic isolation can originate even if ecological speciation occurs in the face of gene flow. For example, Agrawal et al. (2011) present analytical models in which genes subject to divergent natural selection, either directly or via linkage disequilibrium with other loci, also affect intrinsic isolation. Such genes can overcome gene flow to diverge between populations, resulting in the evolution of intrinsic isolation.

There is also empirical evidence that divergent selection can drive the evolution of intrinsic postmating isolation. A classic case is intrinsic hybrid sterility in *Mimulus* monkeyflowers adapted to different soil types (Macnair and Christie 1983, Christie and Macnair 1984). In this example, crossing experiments revealed that a yet unidentified gene tightly physically linked to a gene conferring tolerance to copper interacts with a small number of genes in another population to generate intrinsic hybrid incompatibility. Thus, genes conferring adaptation to copper contribute to the evolution of postmating isolation (Wright and Willis, unpublished). Evidence for a role for divergent adaptation in the evolution of intrinsic reproductive isolation also stems from broad comparative studies showing positive correlations between the degree of ecological divergence and levels of intrinsic postmating isolation between taxon pairs of insects (Funk et al. 2006) and fish (Bolnick et al. 2006). Experimental and genetic studies in yeast also implicate divergent adaptation in the evolution of intrinsic postzygotic isolation (Dettman et al. 2007, Lee et al. 2008). Finally, many studies implicate selection in the evolution of intrinsic isolation (Fitzpatrick 2002, Coyne and Orr 2004, Orr 2005a) but are yet to isolate a role for divergent selection specifically.

An interesting example of ecological divergence generating genetic incompatibilities concerns cytonuclear interactions in plants (Fig. 4.5). Sambetti et al. (2008) found that habitat adaptation of cytoplasmic genomes contributes to reproductive barriers in the hybridizing sunflower species, *Helianthus annuus* and *Helianthus petiolaris*. Transplant experiments were used to measure survivorship of parental genotypes, F1 hybrids, and all possible backcross combinations of nuclear and cytoplasmic genomes in xeric and mesic habitats of the parental species. The results revealed that the parental species cytoplasms were strongly locally adapted. In turn, cytonuclear interactions sometimes generated incompatibilities that reduced the fitness of hybrid genotypes with mismatched cytoplasmic and nuclear genomes (see Woods et al. 2009 for a related example in animals). Cytonuclear incompatibilities in general appear relatively common in plants (Fishman and Willis 2006, Rieseberg and Willis 2007, Rieseberg and Blackman 2010), and it will be of interest to determine how often divergent selection is involved in their evolution.

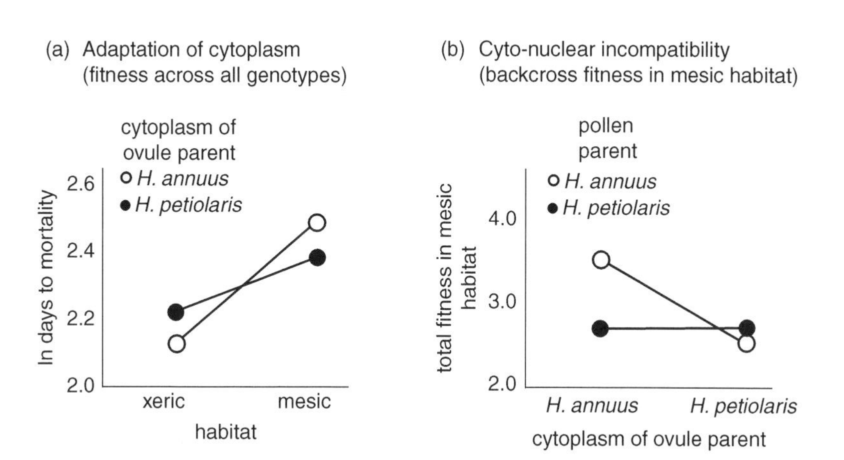

Figure 4.5. Evidence that divergent adaptation contributes to cytonuclear incompatibilities between hybridizing sunflowers. a) Evidence for local adaptation of cytoplasm from a cytoplasm by habitat interaction for survivorship. b) Total fitness of backcross populations in mesic habitat: hybrid genotypes with mismatched cytoplasmic versus nuclear genomes sometimes exhibited reduced fitness. Modified from Sambetti et al. (2008) with permission of John Wiley & Sons Inc.

In conclusion, theoretical and empirical evidence clearly shows that divergent adaptation can generate intrinsic genetic incompatibilities. It remains to be seen how commonly and quickly this occurs. Additionally, it has been argued that intrinsic genetic incompatibilities, although not "ecologically dependent" in the true sense, might vary in their manifestation. For example, genetic incompatibilities might not occur in benign lab environments, manifesting themselves only under harsh or stressful conditions. A test of this idea in killifish (*Lucania goodei* and *L. parva*) found no effect of the environment on existing genetic incompatibilities (Fuller 2008). Further quantification of the effect of environmental conditions on the manifestation of genetic incompatibilities is needed.

4.1.6. Ecologically dependent selection against hybrids

Postzygotic isolation can also arise when hybrid fitness is reduced because of an ecological mismatch between hybrid phenotype and the environment (Rice and Hostert 1993, Rundle and Whitlock 2001, Rundle and Nosil 2005). Basically, hybrids are not well adapted to either parental environment, and in effect, fall between niches. Ecologically dependent postmating isolation is analogous to immigrant inviability except that divergent selection is acting against hybrid instead of parental individuals. As with immigrant inviability, ecologically dependent postzygotic isolation and divergent selection between environments can be considered two sides of the same coin. To the extent that hybrid phenotypes are intermediate and intermediate habitats are lacking, ecologically dependent postzygotic isolation is a necessary consequence of divergent selection. Until recently, ecologically dependent postzygotic isolation has received less attention than intrinsic postzygotic isolation.

There are at least three techniques for demonstrating ecologically dependent postzygotic isolation. In the first, the fitness of hybrids in the wild is compared to that in a benign environment (e.g., Hatfield and Schluter 1999, McBride and Singer 2010). The benign environment is assumed to remove the ecological factors that reduce hybrid fitness, thus permitting an estimate of any intrinsic genetic isolation. Comparison of hybrid fitness in the wild to that in the benign environment yields an estimate of ecologically dependent isolation. Caution is warranted, however, because non-ecological reductions in hybrid fitness may differ between environments, complicating this method.

In the second, backcrosses of F1 hybrids to both parental forms are used in reciprocal transplants between environments. A comparison of the fitness of the two types of backcrosses estimates a component of ecologically dependent isolation while controlling for any genetic incompatibilities (c.f., Rundle and Whitlock 2001). The basic premise is that backcrosses to different parental species share similar amount of alleles contributing to intrinsically "bad" genotypic combinations which generate intrinsic incompatibility, but differ in their phenotypes according to which

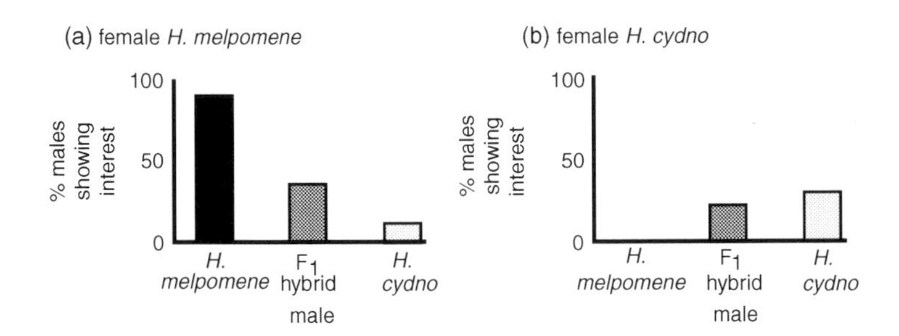

Figure 4.8. Sexual selection against hybrids between two species of *Heliconius* butterflies. Modified from Naisbit et al. (2001) with permission of the Royal Society of London.

1993b, Rowe and Houle 1996, Servedio 2009), and hybrid condition is reduced as a result of ecological mechanisms (van Doorn et al. 2009). Finally, divergent selection is implicated if mate preferences diverge between parental species as a consequence of ecological mechanisms and this renders hybrids unattractive because of their intermediate phenotypes. The above possibilities have received little attention, although the latter situation appears to be involved in the reduced mating success of hybrids between species of *Heliconius* butterflies (Fig. 4.8) and *Pundamilia* cichlids (Stelkens et al. 2008, van der Sluijs et al. 2008). For example, in the butterfly system, color patterns diverged as adaptations to mimic different model taxa, but are also important traits in mate choice. Hybrid color patterns are intermediate and fall largely outside of the range of parental mate preferences (Naisbit et al. 2001). An analogous example from plants is pollinator-based discrimination against hybrid plants possessing intermediate floral traits (Schemske and Bradshaw 1999, Emms and Arnold 2000).

4.1.8. Forms of reproductive isolation: conclusions

There is some evidence that each barrier considered could evolve via divergent selection, and thus contribute to ecological speciation. However, with the exception of numerous reciprocal transplant experiments in nature testing for immigrant inviability, almost all the evidence stems from laboratory studies of reproductive isolation. Thus, further work in nature is needed, particularly to show that the barrier evolved via divergent selection and is actually involved in reductions in gene flow. Even once divergent selection has been implicated, the specific mechanistic causes of evolution (e.g., for habitat isolation, the fitness trade-offs versus information-processing hypotheses) have rarely been addressed.

Given evidence that divergent selection can drive the evolution of diverse types of reproductive barriers, future work could fruitfully focus on three questions: (1) How common are different forms of reproductive isolation during ecological speciation? (2) For a given point in the speciation process, do multiple reproductive barriers act, and what are their relative contributions to total reproductive isolation? (3) Across stages of the ecological speciation process, at what point do different barriers tend to evolve? I now treat each of these questions in turn.

4.2. How common are different forms of reproductive isolation during ecological speciation?

Table 4.4 summarizes the results of four previous reviews of components of reproductive isolation (Funk et al. 2002, Nosil et al. 2005, Lowry et al. 2008a, Matsubayashi et al. 2010). Two major conclusions are evident. First, all barriers appear at least somewhat common. For example, averaged across the four reviews, the least likely barrier to be observed (in this case, intrinsic postmating isolation) was still present in 41% of study systems. Second, there is variation in how common different barriers are. For example, averaged across reviews the most commonly observed barrier (immigrant inviability) was present in 94% of study systems. These

Table 4.4. Summary of the presence of various reproductive barriers in past reviews of multiple components of reproductive isolation. Percentages refer to the percent of study systems examined that showed the presence of a particular barrier. Numbers in parentheses refer to the number of study systems investigated (i.e., sample size).

Organisms	1. HI	2. TI	3. II	4. SI/PI	5. GAI	6. CSP	7. IGI	8. EDPI	9. SSH
1. Herbivorous insects	93% (14)	73% (11)	92% (13)	64% (14)	29% (7)	100% (1)	25% (12)	56% (9)	33% (3)
2. Plants and animals	83% (12)	n/a	100% (20)	73% (11)	n/a	100% (2)	21% (14)	55% (11)	66% (3)
3. Herbivorous insects	91% (11)	60% (10)	n/a	73% (11)	n/a	n/a	n/a	n/a	n/a
4. Flowering plants	100% (4)	73% (30)	89% (18)	76% (21)	68% (28)	71% (14)	78% (18)	50% (10)	n/a
Average	**92%**	**69%**	**94%**	**72%**	**49%**	**90%**	**41%**	**54%**	**50%**

1. Matsubayashi et al. 2010. Presence or absence values taken directly from their Table 1.
2. Nosil et al. 2005. Values of zero and negative values counted as the absence of a barrier. All others counted as the presence. Strong ascertainment bias toward the detection of II because study systems were chosen *a priori* where II was measured. Nonetheless, II could be been found missing once measured.
3. Funk et al. 2002. From their Table 1, plus values taken as presence and minus values as absence.
4. Lowry et al. 2008a. Values of zero, negative values, values below 0.01, and those denoted as "low" by the authors taken as the absence of a barrier. All other values taken as the presence. F_1 seed formation was taken as GAI. Four different forms of IGI were examined (F_1 seed germination, viability, male fertility, and seed set). Presence of any one was counted as the presence of IGI.

HI, habitat isolation; TI, temporal isolation; II, immigrant inviability; SI/PI, sexual or pollinator isolation; GAI, gametic incompatibility; CSP, conspecific sperm or pollen preference; IGI, intrinsic genetic incompatibility; EDPI, ecologically dependent postmating isolation; SSH, sexual selection against hybrids.

reviews stem from studies of reproductive barriers generally, without reference to whether divergent selection drove their evolution.

What is known then about how strongly divergent selection drives the evolution of different types of reproductive isolation? Little is known, but some new insight stems from close examination of a comparative study by Funk et al. (2006). As discussed in Chapter 2, this study added an ecological dimension to comparative studies examining the relationship between reproductive isolation and time. This allowed the authors to statistically isolate the association between ecological divergence (in diet, habitat use, and body size) and reproductive isolation, independent of time. Four of the datasets examined had data on both premating (sexual) isolation and intrinsic postmating isolation (Fig. 4.9). In these datasets, the association between ecological divergence and reproductive isolation was sometimes more strongly positive for premating isolation (e.g., darters and body size, fruit flies and habitat), but sometimes more strongly positive for postmating isolation (e.g., angiosperms and habitat). Paired differences between these different forms of reproductive isolation within datasets were insignificant (paired t-test, $t = 1.18$, d.f. $= 3$, $p = 0.32$), albeit power is very low ($n = 4$ datasets). Thus, although it is premature to draw strong conclusions with such limited data, there is as of yet no conclusive evidence that divergent selection more strongly promotes premating than intrinsic postmating isolation. Thus, both forms of reproductive isolation could be common during ecological speciation, and the relative degree to which divergent selection promotes their evolution warrants particular attention in future studies.

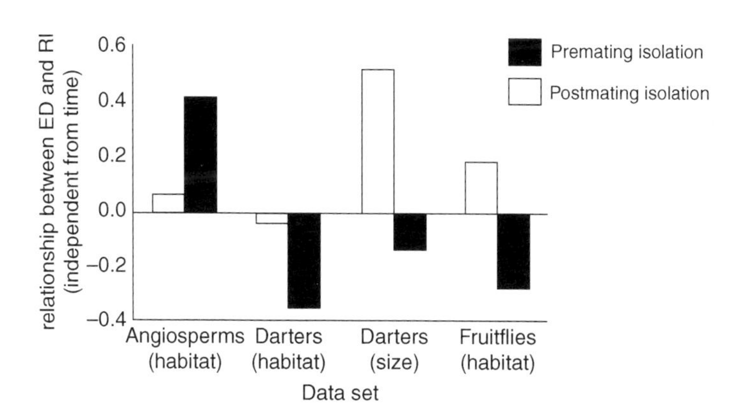

Figure 4.9. The strength and direction of the relationship (i.e., r-values in regression analyses) between reproductive isolation (RI) and ecological divergence (ED), independent from time since divergence, for premating (sexual) versus postmating (intrinsic genetic) isolation. Data depicted is from the four datasets considered by Funk et al. (2006) that contained both forms of reproductive isolation.

4.3. For a given point in the speciation process, do multiple reproductive barriers act, and what are their relative contributions to total reproductive isolation?

Even if most barriers are somewhat common, as argued above, the question remains of whether multiple barriers act simultaneously within a single system. Past reviews (Funk et al. 2002, Nosil et al. 2005, Lowry et al. 2008a, Matsubayashi et al. 2010), as well as detailed case studies, indicate that multiple barriers indeed often act within a single study system. Table 4.5 outlines some cases studies supporting this conclusion, and Fig. 4.10 illustrates an example in which ecological speciation specifically was supported. Another example concerns closely related species of the phytophagous ladybird beetles (*Henosepilachna vigintioctomaculata* and *H. pustulosa*), where six reproductive barriers exist, yet each individual barrier is far from complete. Nonetheless, the joint action of all the different barriers results in almost complete reproductive isolation (Matsubayashi and Katakura 2009), illustrating the ability of multiple, yet incomplete, components of isolation to almost completely prevent gene flow.

An obvious question that emerges if multiple barriers act simultaneously is how strong are each of their contributions to total reproductive isolation? Addressing this question requires looking at paired estimates of different forms of reproductive isolation within single study systems. Some such data exist in reviews of reproductive isolation. Here I focus on comparing sexual/pollinator isolation to intrinsic

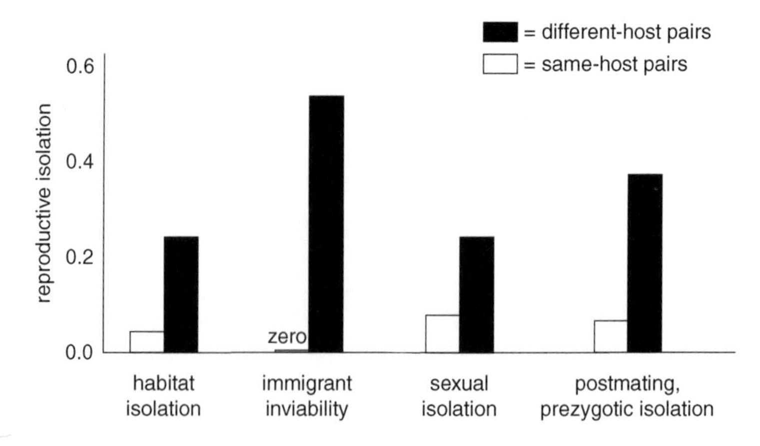

Figure 4.10. The correlation between ecological divergence and reproductive isolation among population pairs of *Timema cristinae* walking-stick insects feeding on different versus the same host plant species. Variation in geographic setting is controlled for as all depicted populations pairs are allopatric. Multiple population pairs of each type were examined and different-host population pairs generally exhibited significantly greater reproductive isolation than same-host pairs. Modified from Nosil (2007) with permission of the University of Chicago Press.

Table 4.5. Some examples of detailed case studies of components of reproductive isolation (RI) where multiple components were detected in the same study system.

Study system	Components of RI observed	reference
Z- and E-pheromone forms of *Ostrinia nubilalis* corn borers	Temporal isolation, sexual isolation, postmating-prezygotic isolation, some forms of intrinsic postzygotic isolation (e.g., F_1 male behavioral dysfunction)	(Dopman et al. 2010)
Henosepilachna vigintioctomaculata and *H. pustulosa* ladybird beetles	Habitat isolation, sexual isolation, postmating–prezygotic isolation (specifically, conspecific sperm precedence), intrinsic postmating isolation	(Matsubayashi and Katakura 2009)
Gryllus firmus and *G. pennsylvanicus* field crickets	Temporal isolation, immigrant inviability, sexual isolation, intrinsic postmating isolation	(Maroja et al. 2009b and references within)
Heliconius erato and *H. himera* mimetic butterflies	Immigrant inviability, sexual isolation, extrinsic postmating isolation	(McMillan et al. 1997)
Timema cristinae host ecotypes	Habitat isolation, immigrant inviability, sexual isolation, postmating–prezygotic isolation	(Nosil 2007)
Mimulus lewisii and *M. cardinalis* monkeyflowers	Habitat isolation, pollinator isolation, postmating–prezygotic isolation (conspecific pollen precedence), intrinsic postmating isolation	(Ramsey et al. 2003)
Loxia curvirostra crossbill call types	Habitat isolation, immigrant inviability, sexual isolation	(Smith and Benkman 2007)
Spodoptera frugiperda fall armyworm host races	Habitat isolation, sexual isolation, possibly some intrinsic postmating isolation and sexual selection against hybrids	(Groot et al. 2010)
Lake Victoria cichlid fish	Habitat isolation, temporal isolation, sexual isolation	(Seehausen et al. 1998)

postmating isolation, because there has been much debate about their relative importance (Coyne and Orr 2004; Mallet 2006). In the review of Nosil et al. (2005), eight study systems had estimates of both these forms of reproductive isolation (Table 4.6). Sexual/pollinator isolation was stronger in six comparisons, with two additional ties ($Z = 2.20$, $p = 0.028$, Wilcoxon's signed ranks test). Lowry et al. (2008a) present a similar review, which also suggests pollinator isolation tends to be stronger than intrinsic postmating isolation (but only three systems had paired estimates, precluding a statistical test). Finally, analyses of multiple taxon pairs of cichlid fish also suggest divergent selection more strongly promotes sexual than intrinsic genetic isolation, particularly in sympatric conditions (Stelkens and Seehausen 2009b, Stelkens et al. 2010).

Table 4.6. Paired estimates of two different forms of reproductive isolation within single study systems. Data from Nosil et al. (2005).

System	Sexual/pollinator isolation	Intrinsic postmating isolation
1. *Agelenopsis aperata*	0.06	–0.15
2. *Eurosta solidaginis*	0.37	0.00
3. *Galerucella nymphaeae*	0.00	0.00
4. *Gasterosteus aculeatus*	0.63	0.00
5. *Heliconius erato*	0.00	0.00
6. *Iris* spp. 1	0.00	–0.62
7. *Littorina saxatilis*	0.52	0.00
8. *Mimulus* spp.	0.98	0.42

Although more data are clearly needed, the results suggest that sexual/pollinator isolation within single systems tends to be stronger than intrinsic postmating isolation. A review of hybrid zones also reached the conclusion that prezygotic factors are of particular importance for speciation (Jiggins and Mallet 2000). This review compared bimodal versus unimodal hybrid zones, under the assumption that bimodal zones represent greater reproductive isolation. Bimodal hybrid zones tended to exhibit prezygotic isolation whereas unimodal zones do not. In contrast, bimodal versus unimodal zones did not differ in postzygotic isolation. The collective results suggest prezygotic isolation is more important, or perhaps even necessary, for speciation.

4.4. Across the ecological speciation process, at what point do different barriers evolve?

A final and critical issue for understanding ecological speciation is the point in the process that different forms of reproductive isolation arise. In other words, do rates of evolution vary predictably among reproductive barriers? Do some barriers tend to evolve earlier than others? In the context of ecological speciation specifically, these questions cannot yet be answered because almost nothing is known. This is an oversight, because these questions are important—for example, because reproductive barriers that arise before the speciation process is complete are the most causally important for speciation (Coyne and Orr 2004, Via 2009). When divergence occurs with gene flow, the barriers that arise earliest in the process might be especially important, even if they are weak, because any reduced gene flow they cause could facilitate further divergence (Orr 1995, Hendry et al. 2001, van Doorn and Weissing 2001, Feder and Nosil 2010). Addressing the timing of evolution (i.e., early or late) of different reproductive barriers requires "snapshots" of different stages of the speciation process within a group of closely related organisms. Several comparative studies examining numerous species pairs have examined how reproductive isolation builds up with increasing time since divergence (Coyne and Orr 2004,

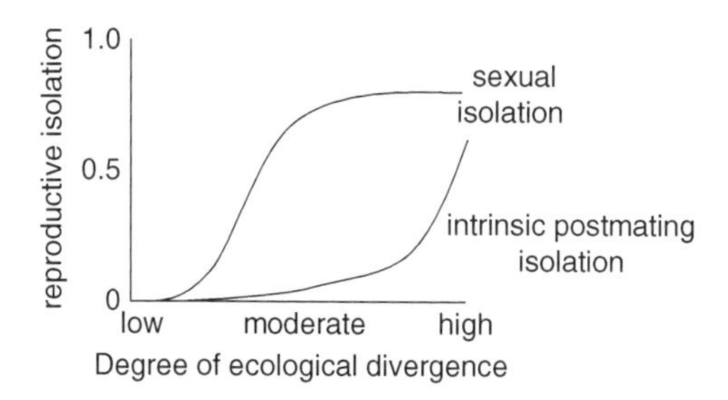

Figure 4.11. A hypothetical example examining how different forms of reproductive isolation build up as ecological divergence increases. Data such as that depicted can yield insight into the point in the speciation process that different reproductive barriers evolve. For example, in the example depicted sexual isolation is stronger than intrinsic postmating isolation for any given level of ecological divergence, and sexual isolation first evolves at lower levels of ecological divergence.

Fitzpatrick 2004, Mendelson et al. 2004, Gourbiere and Mallet 2010, Stelkens et al. 2010). However, studies that examine a role for divergent selection, and thus pertain to ecological speciation specifically, are lacking. Hence, studies examining how different forms of reproductive isolation build up with increasing levels of ecological divergence are sorely needed (Fig. 4.11).

4.5. Forms of reproductive isolation: conclusions and future directions

There is at least some evidence that every form of reproductive isolation can be part of ecological speciation. Critical remaining questions concern the relative importance and timing of evolution of different reproductive barriers. It seems logical that inherently ecological forms of reproductive isolation, such as immigrant inviability and ecologically dependent postmating isolation, can be the first barriers to evolve; these barriers presumably evolve at the same rate as adaptive divergence itself. However, strong tests of this assumption are few. There is some evidence that sexual isolation is stronger and evolves earlier than intrinsic postmating isolation during ecological speciation (Mendelson 2003), but exceptions to this trend exist, and outcomes might depend on the geography of divergence (Stelkens et al. 2010). Thus, it is premature to conclude that intrinsic postmating isolation is a less important part of ecological speciation than other forms of reproductive isolation, although such intrinsic isolation might be somewhat weaker in the early stages of speciation.

Explicit theory concerning the timing of evolution of different reproductive barriers, and how different barriers might affect each other, is scarce. Thus, it is difficult to predict which barriers are most likely to drive ecological speciation.

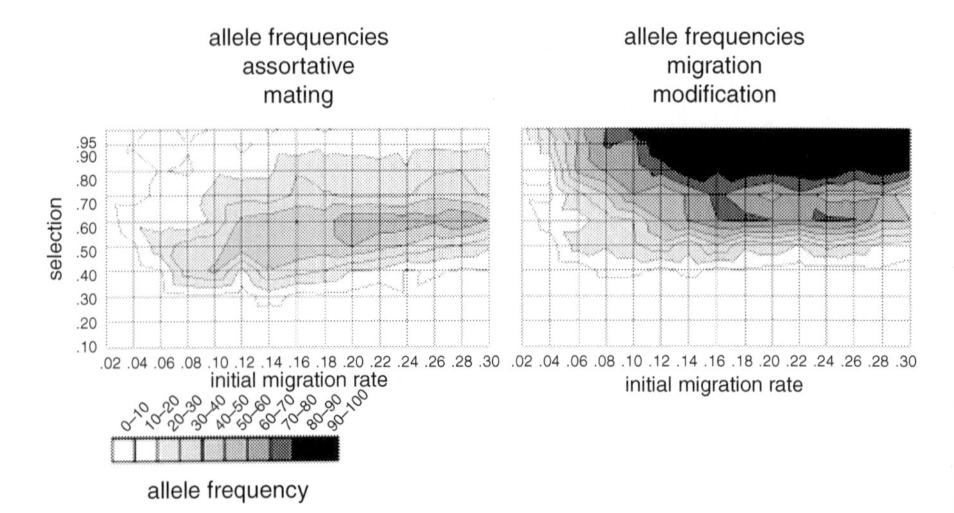

Figure 4.12. Allele frequencies for assortative mating (divergent mate preferences) versus migration modification (divergent habitat preferences) in simulations where both forms of isolation evolve simultaneously. Under strong selection (upper values of *y*-axis), the evolution of migration modification generally interfered with the evolution of assortative mating, by decreasing migration between populations and thereby reducing selection for assortative mating. Modified from Nosil and Yukilevich (2008) with permission of John Wiley & Sons Inc.

Some exceptions to this lack of theory exist. For example, Nosil and Yukilevich (2008) used simulation models of ecologically driven reinforcement to study the joint evolution of two forms of premating isolation: namely, divergent mating preferences and divergent habitat preferences (Fig. 4.12). In their models, the two different forms of reproductive isolation evolved simultaneously, potentially facilitating or interfering with each other's evolution. Under weak selection, the evolution of either form of reproductive isolation was unlikely. Under intermediate selection, the conditions favoring the rise of one form of reproductive isolation favored the rise of the other. Under strong selection, habitat preference evolved rapidly and then interfered with the evolution of assortative mating. This occurred because once habitat preference evolved, it reduced migration between populations, thereby decreasing the production of hybrids and reducing reinforcing selection for assortative mating. Relatedly, models by Agrawal et al. (2011) revealed that the evolution of intrinsic postmating isolation can constrain genetic divergence in more ecologically based reproductive barriers. These results generate an "interference hypothesis" in which the evolution of one barrier constrains or prevents the evolution of other barriers. Further work on the interaction of different barriers is sorely needed.

5

A genetic mechanism to link selection to reproductive isolation

During ecological speciation, divergent selection drives the evolution of reproductive isolation. Thus, the third and final component of ecological speciation is the genetic mechanism by which divergent selection on ecological traits is transmitted to the traits causing reproductive isolation. There are two ways this can occur, distinguished by the relationship between the genes under divergent selection (i.e., those affecting ecological traits) and those causing reproductive isolation (Rice and Hostert 1993, Kirkpatrick and Ravigné 2002). In the first, these genes are one in the same. In this case, a single gene has effects on two phenotypic traits: an ecological trait and reproductive isolation (which itself is considered a "trait"). Reproductive isolation thus evolves by direct selection because it is the pleiotropic effect of the genes under selection (Kirkpatrick and Barton 1997). In the second, genes under divergent selection are physically different from those causing reproductive isolation (e.g., the genes causing reproductive isolation might themselves be neutral). In this case, reproductive isolation evolves by indirect selection arising from the nonrandom association (linkage disequilibrium) of the genes for reproductive isolation and those for ecological traits. In essence, linkage disequilibrium between loci causes divergent selection on fitness loci to "spill over" onto reproductive isolation loci. In turn, this causes the evolution of reproductive isolation.

Understanding the nature of these genetic relationships is important, because they affect the strength of selection transmitted to the genes affecting reproductive isolation. In general, indirect selection transmitted via linkage disequilibrium is thought to be less effective than direct selection due to pleiotropy in the evolution of reproductive isolation (Felsenstein 1981, Kirkpatrick and Ryan 1991, Kirkpatrick and Barton 1997). This is for two reasons. First, under the scenario of linkage disequilibrium the statistical association between the genes under selection and those causing reproductive isolation tends to be imperfect, thus weakening the transmission of selection onto genes causing reproductive isolation. Second, the association between genes due to linkage disequilibrium is vulnerable to being weakened even further via recombination. Figure 5.1 illustrates these genetic scenarios of pleiotropy and linkage disequilibrium and depicts how tight physical linkage between loci on a chromosome can promote the maintenance of linkage disequilibrium. This chapter thus explores the conditions under which pleiotropy versus linkage disequilibrium can drive ecological speciation, including consideration of just how tight physical linkage needs to be for linkage disequilibrium to be an effective mechanism for speciation. The chapter concludes by considering the

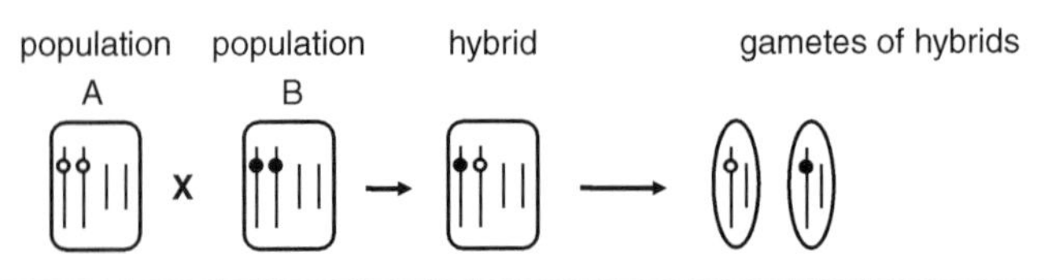

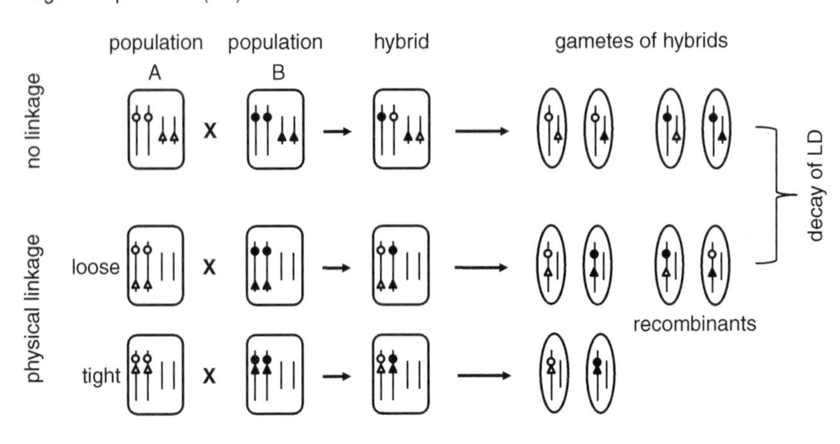

Figure 5.1. Genetic mechanisms for linking divergent selection to reproductive isolation. The figure shows individual genotypes in hybridization scenarios between two populations (both 2n = 4). Two loci are shown on each chromosome. Circles are divergently selected loci and triangles are reproductive isolation loci. Open circles are alleles conferring an advantage in environment A and solid circles are alleles conferring an advantage in environment B. a) Pleiotropy: the selected gene and the reproductive isolation gene are one and the same, resulting in divergent adaptation and reproductive isolation between populations. b) Linkage disequilibrium (LD): the selected gene and the reproductive isolation gene are different. A lack of physical linkage, or loose linkage, can result in the decay of LD. Tight physical linkage can maintain LD, resulting in patterns similar to pleiotropy. Modified from Matsubayashi et al. (2010) with permission of John Wiley & Sons Inc.

number of loci underlying ecological speciation and the specific genes driving the process.

5.1. Genetics of ecological speciation: the theory of divergence hitchhiking

"... what is required for divergence hitchhiking to be important is that effective recombination is significantly reduced locally in the genome without being substantially reduced globally" anonymous reviewer.

Pleiotropy is more effective than linkage disequilibrium in driving speciation (Rice and Hostert 1993, Kirkpatrick and Ravigné 2002, Gavrilets 2004). However, reduced recombination between loci, for example due to tight physical linkage, can result in different loci having effects similar to a single, pleiotropic locus (Hawthorne and Via 2001, Kirkpatrick and Ravigné 2002, Via and Hawthorne 2002, Gavrilets 2004, Servedio 2009). A recent verbal theory of "divergence hitchhiking" relies on this idea to generate a mechanism by which speciation in the face of gene flow may be easier than previously thought (Via and West 2008, Via 2009). The premise is that divergent selection reduces gene flow and interpopulation recombination between populations in different habitats, and even if recombination occurs, selection reduces the frequency of immigrant alleles in advanced generation hybrids (Gavrilets 2004, Nosil et al. 2005, Via and West 2008, Via 2009, Kobayashi and Telschow 2011). This reduction in effective gene flow might allow large regions of genetic differentiation to build up in the genome around the few loci subject to divergent selection. In a metaphorical sense, a focal gene under selection is in the "driver's seat" and loci physically linked to this gene on a chromosome (i.e., genes in the same car as the driver) "hitch a ride" to differentiation (Smadja et al. 2008). This general idea rests on the assumption that a site under divergent selection will create a relatively large region of reduced gene flow around it, enhancing the potential to accumulate differentiation at linked sites. These linked sites might contain genes affecting reproductive isolation, promoting speciation via the mechanism of linkage disequilibrium.

The extent to which divergence hitchhiking allows for linkage disequilibrium to act as an effective mechanism for linking genes under selection to those causing reproductive isolation will depend, in part, on the size of the affected genomic region (i.e., the size of the region in linkage disequilibrium): the larger the region, the larger the proportion of the genome resistant to gene flow, and the greater the possibility that genes within the region contribute to further reproductive isolation. As we will see in more detail in a subsequent chapter on genomics, empirical evidence concerning the size of differentiated gene regions during speciation is mixed, with evidence for regions of differentiation that are both large (e.g., Emelianov et al. 2004, Harr 2006, Rogers and Bernatchez 2007a, Via and West 2008) and small (e.g., Dopman et al. 2005, Turner et al. 2005, Geraldes et al. 2006, Machado et al. 2007, Noor et al. 2007, Turner and Hahn 2007, White et al. 2007, Yatabe et al. 2007, Makinen et al. 2008a, Makinen et al. 2008b, Storz and Kelly 2008, Wood et al. 2008, Strasburg et al. 2009, Scascitelli et al. 2010). Theoretically then, what are the factors that predict the size of differentiated regions within the genome?

Feder and Nosil (2010) developed the beginnings of a formal mathematical theory of divergence hitchhiking to start to address this question. Specifically, they used a combination of analytical and simulation approaches to expand the single-locus models of Charlesworth et al. (1997) to any number of loci under selection and to a wider range of parameter values. Feder and Nosil (2010) considered two demes subject to divergent selection and exchanging migrants at a gross rate m and quantified the ability for reduced gene flow surrounding selected sites to maintain neutral differentiation at regions of increasing recombination distance from the selected site. Specifically, effective gene flow at different genomic regions was quantified under scenarios where selection against immigrant alleles results in the effective migration rate being lower than the gross migration rate (Bengtsson 1985, Barton and Bengtsson 1986, Pialek and Barton 1997, Gavrilets and Cruzan

1998). The strength of selection, recombination rate between neutral and selected loci, gross migration rate, and numbers of loci under selection were varied. Thus, for each locus and set of parameter values, the effective migration rate was calculated, and then converted to F_{ST} using population genetic formula. The results allowed visualization of the decay of F_{ST} along a chromosome as one moves away from a divergently selected site (Fig. 5.2).

The main finding of Feder and Nosil (2010) was that divergence hitchhiking around a single locus can generate large regions of differentiation under some conditions, but that these conditions are limited. For a single locus under selection, regions of differentiation do not extend far along a chromosome away from a selected site unless selection is very strong and both effective population sizes and migration rates are low. When multiple unlinked loci under selection are considered, regions of differentiation can be larger. However, with many loci under selection, effective migration rates become so low that genome-wide divergence occurs and isolated regions of divergence do not occur. In essence, what is required for divergence hitchhiking to be important is that effective recombination is

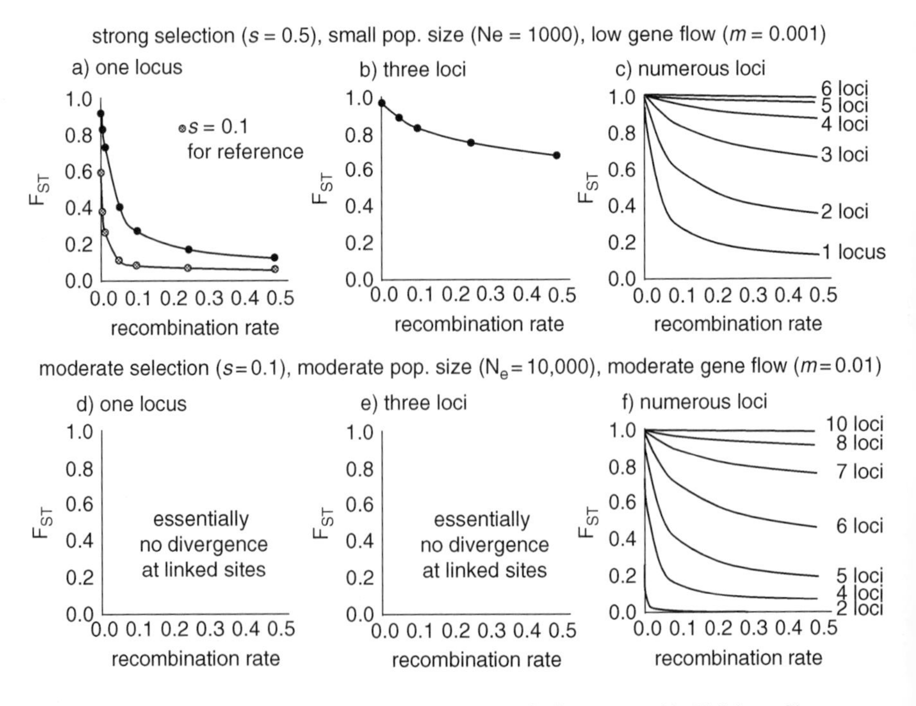

Figure 5.2. Some results concerning the theory of divergence hitchhiking. Shown are equilibrium levels of divergence (F_{ST}) at neutral sites linked to those under selection at various recombination distances. Divergence hitchhiking can generate regions of neutral differentiation extending away from a selected site, but only under certain conditions. Note that when numerous loci are under selection, genome-wide divergence can occur such that isolated regions of divergence are erased (e.g., in panel C compare scenarios with 1–3 loci to those with 4–6 loci under selection). Abbreviation: pop. = population. Reprinted from Feder and Nosil (2010) with permission of John Wiley & Sons Inc.

significantly reduced locally in the genome without being substantially reduced globally. The findings of Feder and Nosil (2010) concerning divergence hitchhiking specifically are consistent with general theory concerning the maintaining of linkage disequilibrium in the face of gene flow (Felsenstein 1981, Kirkpatrick and Ravigné 2002, Gavrilets 2004, Servedio 2009).

The conclusion is that linkage disequilibrium can be an effective genetic mechanism of ecological speciation, but this is most likely when: (1) physical linkage between selected and reproductive isolation is extremely tight; or (2) multiple loci are under divergent selection such that effective gene flow between populations is extensively reduced. On the latter point, gene flow reductions due to "multifarious" selection acting on many different and physically isolated genes may thus be an important promoter of speciation. Feder et al. (2011a) recently defined the term "genome hitchhiking" to describe the process by which genetic divergence across the genome is facilitated, even for neutral loci unlinked to those under selection, by the reductions in gene flow that selection causes (i.e., the genetic barrier to gene flow). To extend the metaphor of hitchhiking, loci unlinked to those under divergent selection are not in the same car as a focal gene under divergent selection, but nonetheless have their movement (i.e., their divergence) facilitated from the lower gene flow caused by selection. Genome hitchhiking is facilitated by multifarious selection and, unlike divergence hitchhiking, does not rely on physical linkage, although it might be aided by it.

The discussion here has focused on equilibrium levels of divergence at neutral sites linked to those under selection. A subsequent chapter considers the potential for existing regions of divergence to aid the build-up of further divergence, for example via affecting the establishment probabilities of new mutations. Future work could also incorporate the role of habitat preference and assortative mating in aiding genomic divergence. I now turn to reviewing the empirical data on the genetics of ecological speciation, first considering pleiotropy and then linkage disequilibrium, and in both cases focusing on the fit of the data to theoretical predictions.

5.2. Linking selection to reproductive isolation via pleiotropy

"The available evidence indicates that pleiotropy is virtually universal"

Wright (1968, p. 61)

Although pleiotropy in general may be common, is it of the specific form that is expected to promote ecological speciation? In other words, how might divergently selected loci cause reproductive isolation pleiotropically? There are numerous ways (note that in such cases, one can consider reproductive isolation as the second "trait" affected by a gene that also affects an ecological trait). For example, habitat isolation will evolve as a direct consequence of selection on habitat preference genes if individuals mate in their preferred habitat. This is the route by which sympatric speciation is thought to be most likely (Johnson et al. 1996) and has been experimentally demonstrated (Rice and Salt 1990). Sexual isolation can evolve owing to changes in mate preferences that arise as a pleiotropic consequence of the adaptive divergence of mating systems during sensory drive (Boughman 2002). Such changes in mate preferences may also cause sexual selection against hybrids as a direct consequence (Liou and Price 1994). In plants, pollinator isolation is a direct consequence of adaptation to different pollinators (e.g., Schemske and Bradshaw 1999), and temporal isolation, caused by differences in flowering time, may arise as the pleiotropic effect of adaptation to different environments (Lowry et al. 2008b). Intrinsic postzygotic isolation can arise pleiotropically if alleles favored by selection within each population contribute to incompatibilities between them (Dettman et al. 2007, Lee et al. 2008, Agrawal et al. 2011).

A perhaps less glamorous, but equally legitimate, example concerns loci under divergent selection generally, which can automatically cause immigrant inviability and extrinsic postmating isolation. Whenever such loci are detected, they represent scenarios of pleiotropy driving the evolution of (i.e., extrinsic) reproductive isolation. For example, quantitative trait locus (QTL) mapping in pea aphids identified loci with opposite effects on fecundity on the two hosts used by different host races (Hawthorne and Via 2001, Via and Hawthorne 2002). Such a fundamental genetic trade-off in fecundity probably contributes to extrinsic isolation between the host races. Because these forms of reproductive isolation appear common in nature, this may be the most general way that pleiotropy contributes to ecological speciation. However, explicit genetic studies testing this assumption are few.

In practice, separating pleiotropy from close physical linkage will often be a difficult task, because both generate strong associations between ecological traits and reproductive isolation. Pleiotropy might be demonstrated using breeding and experimental approaches that rule out a role for regions physically linked to those under selection (Conner 2002). For example, "near-isogenic lines" (NILs) can be created via repeated backcrossing in the lab, thereby generating breeding lines that differ primarily (or only) in the gene of interest. An example that pertains to ecological speciation concerns two species of monkeyflower that differ in flower color, an important trait contributing to pollinator isolation. Flower color itself is controlled in large part by a single QTL locus (*YUP*). Bradshaw and Schemske (2003) created NILs that differed phenotypically in flower color, but on average differed genetically only in the small proportion of the genome (3%) containing the *YUP* locus. In the predominately bumblebee-pollinated *Mimulus lewisii*, substitution of the *YUP* allele from the hummingbird-pollinated *M. cardinalis* increased its attractiveness to hummingbirds and pleiotropically decreased its attraction to bumblebees, facilitating the evolution of pollinator isolation (Fig. 5.3). In contrast, introgression of the *M. lewisii YUP* allele into *M. cardinalis* increased its attractiveness to bumblebees. The results strongly suggest that a single "*YUP* gene" has pleiotropic effects on both color and pollinator isolation, although physical linkage between different genes in the small introgressed region, one affecting color the other pollinator isolation, remains a possibility.

Other potential examples of genes under selection with pleiotropic effects on reproductive isolation are the wingless gene, which might affect both wing color and sexual isolation between species of *Heliconius* butterflies (Kronforst et al. 2006), and QTL affecting both host preference and performance in pea aphids (Hawthorne and Via 2001) (Fig. 5.4). Finally, correlations between host preference

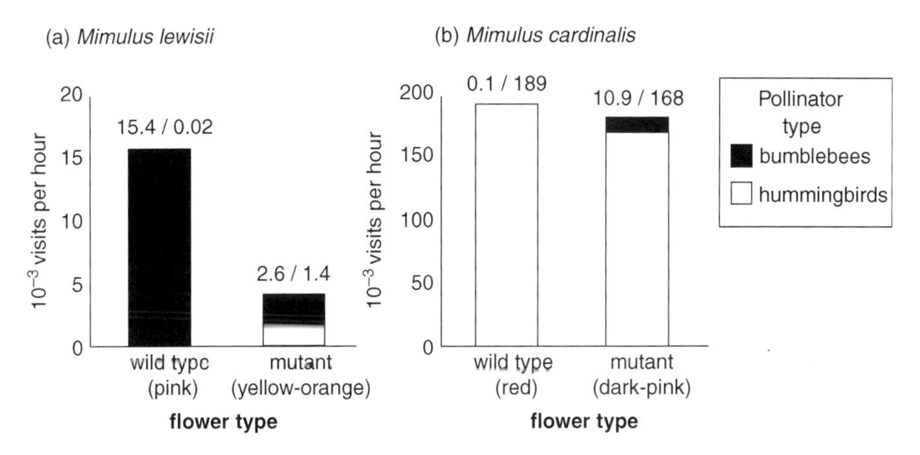

Figure 5.3. An example of pleiotropy contributing to ecological speciation: a QTL affecting flower color also affects rates of pollinator visitation, and thus pollinator isolation. Shown here are pollinator visitation rates to wild-type and near-isogenic lines (mutant) of *M. lewisii* and *M. cardinalis* monkeyflowers. Numbers above bars represent the ratio of bumblebee to hummingbird visitation. Pollinator preferences were shifted in the predicted direction, albeit only slightly, via substitution of the single *YUP* QTL. Data from Bradshaw and Schemske (2003).

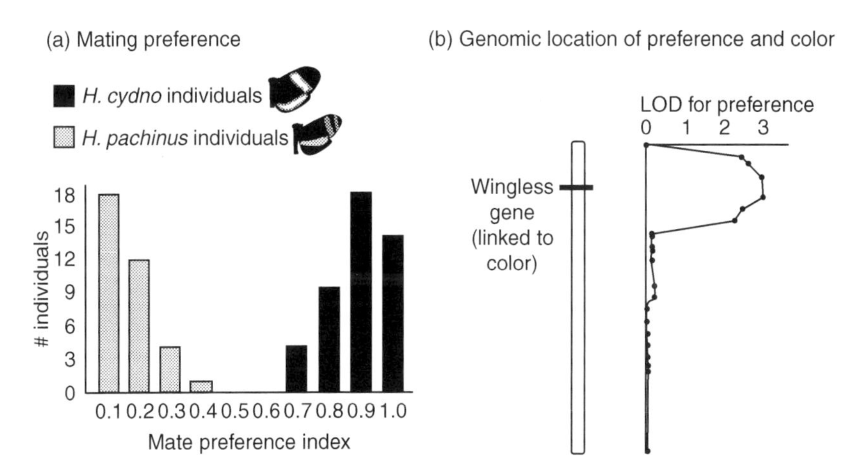

Figure 5.4. An example of pleiotropy or tight physical linkage contributing to ecological speciation. a) Two species of *Heliconius* butterflies differ in color pattern and also mate assortatively by color pattern. Mate preference index values are shown for each species (black versus gray bars), where 0.0 represents complete preference for *H. pachinus*, 0.5 represents no preference, and 1.0 represents complete preference for *H. cydno*. b) Preference and color pattern both map to the same genomic location, which harbors the *wingless* gene. 'LOD' score indicates strength of support for linkage of mate preference to a genomic location. Modified from Kronforst et al. (2006) with permission of the National Academy of Sciences USA.

and performance are relatively widespread in phytophagous insects, but further work is needed to test conclusively how often this involves pleiotropy (Gripenberg et al. 2010).

5.3. Linking selection to reproductive isolation via linkage disequilibrium

Theory predicts that maintenance of linkage disequilibrium between unlinked loci can sometimes be difficult (Felsenstein 1981, Servedio 2009, Feder and Nosil 2010). However, a number of factors might promote the maintenance of linkage disequilibrium between different genes under divergent selection versus those causing reproductive isolation, including very tight physical linkage (Hawthorne and Via 2001), structural features of the genome that reduce recombination (Noor et al. 2001, Rieseberg 2001), one-allele assortative mating mechanisms (Felsenstein 1981, Ortíz-Barrientos and Noor 2005), and strong selection (Charlesworth et al. 1997, Via and West 2008). I treat each in turn.

5.3.1. Tight physical linkage

There are few definitive examples of tight physical linkage between genes under divergent selection and those causing reproductive isolation, because most studies cannot distinguish tight linkage from pleiotropy. Perhaps the best example in which physical linkage appears most likely concerns the classic case of copper tolerance in *Mimulus* monkeyflowers. This has long been touted as an example in which adaptation (in this case to copper in the soil) has a pleiotropic effect on reproductive isolation (in this case intrinsic postmating isolation) (Macnair and Christie 1983, Christie and Macnair 1984). However, recent genetic crossing and QTL studies demonstrate that copper tolerance and postmating isolation are actually controlled by two different, yet tightly linked, loci (Wright and Willis unpublished).

5.3.2. Linkage disequilibrium and factors that reduce recombination: theory

Factors that reduce recombination between loci might help maintain linkage disequilibrium between selected and reproductive isolation loci (see Ortiz-Barrientos et al. 2002 for a review, Servedio 2009). For example, a number of verbal models propose that the recombination-suppressing effects of chromosomal inversions facilitate the maintenance of differences between interbreeding populations (Noor et al. 2001, Rieseberg 2001). The basic premise is that inversions reduce introgression for large regions of the genome and protect favorable genotypic combinations within these regions from being broken up by recombination (Butlin 2005, Hoffmann and Rieseberg 2008). In turn, each gene within these favorable genotypic combinations exhibits a greater composite selective differential between environments than it would individually (Rieseberg 2001). Additionally, inversions might tie up genes that confer divergent adaptation to those that affect assortative mating, thereby doubly promoting the maintenance of genetic differences between hybridizing populations (Butlin 2005). Finally, genotypic combinations within inversions that contribute to intrinsic genetic incompatibilities in hybrids can be protected from invasion by ancestral, compatible genotypic combinations (Noor et al. 2001). These arguments apply to the maintenance of genetic differences within inverted genomic regions, once an inversion has spread through a population. How inversions spread to high frequency in the first place is treated in the next chapter.

Formal mathematical models of the verbal arguments above emerged only recently and are important because they consider quantitatively the fact that low levels of recombination do occur within inversions (Hoffmann and Rieseberg 2008). What are the effects of these low levels of recombination? For genes under selection or causing reproductive isolation, Feder and Nosil (2009) compared the maintenance of genetic differentiation between formerly allopatric populations upon secondary contact and hybridization in three types of genomic regions: (1) inverted regions with no recombination within them; (2) inverted regions with low recombination within them; and (3) collinear regions with free recombination (Fig. 5.5). The most general finding was that even low levels of recombination result in the eventual loss of accentuated divergence in inverted regions, relative to that observed in collinear regions. Nonetheless, after secondary contact there was a transient period in which divergence was higher in inverted regions, and this period can last thousands of generations. The conclusion is that inversions can facilitate the maintenance of adaptive divergence under some conditions, but that large or permanent differences between inverted and collinear regions need not always occur. Rather than discounting the effects of inversions, the results generate explicit predictions about when they are likely to play a role in speciation: (1) when recombination within them is extensively reduced, as might especially occur near breakpoints of inversions; (2) when multiple loci within an inversion are subject to divergent selection; and (3) when hybridization is recent in origin. Future work could examine the effects of low rates of recombination on the build-up, rather than the maintenance, of genetic differentiation (c.f., Navarro and Barton 2003) and might consider neutral (rather than selected) differentiation (Guerrero et al. 2011). Finally, although this section focused on inversions, the role of recombination rate more generally warrants attention in research on the role of linkage disequilibrium in speciation (Nachman and Payseur 2011).

5.3.3. Linkage disequilibrium and factors that reduce recombination: data

Direct empirical evidence that factors that reduce recombination promote ecological speciation is sparse. Perhaps the only clear example that inversions can be involved in ecological speciation stems from field experiments with ecotypes of *Mimulus guttatus*. Specifically, Lowry and Willis (2010) showed that traits involved in local

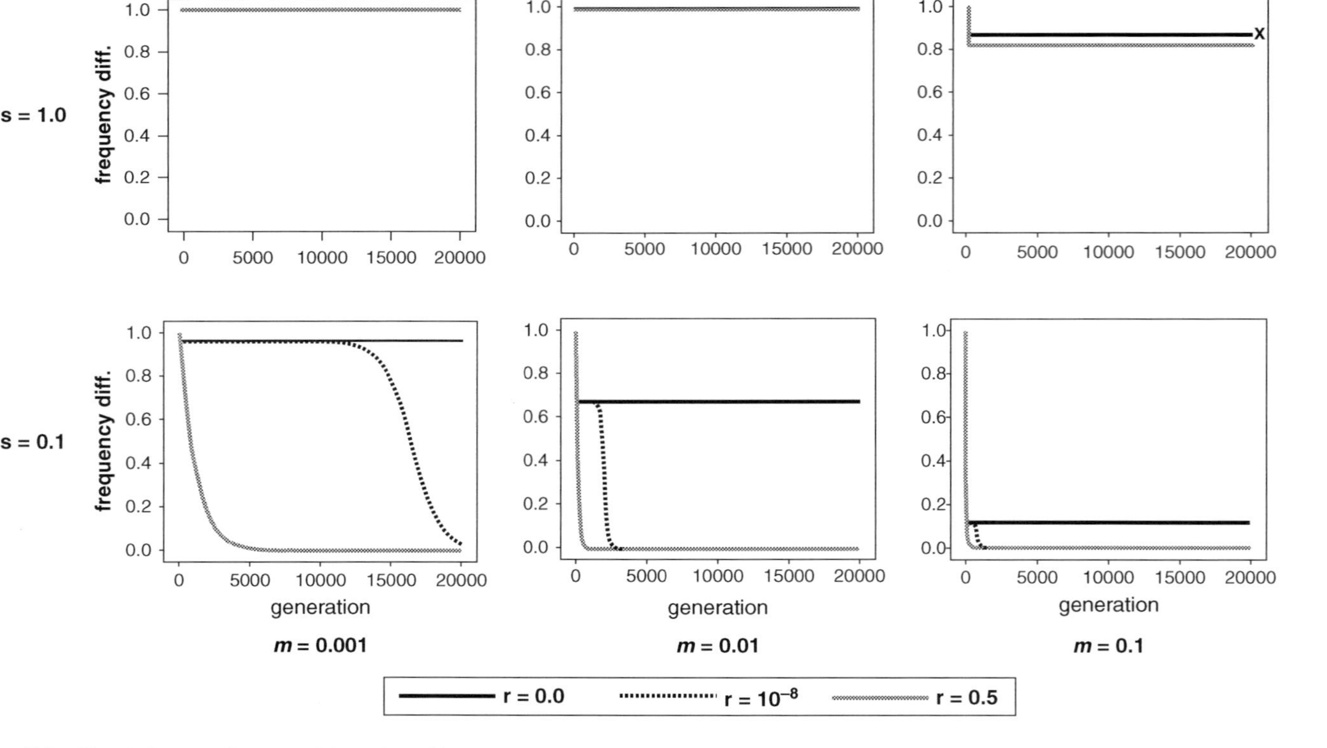

Figure 5.5. Simulation results examining the efficacy of chromosomal inversions in maintaining genetic divergence between hybridizing populations at two divergently selected loci (A and B). Shown is the difference in allele frequency between populations at locus B (frequency diff.) as a function of the number of generations since secondary contact between populations, for different selection (*s*) and migration (*m*) regimes. Each panel contrasts scenarios where the two loci reside in genomic regions with no recombination between loci (inversion with no recombination), versus low recombination between loci ($r = 10^{-8}$ = inversion with low recombination) versus free recombination between loci (collinear genomic region). Frequency differences between populations were maintained at all genomic regions when selection was strong relative to migration and decayed at all genomic regions when migration was high relative to selection. "X" denotes identical results for different recombination rates. Modified from Feder and Nosil (2009) with permission of John Wiley & Sons Inc.

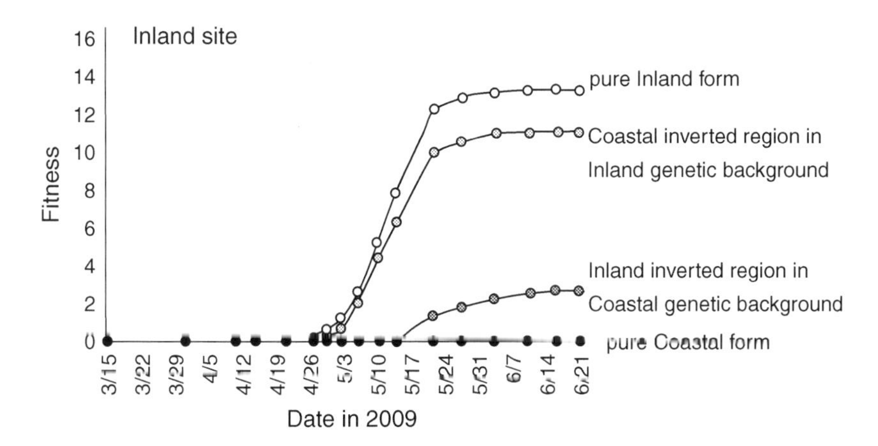

Figure 5.6. Evidence from a reciprocal-transplant field experiment that a chromosomal inversion contributes to divergent adaptation and reproductive isolation in the wild. Repeated backcrossing was used to create NILs that differed primarily in a genomic region harboring the chromosomal inversion. Shown here are the expected fitness of each parental type (coastal and inland pure forms) and two types of backcrosses in the inland environment. Controls for backcrossing were also created (e.g., a coastal inverted region backcrossed into a coastal genetic background; not shown for simplicity). Although the inversion has some effect on fitness, other genomic regions clearly also contribute. See Lowry et al. (2010) for details. Reprinted with permission of the Public Library of Science.

adaptation and reproductive isolation between coastal and inland ecotypes map to a chromosomal inversion. They then established NILs using repeated backcrossing and used them in a field experiment to demonstrate that the inversion contributes to divergent adaptation and reproductive isolation in the wild (Fig. 5.6).

Nonetheless, several empirical patterns provide indirect evidence that inversions promote speciation. First, inversions sometimes more commonly distinguish sympatric versus allopatric taxon pairs (Noor et al. 2001). This suggests that secondary contact between populations lacking inversions results in the fusion of the two populations such that most sympatric taxa observed in nature exhibit inversion differences. Indeed, a study comparing a sympatric to an allopatric taxon pair of *Drosophila* demonstrated that hybrid sterility factors are present in collinear regions in the allopatric pair, whereas virtually all of the sterility factors in the sympatric pair are associated with inverted regions, as expected if inversions facilitate speciation with gene flow (Brown et al. 2004). Second, QTL studies of hybridizing taxa have shown genes affecting adaptive divergence and reproductive isolation to reside within inversions (Rieseberg et al. 1999b, Noor et al. 2001, Feder et al. 2003a, Feder et al. 2003b, Hoffmann and Rieseberg 2008, Manoukis et al. 2008). Third, studies in *Rhagoletis pomonella* host races (Michel et al. 2010), *Drosophila* species pairs (Machado et al. 2007, Noor et al. 2007, McGaugh et al. 2011), and *Anopheles gambiae* mosquitoes (White et al. 2007) supported the prediction that genetic divergence will be accentuated for genes that reside within inversions, relative to collinear regions. However, divergence in these examples was only mildly elevated within inverted regions, and other studies report a limited role for inversions in maintaining elevated genetic divergence (Yatabe et al. 2007, Hoffmann and Rieseberg 2008).

A suite of recent population genomic studies (i.e., genome scans) seems to contradict the importance of inversions. In this literature, genomic regions involved in adaptive divergence or reproductive isolation are inferred via their accentuated genetic differentiation between populations (see Beaumont and Nichols 1996, Beaumont 2005, Nielsen 2005, Storz 2005, Noor and Feder 2006, Stinchcombe and Hoekstra 2008 for reviews). Studies using such approaches often report that regions of exceptionally high differentiation ("outlier loci") are widely distributed across the genome, for example showing little or no sign of linkage disequilibrium with one another and mapping to numerous different chromosomes (Scotti-Saintagne et al. 2004, Achere et al. 2005, Grahame et al. 2006, Rogers and Bernatchez 2007a, Egan et al. 2008, Nosil et al. 2008, Wood et al. 2008, see Nosil et al. 2009a for a review). This suggests that genes affecting adaptive divergence are not always clustered within an inversion and that divergence of collinear regions can readily occur (although it is possible that different outlier loci each lie within different, genomically dispersed inversions). In short, inversion models have some, but not overwhelming, empirical support. As noted by Hoffman and Rieseberg (2008), "*the relative importance of inversions in the adaptive evolution of traits has rarely been addressed. It is therefore usually not clear if inversions play a critical role in adaptive shifts or if they only have a minor effect.*" (p. 31).

It seems reasonable thus to propose that factors that reduce recombination facilitate ecological speciation, but only when they reduce recombination to a very low level, such as occurs near chromosomal breakpoints or centromeres (Turner et al. 2005, Geraldes et al. 2006, Carneiro et al. 2009). A study by Strasburg et al. (2009) provides insight into this unresolved issue by examining divergence between hybridizing *Helianthus* sunflower species at 77 loci distributed across the genome. They report widespread adaptive divergence in collinear regions and find that divergence is not accentuated within inversions, except perhaps near chromosomal breakpoints (Fig. 5.7). In sum, future work should examine just how reduced recombination needs to be for linkage disequilibrium to act as a genetic mechanism of speciation.

5.3.4. One-allele assortative mating mechanisms

The genetic basis of reproductive isolation itself also affects the likelihood of speciation via the mechanism of linkage disequilibrium. In a classic and highly influential paper, Felsenstein (1981) pointed out that there are two distinct possibilities, which he modeled theoretically and termed "one-allele" and "two-allele" mechanisms. In a one-allele mechanism, reproductive isolation is caused by the

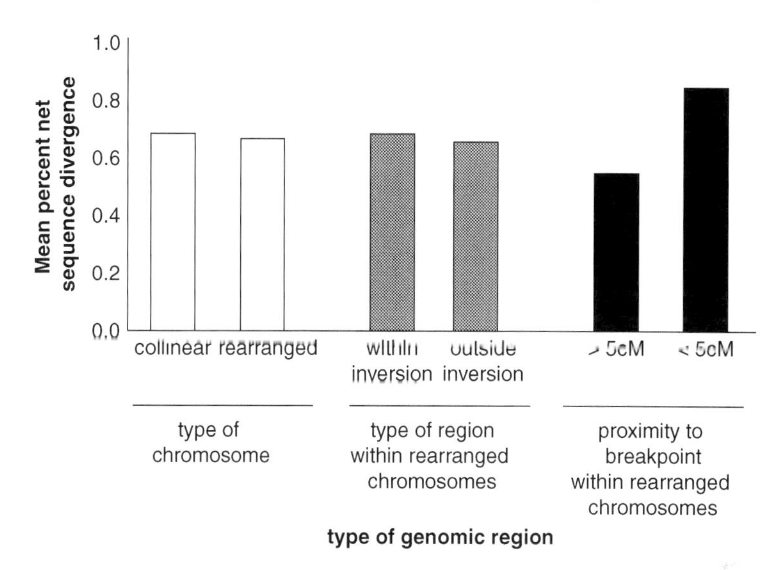

Figure 5.7. Genetic divergence between hybridizing species of *Helianthus* sunflowers in different types of genomic regions. Only differences associated with proximity to breakpoints within rearranged chromosomes approached or achieved statistical significance. Abbreviation: cM, centimorgans. Data from Strasburg et al. (2009).

same allele for assortative mating fixing in both populations (e.g., an allele causing individuals to prefer mates phenotypically similar to themselves). In a two-allele mechanism, different alleles fix in each population (e.g., a preference allele for large individuals in one population and a different preference allele for small individuals in the other). This distinction is important when considering the effects of recombination. Recombination in a two-allele mechanism breaks down linkage disequilibrium between genes under selection and those causing assortative mating, weakening or erasing the genetic association between them. In contrast, this problem is completely alleviated for a one-allele mechanism. Thus, one-allele mechanisms are a powerful genetic mechanism of speciation. The strongest evidence for the existence of a one-allele mechanism stems from studies in *Drosophila*. Females from populations of *D. pseudoobscura* that co-occur with its sibling species, *D. persimilis*, exhibit high reluctance to mate with *D. persimilis* males (i.e., reinforcement) (Noor 1995). In contrast, females from allopatric populations of *D. pseudoobscura* are more inclined to hybridize. Ortíz-Barrientos and Noor (2005) introgressed alleles for mating discrimination from *D. pseudoobscura* into the genetic background of *D. persimilis*. They showed that alleles conferring high or low mating discrimination in *D. pseudoobscura* (i.e., those stemming from sympatric versus allopatric parts of the species range, respectively) produce the same effects when inserted into *D. persimilis*. This suggests that the same assortative

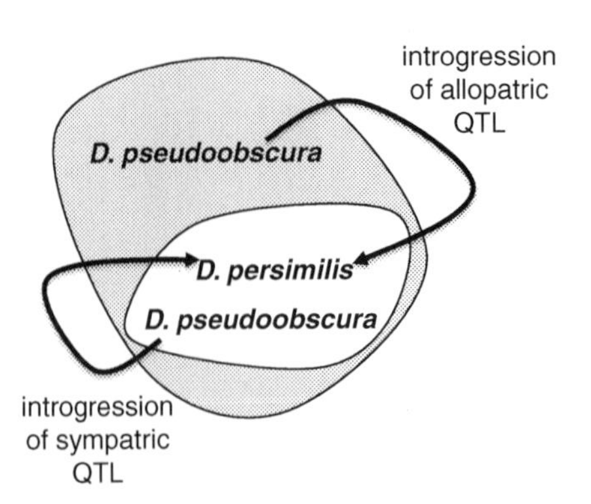

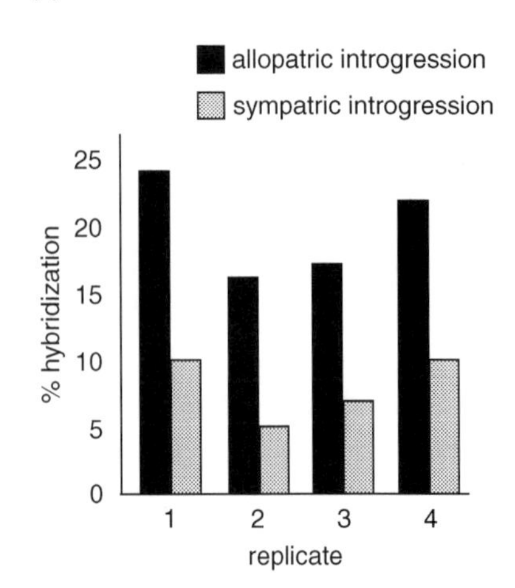

Figure 5.8. An example of a one-allele assortative mating in *Drosophila*. See text for details. Data from Table 1 of Ortíz-Barrientos and Noor (2005).

mating allele occurs in both species (Fig. 5.8). Further such tests are sorely needed, because the frequency of one-allele mechanisms in nature is unknown and because the *Drosophila* example is not known to represent ecological speciation.

5.3.5. Strong selection

A final consideration is the strength of divergent selection. Theory indicates that effective migration rates between populations will decrease, and thus that divergence hitchhiking will increase, with increasing selection strength (Charlesworth et al. 1997, Gavrilets and Cruzan 1998, Barton 2000, Gavrilets 2004, Feder and Nosil 2010). Specifically, as selection gets stronger, regions under divergent selection will become more differentiated, and hitchhiking in adjacent chromosomal regions can initially span larger areas than when selection is weaker. Indeed, in the models discussed in this chapter, strong selection often played roles analogous to, and sometimes more powerful than, reduced recombination in promoting genetic differentiation (Feder and Nosil 2009, 2010, Feder et al. 2011a). As discussed in more detail in Chapter 7, other factors such as epistasis and the recency of selective sweeps will interact with the strength of selection to determine levels of genetic divergence (Sinervo and Svensson 2002, Gavrilets 2004, Barrett and Schluter 2008). These theoretical issues cannot yet be properly tested because most data on selection strength in nature at the genetic level are observational and based on indirect inferences from patterns of differentiation (Linnen et al. 2009, Nosil et al. 2009a), rather than on direct experimental measurements of selection (but see Barrett et al. 2008, Michel et al. 2010). Sorely needed now are experimental data on how selection strength varies across the genome.

5.3.6. Genetics mechanisms linking selection to reproductive isolation: conclusions

Table 5.1 summarizes theoretical predictions, empirical questions, and current knowledge concerning the role of different genetic factors in ecological speciation. Perhaps the main conclusion is that pleiotropy is the most effective genetic mechanism of ecological speciation. Some examples of the types of pleiotropy conducive

Table 5.1. A summary of the role of four genetic factors in promoting ecological speciation.

General factor	Theoretical prediction	Empirical question(s)	Existing data and what is missing
1. Pleiotropy	Pleiotropy is the most effective mechanism for linking divergent selection to genes causing reproductive isolation.	Do examples of such pleiotropy exist?	Yes, a few explicit examples exist. All cases of immigrant inviability and extrinsic postmating isolation probably represent pleiotropy, but genetic evidence is scarce.
2. Physical linkage, perhaps facilitated by regions of reduced recombination in the genome (e.g., chromosomal inversions)	Physical linkage can allow linkage disequilibrium to be an effective genetic mechanism of speciation, but only under specific conditions (e.g., strong selection, very tight linkage).	How large are regions of differentiation in the genome created by divergence hitchhiking? How tight does linkage need to be for substantial hitchhiking? How effective are factors that reduce recombination in promoting genetic divergence?	Mixed, with examples of large and small regions of differentiation—many cases suggest that physical linkage needs to be very tight and recombination extensively reduced for speciation to be promoted.
3. One-allele assortative mating mechanism	A highly effective genetic mechanism of speciation.	Are there examples of one-allele assortative mating mechanisms?	One good example exists, but links to ecological speciation specifically are lacking.
4. Strong selection	Strong selection can promote speciation in a manner analogous to reduced recombination.	Are there examples of strong selection playing roles analogous to favorable genetic architectures?	Strong selection can occur, but has almost never been directly measured at a genetic level.

for ecological speciation have emerged, but the critical question of how common they are remains. The more common it is, the more likely ecological speciation will be a widespread mechanism of speciation. In comparison to pleiotropy, linkage disequilibrium is a less-effective mechanism of speciation. For example, divergence hitchhiking around specific individual loci is expected to cause large genomic regions of differentiation only under relatively limited conditions. Nonetheless, we can generate predictions about when linkage disequilibrium will be an effective genetic mechanism of speciation: when physical linkage is tight and selection strong, when selection acts on many loci across the genome to reduce genome-wide rates of effective gene flow, when other factors such as chromosomal inversions strongly reduce recombination, and when the same assortative mating allele fixes in both of two diverging populations.

5.4. Genetic constraints on ecological speciation

The discussion thus far has focused on how different genetic scenarios promote, rather than constrain, speciation. However, two main types of genetic constraints might hinder divergence in a selected trait, thereby constraining ecological speciation: (1) constraints due to lack of genetic variation in the selected trait (Lande 1979); and (2) constraints due to genetic correlations between traits (Lande 1979, Barton 1995, Schluter 1996a, Orr 2000, Otto 2004). On the first possibility, persistent directional selection is expected to occur during ecological speciation and can deplete genetic variation, potentially resulting in constraints on adaptive divergence. For example, Futuyma et al. (1995) examined genetic constraints related to host-plant adaptation in multiple species of leaf beetles in the genus *Ophraella*. In some species, they report a complete lack of genetic variation in the ability to use host plants of congeners, and in other cases, only limited variation was found. They conclude that host shifts and adaptation to a different host could be strongly constrained by lack of suitable genetic variation. Further such studies are warranted.

On the second possibility, it is well known that traits do not evolve independently, owing to genetic covariance between traits arising from pleiotropy and linkage disequilibrium (Arnold 1992, Lynch and Walsh 1998, Chenoweth et al. 2010). Thus, evolutionary divergence in any single trait might be influenced by correlations with other traits. For example, adaptive divergence in floral traits in *Ipomoea hederacea* is constrained by correlations between traits (Smith and Rausher 2008). More generally, the between-generation evolutionary change in trait means is a function of the matrix of genetic variances and covariances ("the G-matrix"), and the vector of selection gradients (Lande 1979). As a consequence, it has been proposed that adaptive divergence will be biased in the direction of maximal multivariate genetic variation, popularized as the "genetic lines of least resistance" (Schluter 1996a). Although examples of such biased divergence exist (Schluter 2000b, Chenoweth et al. 2010), a systematic review of the effects of the G-matrix on evolution emerged only recently. Agrawal and Stinchcombe (2009) measured the impact of genetic covariances on the rate of adaptation using 45 datasets spanning a range of plant and animal taxa. They compared the rate of fitness increase given the observed G-matrix to the expected rate if all the covariances in the G-matrix were set to zero. No net tendency for covariances to constrain the rate of adaptation was found. There were some examples in which covariances strongly constrained the rate of adaptation, but these were balanced by counterexamples in which covariances facilitated the rate of adaptation, and cases in which covariances had little or no effect. Some of this variability might be due to the underlying causes of genetic covariances. For example, genetic covariances caused by pleiotropy could be more likely to constrain divergence than those caused by linkage disequilibrium, because the latter will be weaker and can be altered via recombination (Otto 2004).

A final consideration concerns factors that reduce recombination, which are generally argued to promote speciation. Reduced recombination, however, may decrease the chance of gene combinations favorable for ecological speciation being brought together in the first place (Kirkpatrick et al. 2002, Ortiz-Barrientos et al. 2002, Feder et al. 2011a). Therefore, whether reduced recombination promotes or constrains ecological speciation will depend on the relative importance of building up versus maintaining appropriate forms of linkage disequilibrium (Servedio 2009). In turn, this might vary across the speciation process, with the build-up of linkage disequilibrium being important for initiating speciation, and its maintenance being critical for completing the process. Finally, almost all work to date on genetic constraints has focused on adaptive divergence, rather than reproductive isolation, and thus studies of the latter are especially needed to better understand the role of genetic constraints in ecological speciation.

5.5. The individual genetic basis of traits under selection and traits conferring reproductive isolation

Independent of how selected and reproductive isolation loci are associated with each other, their individual genetic bases are of interest for understanding speciation. Here I review the individual genetic bases of adaptive divergence and reproductive isolation, focusing on the number of genes affecting a specific trait under selection or a specific form of reproductive isolation. A subsequent chapter on genomics considers the related issue of the number of loci that diverge across the entire genome. There are two extremes under which the genetic basis of ecological speciation may fall: (1) many loci with a small effect (polygenic control), as opposed to (2) one or a few loci with a large effect (simple genetic control).

5.5.1. Theoretical relevance of number of loci under selection for speciation

In general, genetic divergence at more loci promotes speciation, for example by increasing the probability of divergence at loci that are incompatible with one another (Orr 1995, Gavrilets 2004). However, the number of loci affecting a particular trait can influence the rate and likelihood that population divergence actually occurs, for example by affecting per locus selection coefficients (Gavrilets 2004, Gavrilets and Vose 2005), and the manner in which genetic divergence is distributed across the genome (i.e., how localized it is). Thus, having more loci affecting a specific trait under divergent selection does not always promote speciation.

For a given total strength of divergent selection, the fewer the genes subject to selection the higher the per-locus selection coefficients. Divergence for a given locus is a function of its selection coefficient (s) and rates of gene flow (m) (Gavrilets 2004, Mallet 2006), with divergence generally occurring so long as $s > m$. Thus, it may be easier for selection to overcome gene flow and generate adaptive divergence when few loci are involved (Gavrilets and Vose 2005, Yeaman and Otto 2011, Yeaman and Whitlock 2011). In addition, genes with large selection coefficients, if they happen to cause reproductive isolation, may cause greater reproductive isolation than more weakly selected loci.

However, if only a few genes diverge, then it may be unlikely overall that forms of reproductive isolation that are not direct consequences of selection will arise via the fortuitous pleiotropic effects of divergence (Rice and Hostert 1993, Nosil et al. 2009b). In contrast, when divergent selection acts on many genomic regions, differentiation across the genome is "sampled" and might incidentally involve genes yielding additional isolation (e.g., genes responsible for mating preference or intrinsic hybrid sterility). At the genetic level, as more and more loci diverge, the number of pairwise interactions that can lead to reproductive incompatibilities accumulates rapidly, and this should lead to a "snowball" of the accumulation of genetic incompatibilities (Orr 1995, Orr and Turelli 2001, Gourbiere and Mallet 2010). In essence, simple genetic control can facilitate divergence in the face of gene flow, but polygenic control can sometimes result in a greater correlated evolutionary response and associated reproductive isolation (Rice and Hostert 1993, Nosil et al. 2009b) (Fig. 5.9). Given these theoretical considerations, what is the genetic basis of adaptive divergence and reproductive isolation in nature?

5.5.2. Genetic basis of adaptive divergence and reproductive isolation during ecological speciation

I focus here on an article that reviewed the genetic basis of adaptation to, and behavioral preference for, different host-plant species by phytophagous insects (Matsubayashi et al. 2010). Although many other studies exist, a full review of the genetics of adaptation and speciation is beyond the scope of this book, and the results discussed here pertain to ecological speciation specifically and illustrate a

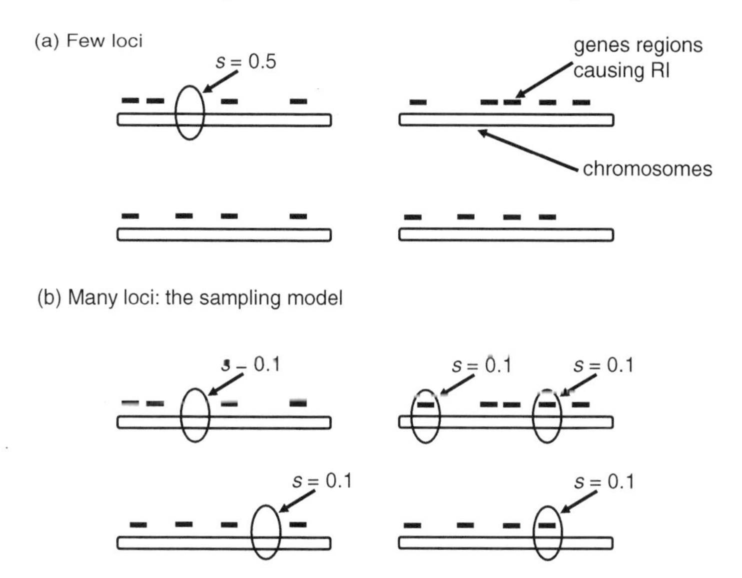

Figure 5.9. Theoretical relevance of the number loci affecting a trait under divergent selection on ecological speciation. Black bars above chromosomes represent genomic regions harboring genes which cause reproductive isolation (RI), where RI refers to forms of RI that are not direct consequences of divergent selection. "s" represents per-locus selection coefficients, assuming a finite amount of total selection (here = 0.5). Black ovals represent genes under divergent selection and divergence in these genes is assumed to occur so long as $s > m$. a) Few loci subject to selection results in high per locus s, which can overcome strong gene flow to cause divergence, but may be unlikely to incidentally cause RI via pleiotropy. b) Many loci subject to divergent selection. Countering gene flow is more difficult, but if divergence across more of the genome can occur (i.e., is "sampled"), then the probability of divergence in regions causing RI is increased.

few key points (see Coyne and Orr 2004, Lowry et al. 2008a, Schluter and Conte 2009 for more thorough reviews). The review of Matsubayashi et al. (2010) considered studies that used experimental crosses between closely related populations or species of insects spanning numerous orders. For adaptation, ten relevant studies were identified. Six studies supported simple genetic control, but the others supported polygenic control. In all studies examined, performance loci were to some extent located on autosomes. Seven studies revealed complete dominance, indicating that F_1 hybrids would survive only on one of the parental hosts. For divergent host-plant feeding or ovipositing preferences, 14 relevant studies were identified. Of these, ten revealed that host preference is determined by a few loci and the others reported more polygenic control. Eight of the studies reviewed revealed that preference for one host was dominant over preference for the other. Although the number of available studies is few, what can be said is that host adaptation and preference can involve only a few loci of large effect, potentially contributing to the propensity for herbivorous insects to undergo speciation with gene flow (Berlocher and Feder 2002, Drès and Mallet 2002). However, there are certainly examples of more polygenic control of reproductive isolation (Presgraves 2003, Coyne and Orr 2004, Ellison et al. 2011), and a review of the genetic basis of speciation in flowering plants reported that the number of loci involved in reproductive isolation varied among systems and traits (range = 1–17), with no clear differences between pre- and postmating isolation (Lowry et al. 2008a). Thus, much work remains to be done before theoretical expectations can be connected to empirical patterns.

5.5.3. Opposing genetic dominance as a form of reproductive isolation

An interesting finding reported by Matsubayashi et al. (2010) is that roughly half of the studies reviewed reported evidence for genetic dominance. Additionally, in some systems, such as *D. melanogaster* species complex, *Acrocercops transecta* moths, and *Mitoura* butterflies, the direction of dominance was opposite between preference and performance traits. This finding has implications for speciation, because opposing dominance acts to reduce gene flow. Consider an illustrative example. In *A. transecta*, performance on *Juglans* plants was completely dominant over that on *Lyonia*, but preference for *Lyonia* was completely dominant over that for *Juglans* (Ohshima 2008, 2010). Because of the dominant performance on *Juglans*, F_1 larvae can survive only on *Juglans*. This suggests that gene flow will occur (i.e., asymmetrically) from the *Lyonia* race to the *Juglans* race. However, eclosed F_1 females avoid ovipositing on *Juglans* owing to the expression of the dominant *Lyonia* preference. This results in the genetic components that came from the *Lyonia* race being removed from the *Juglans* race. Thus, differences in the mode of dominance inheritance between preference and performance itself functions as a barrier to gene flow between populations adapted to different hosts (Forister 2005, Ohshima 2008, 2010). Differences in the chromosomal location of genes (autosomes versus Z chromosomes) could also contribute to this type of reproductive isolation (Nygren et al. 2006). Finally, the observed differences in dominance direction between performance and preference traits suggest that genes responsible for the two traits are likely to be physically different from one another. This means that genetic covariance between the two traits is not caused by pleiotropy, and thus that divergence involves multilocus evolution.

5.6. Ecological speciation genes

The longstanding goal of finding specific genes causing reproductive isolation is being achieved (Coyne and Orr 2004, Orr 2005a, Noor and Feder 2006, Rieseberg and Blackman 2010). However, this chapter on genetics has thus far been void of much discussion of actual genes. Which genes contribute to ecological speciation?

What are their effect sizes? Are the same genes involved repeatedly in independent speciation events? How prevalent are Dobzhansky–Muller incompatibilities compared with additive genetic effects? Are changes at regulatory sites or coding regions more likely to underlie speciation? The solutions to such questions are of broad interest because they will help us clarify the process of speciation, including the roles of different forms of selection versus drift in species formation and the importance of genetic architecture for speciation (Felsenstein 1981, Coyne and Orr 2004, Schluter 2009).

The genes to help us answer these questions are "speciation genes," which Nosil and Schluter (2011) recently defined as "*those genes whose differentiation between populations made a significant contribution to the evolution of reproductive isolation.*" Under this definition, divergence at a speciation gene increases the total amount of reproductive isolation between populations. It thus follows that divergence must occur before reproductive isolation is complete. This contrasts with most previous definitions that considered a speciation gene to be any gene contributing to a contemporary component of reproductive isolation (Orr and Presgraves 2000, Orr et al. 2004, Wu and Ting 2004, Orr 2005a). However, not all genes affecting reproductive isolation today played a role in the process of speciation (Coyne and Orr 2004, Via 2009). Instead, many "post-speciation" genes will have diverged after reproductive isolation was complete, thereby duplicating the effects of genes that diverged earlier in the process. In contrast, speciation genes contribute to the evolution of reproductive isolation. In such cases, one cannot predict whether speciation will eventually be completed, but we nevertheless learn about the process of speciation (i.e., the evolution of reproductive isolation) in the meantime.

Beyond the timing of divergence, we might consider the magnitude of a gene's contribution to the evolution of reproductive isolation at the time it diverged, referred to as its "effect size." The concept of effect size is conceptually straightforward when applied to genes that made standalone increments to total reproductive isolation. However, sometimes (or perhaps often) only the later-fixing gene of a pair (or set) of interacting genes led to an immediate increase in reproductive isolation, whereas the earlier-fixing gene had little or no effect at the time of its divergence. In such cases the effect size of the later-fixing gene is assigned retroactively to both members of the pair. Effect size is difficult to measure, but there is little option if we wish to distinguish the genes that mattered to speciation from those that did not. These considerations lead to the following criteria to identify a speciation gene (following Nosil and Schluter 2011):

1) demonstrating that the gene has an effect on a component of reproductive isolation;
2) demonstrating that divergence at the locus occurred before speciation was complete;
3) quantifying the degree to which divergence at the locus increased total reproductive isolation.

Here I focus on "ecological speciation genes," which I define as above, but more specifically as genes *subject to divergent selection* whose differentiation between populations made a significant contribution to the evolution of reproductive isolation. Thus a fourth criterion is added:

4) the gene is subject to divergent natural selection.

This fourth criterion is automatically met for genes causing extrinsic reproductive isolation (immigrant inviability and ecologically dependent postmating isolation). Genes causing other types of reproductive isolation might also meet this criterion, but further evidence is required to demonstrate this. Nosil and Schluter (2011) reviewed candidate speciation genes and found no clear examples of genes known to cause extrinsic isolation. To compensate, they tabulated some cases of genes known or suspected of being under divergent natural selection, on the assumption that these genes contribute to extrinsic isolation. I focus below on these candidate ecological speciation genes (Table 5.2). Examples other than those considered here exist, but those considered here span a range of reproductive barriers, illustrate my main points, and provide a starting point for evaluating speciation genes.

5.6.1. Genes that affect reproductive isolation today

The strength of evidence that a given gene is the cause of reproductive isolation varies greatly among candidate speciation genes (Table 5.2). The strongest evidence comes from studies that have mapped reproductive isolation to a candidate gene and then used experimental methods such as positional cloning, gene replacement/knockout, gene expression assays, and transgenic manipulations to confirm the gene involved. The examples most clearly pertaining to ecological speciation are the odorant binding protein genes, *OBP57d* and *OBP57e*, which affect taste perception and host-plant preference, and thus probably contribute to habitat isolation between *D. melanogaster and D. sechellia* (Matsuo et al. 2007). In other cases, reproductive isolation has been mapped to a narrow genomic region and a candidate gene within the QTL has been identified, but the evidence for the gene's involvement in reproductive isolation remains correlative. In these cases, the effects of the candidate genes are yet to be disentangled from those of physically linked genes. In a third class of examples, the evidence is even more indirect, where a specific gene has been found to affect a phenotypic trait known to be under divergent natural selection between the contrasting environments of the parent species, but effects on extrinsic reproductive isolation have not yet been measured. An example is the Ectodysplasin (*Eda*) gene in threespine stickleback fish, which is known to affect differences between wild marine and freshwater stickleback populations in their number of bony lateral plates (Colosimo et al. 2005) (Fig. 5.10). *Eda* thus probably contributes to immigrant inviability and extrinsic postmating isolation, but this is yet to be

Table 5.2. Examples of candidate ecological speciation genes and their fit to the criteria outlined in the text. In many cases, the gene in question has yet to be directly shown to contribute to reproductive isolation, but by affecting divergent adaptation it probably contributes to extrinsic isolation (text for details). The list is far from exhaustive, but illustrates the main points covered in the text, including how each of the criteria for an ecological speciation gene might be tested. I refer readers to past reviews for a more thorough treatment (Coyne and Orr 2004, Rieseberg and Blackman 2010, Nosil and Schluter 2011).

Gene	Details of gene or phenotype affected	Study system	Putative form of RI	Criterion 1: gene affecting RI	Criterion 2: timing of divergence	Criterion 3: increase in total RI	References
1. *OBP57d* and *OBP57e*	Odorant binding proteins that affect test perception and host plant preference	*D. melanogaster* and *D. sechellia*	HI	**Gene affects RI** (affects host plant preference)	**Unknown**	**Unknown**	(Matsuo et al. 2007)
2. *Eda*	Lateral plate number (in part, adaptation to predation regimes)	*Gasterosteus aculeatus* freshwater and marine stickleback fishes	II, EDPI	**Gene known, RI only inferred**	**Before complete RI** (RI between forms incomplete)	**Unknown**	(Colosimo et al. 2005, Barrett et al. 2008, Marchinko 2009)
3. *Pitx1*	Pelvic spine apparatus (in part, adaptation to predation regimes)	*Gasterosteus aculeatus* freshwater and marine stickleback fishes	II, EDPI	**Gene known, RI only inferred** (effect on RI could be small)	**Before complete RI** (as for *Eda*)	**Unknown**	(Shapiro et al. 2004, Chan et al. 2010)
4. *Mc1r*	Pigmentation (adaptation to be camouflaged against visual predation)	*Peromyscus poliontos* mainland and beach oldfield mouse subspecies	II, EDPI	**Gene known, RI only inferred** (effect on RI could be small and effects on phenotype vary among traits)	**Before complete RI**	**Unknown**	(Hoekstra et al. 2006)

Table 5.2. *Cont.*

Gene	Details of gene or phenotype affected	Study system	Putative form of RI	Criterion 1: gene affecting RI	Criterion 2: timing of divergence	Criterion 3: increase in total RI	References
5. *Agouti*	Pigmentation (adaptation for crypsis via visual predation)	*Peromyscus poliontos* beach mouse subspecies	II, EDPI	**Gene associated with traits, linked genes could contribute, RI only inferred**	**Before complete RI**	**Unknown**	(Steiner et al. 2007)
6. *LWS Opsin*	Vision (visual adaptation to different light environments)	*Pundamilia pundamilia* and *P. nyererei* cichlid fishes	EDPI, possibly HI and SI	**Gene associated with RI**	**Before complete RI** (common hybridization known)	**Unknown**	(Seehausen et al. 2008b, Seehausen personal communication)
7. *ROSEA* (also *ELUTA, SULFUREA*)	Flower color (attractiveness and adaptation to different pollinators)	*Antirrhinum majus striatum* and *Antirrhinum m. pseudomajus* flower morphs	II, EDPI	**Gene known, RI only inferred**	**Before complete RI** (hybridization known)	**Unknown**	(Whibley et al. 2006)
8. *Wingless* (QTL)	Mimetic wing color and mate preference (adaptation to visual predators)	*Heliconius cydno* and *H. pachinus* mimetic butterflies	II, EDPI, SI	**Gene associated with RI** (gene is within a QTL for RI)	**Before complete RI** (hybridization known)	**Unknown**	(Kronforst et al. 2006)
9. *CDPK* (QTL)	Salt tolerance (adaptation to different soil salinities)	*Helianthus* sunflower species	II	**Gene associated with RI** (gene is within a QTL for RI)	**Unknown**	**Unknown**	(Lexer et al. 2004)

Gene	Details of gene or phenotype affected	Study system	Putative form of RI	Criterion 1: gene affecting RI	Criterion 2: timing of divergence	Criterion 3: increase in total RI	References
10. *Tpi* (QTL)	Perhaps diapause development timing differences	Pheromone and diapause races of *Ostrinia nubilalis* corn borers	TI	**Gene associated with RI** (gene is within a QTL for RI)	**Before complete RI** (hybridization known)	**Unknown**	(Dopman et al. 2005)
11. *YUP* (QTL)	Flower color (attractiveness and adaptation to different pollinators)	*Mimulus lewisii* and *M. cardinalis* monkeyflower species	II, EDPI, SI	**Gene unknown, RI known**	**Unknown** (RI is strong but incomplete, hybrids are fertile)	**Unknown**	(Schemske and Bradshaw 1999, Bradshaw and Schemske 2003, Ramsey et al. 2003)
12. *Pe*/*Pr* (QTL)	Performance on (Pe) and preference for (Pr) different hosts plant species	*Acyrthosiphon pisum* host races on alfalfa versus clover	II, EDPI, HI	**Gene unknown, RI known**	**Before complete RI**	**Large** (as barriers other than Pe/Pr are weak)	(Hawthorne and Via 2001, Via and West 2008)

[1]**Criterion 1.** Evidence that a gene affects RI. The four classes of evidence discussed in the text are evaluated. **Gene affects RI.** Includes cases in which there is causal evidence that gene in question causally affects RI. **Gene associated with RI.** A known gene is associated with RI, but the genetic evidence is correlative such that another gene could be causing the association. **Gene known, RI only inferred.** Strong evidence is available that a gene affects a phenotypic trait under divergent selection, but effects on RI are indirectly inferred. **Gene unknown, RI known.** Reproductive isolation maps to a genomic region, but the specific gene(s) are yet to be identified.

[2]**Criterion 2.** Evidence that divergence in the gene occurred before the evolution of complete reproductive isolation. **Before complete RI.** Includes intraspecific comparisons and cases in which recurrent hybridization occurs between species.

[3]**Criterion 3.** Effect size of the gene. **Large:** Includes cases in which contributions of the gene to current day reproductive isolation are detectable and considerable, and speciation is not yet complete.

Abbreviations: II, immigrant inviability; EDPI, ecologically dependent postmating isolation; SI, sexual isolation (= pollinator isolation in plants); HI, divergent habitat preferences; TI, temporal isolation.

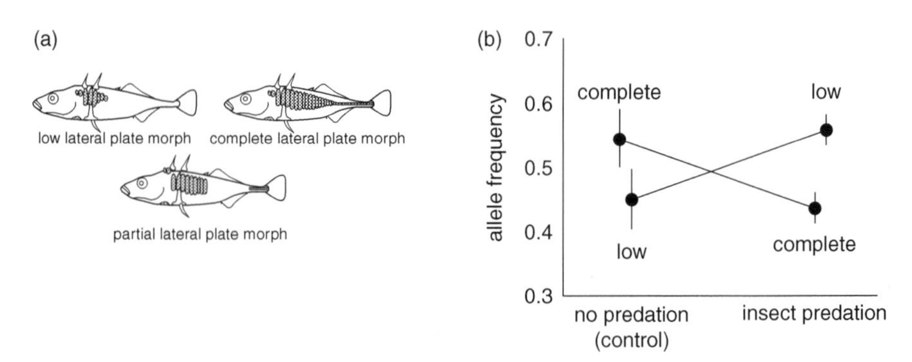

Figure 5.10. *Ectodysplasin* (*Eda*) in threespine stickleback fish is an example of a gene affecting a phenotypic trait (lateral plate number) under divergent selection between populations in freshwater versus marine environments (Reimchen 1983, 2000). To the extent that divergence in *Eda* causes extrinsic isolation, it constitutes an ecological speciation gene. a) Lateral plate number differs between marine and freshwater populations, with marine fish usually exhibiting a greater number of plates than freshwater populations. b) Evidence for divergent selection on the genomic region containing *Eda* from a pond enclosure experiment on young F2 hybrids between a high-plated and a low-plated species fixed for alternative alleles at *Eda*. In the presence of insect predation, which is more common in freshwater environments, genotypes carrying the low *Eda* allele had higher fitness relative to the control. The effect might be due to *Eda* itself, but other phenotypic differences map to the same region of the genome in the F2s. Modified from Marchinko (2009) with permission of John Wiley & Sons Inc.

shown (Table 5.2 for other examples). Finally, we have several examples in which reproductive isolation has been mapped to a QTL, but a candidate gene has yet to be identified.

5.6.2. Timing of divergence

The criterion that divergence takes place before the completion of speciation is automatically met in all examples in which reproductive isolation is not yet complete. Evidence is more difficult to obtain when species pairs are already completely reproductively isolated. In such cases, approaches are available to help establish timing of divergence retroactively. One approach uses phylogenetic methods to map nucleotide substitutions of candidate speciation genes onto a known phylogeny, and tests whether divergence of the gene occurred before the split of the species pair (Fig. 5.11). A second approach compares gene genealogies of candidate speciation genes to those reconstructed from genes not involved in the speciation process, such as unlinked neutral loci. The logic is that genes affecting reproductive isolation do not flow readily between the species, and if they cease to flow before time of gene-flow cessation of neutral loci, then their divergence must have occurred before

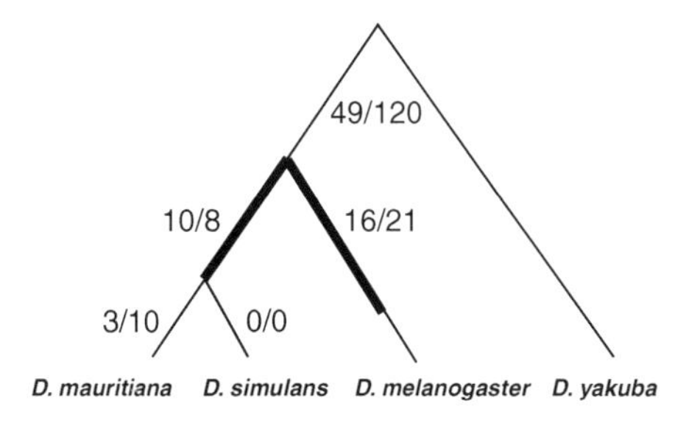

Figure 5.11. Inferred timing of divergence at *Nup96*, a gene that causes intrinsic postmating isolation between *Drosophila melanogaster* and *D. simulans*. The species, which currently exhibit complete reproductive isolation, split millions of years ago (Powell 1997). Depicted are the ratios of replacement to silent substitutions mapped onto the phylogeny of the *D. melanogaster* group of species (results shown are for fixed differences; an indel and an insertion are not depicted). Bold branches indicate those in which *Nup96* experienced significant positive selection. All substitutions present in *D. simulans* took place before the split with *D. mauritiana*, indicating that divergence at *Nup96* was not recent, although the exact timing of divergence is unknown. Modified from Presgraves et al. (2003) with permission of the Nature Publishing Group.

speciation was complete (Ting et al. 2000, Wu 2001b, Via 2009). The pattern predicted in such a case is discordance between the genealogies of speciation genes and other loci (Fig. 5.12) (Hey and Nielsen 2004, Hey 2006, Via and West 2008, Wakeley 2008b). Speciation genes should often "reflect species boundaries," whereas loci not involved in speciation may, for example, show little phylogenetic resolution, or group taxa by geographic location (Via 2001, Maroja et al. 2009a, see Nosil et al. 2009a for review). An example is the pheromone races of *Ostrinia nubilalis*, in which only genealogies for a gene lying within a QTL for reproductive isolation showed pheromone race exclusivity (Dopman et al. 2005). Although this approach is promising, the degree of genealogical discordance required to be confident that a gene diverged before gene flow ceased requires further study. In general, coalescent-based approaches are now commonly employed to tease apart the effects of divergence time and gene flow on patterns of genetic divergence at neutral loci (Hey 2006). Such approaches could be applied to speciation genes, but their efficacy is complicated by potential selection acting on such genes (Hey 2006, Strasburg and Rieseberg 2011).

5.6.3. Speciation effect sizes

The amount of reproductive isolation that already exists between populations limits the degree to which subsequent divergence at a gene can increase reproductive isolation (i.e., it limits the effect size of a gene). For example, a gene that completely

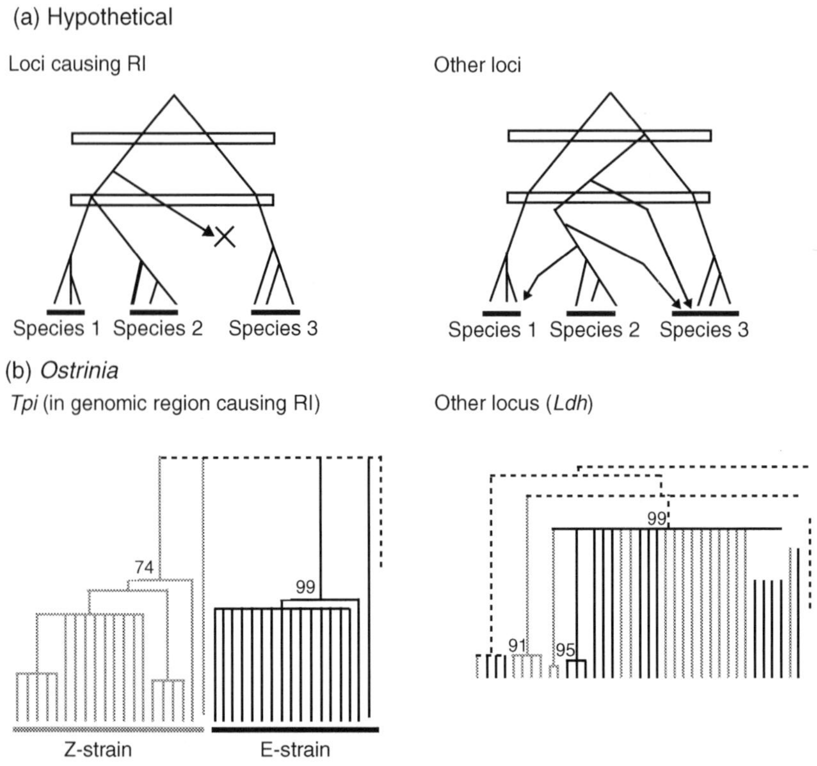

(a) Hypothetical

Loci causing RI

Other loci

Species 1 Species 2 Species 3

Species 1 Species 2 Species 3

(b) *Ostrinia*

Tpi (in genomic region causing RI)

Other locus (*Ldh*)

74

99

91

95

99

Z-strain

E-strain

Figure 5.12. One method to infer that divergence at a candidate speciation gene took place before the evolution of complete reproductive isolation (RI) (following Ting et al. 2000). Depicted are gene trees for different types of loci (e.g., candidate speciation genes versus other loci, such as neutral genes). a) Gene flow between species (depicted by arrows) is prevented (large "X") at speciation genes. Horizontal bars represent speciation events. b) The candidate speciation gene *Tpi* resides in a QTL underlying temporal reproductive isolation between two strains of *Ostrinia nubilalis* corn borers. A phylogeny based on this gene clearly sorts the two strains. Gray lines = Z-strain individuals. Black lines = E-strain individuals. Dotted lines = unresolved. However, a phylogeny based on *Ldh*, which has no known effects on reproductive isolation in these strains, fails to sort the strains. A possible explanation is that gene flow occurred frequently at *Ldh* since the time of divergence at *Tpi*, implying that reproductive isolation was not yet complete when *Tpi* diverged. Numbers by nodes represent bootstrap support. Modified from Dopman et al. (2005) with permission of the National Academy of Sciences USA.

sterilizes hybrids has an effect size of only 1% if, at the time of its divergence, it brought total reproductive isolation from 99 to 100%.

Other than zero, is there a minimum effect size for a gene to be considered a speciation gene? It is too soon to definitively answer this question, as we currently lack information on the distribution of effect sizes, and even small effects might be important. For example, reproductive isolation at the level of 95% will often enable ecologically divergent populations to coexist and maintain their distinctiveness, as is seen in many host races of phytophagous insects (Drès and Mallet 2002, Nosil et al. 2005) and in sympatric form of fish (Gow et al. 2006, Seehausen et al. 2008a, Seehausen and Magalhaes 2010). A gene that raised the total to 100% reproductive isolation (i.e., effect size = 5%) would staunch all gene flow, enable populations to diverge further by stochastic processes, and perhaps ensure speciation is not reversed (Price 2007, Seehausen et al. 2008a). Conversely, a gene that brings total reproductive isolation from 0 to 5% will not by itself have as much impact, though it can make it slightly easier for later genes to diverge (Coyne and Orr 2004, Mallet 2006). These examples suggest that the effect size of a speciation gene is related, but not identical, to its importance in the speciation process.

How does one estimate effect sizes? This task may not be overly difficult when current reproductive isolation is far from complete, as the effect size of a gene can be very similar to its contribution to current reproductive isolation. In some other instances, for example if divergence occurred long ago and many different barriers to gene flow have evolved, estimating effect size will be difficult. I refer readers to Nosil and Schluter (2011) for detailed treatment of this issue, and focus here on the particularly informative example of the QTL locus *YUP* in *Mimulus* monkeyflowers. *YUP* strongly affects both flower color and pollinator isolation between *M. cardinalis* and *M. lewisii* (Schemske and Bradshaw 1999, Bradshaw and Schemske 2003). Current total reproductive isolation between these species is very strong (> 0.99) and arises via the combined effects of multiple components (Ramsey et al. 2003). Although pollinator isolation on its own is strong (= 0.976), total reproductive isolation between species would remain very high even in its absence, because of the effects of the other barriers. Specifically, from Table 2 of Ramsey et al. (2003), it can be estimated that total isolation in the absence of pollinator isolation would be 0.957 (averaging across estimates of reproductive isolation in each direction). If pollinator isolation is completely due to the *YUP* locus, and if *YUP* is the most recent speciation gene to diverge, the effect size of *YUP* would be $\approx 0.99 - 0.957 = 0.033$, about 3%. However, if *YUP* was the first gene to diverge—an unlikely scenario (Ramsey et al. 2003)—its effect size could be as large as 0.976. These calculations demonstrate that estimating effect size is challenging, but that upper and lower bounds can be obtained.

5.6.4. Ecological speciation genes: conclusions

To summarize, specific genes unambiguously known to be subject to divergent selection and to have caused increases in reproductive isolation have not yet been identified. Thus, there are few or no clear examples of ecological speciation genes, although the list of candidates is growing and some might be confirmed as speciation genes before this book comes to print (van't Hof et al. 2011). In addition, even for candidate ecological speciation genes, further information on timing of divergence and effect sizes is needed. If many genes with small speciation effect sizes exist, a more "genomic" view that considers how individual genes are arrayed in the genome and how they interact might best characterize the speciation process (Johnson and Kliman 2002). Thus, there is much work to be done in the hope that improved characterization of speciation genes will clarify how numerous they are, their properties, and how they affect genome-wide patterns of divergence.

5.7. Genetic mechanisms: conclusions and future directions

Recent years have seen great advances in our theoretical and empirical understanding of the genetics of ecological speciation (Schluter and Conte 2009). Pleiotropy is a very effective mechanism to link selection to reproductive isolation, and examples of types of pleiotropy appropriate for ecological speciation exist. A number of other factors have been identified that might make linkage disequilibrium also an effective mechanism of speciation, including physical linkage, structural features that reduce recombination, one-allele assortative mating mechanisms, and strong selection. Nonetheless, the empirical and theoretical data concerning the effectiveness of these mechanisms are mixed, and outcomes often depend on particulars such as just how reduced recombination is. This could be an area of speciation research where details really matter. Not much is yet known about genetic constraints on speciation, and further work on the individual genetic basis of traits involved in adaptive divergence and reproductive isolation is required. In particular, is speciation more likely when traits are controlled by few or many loci? Finally, the search for ecological speciation genes is underway, but many conclusive examples are yet to emerge. Future work should pay attention to meeting the criteria for speciation genes, and consider what such genes tell us about the mechanisms of ecological speciation.

Part III

Unresolved issues

6

The geography of ecological speciation

"It is very unlikely that the purely genetic processes of mutation pressure and random fixation cause changes of a sufficiently high order. . . to hold any sizable gene flow in check. It may be different with selection pressure . . . " Mayr (1947, p. 268)

This book focuses on the role of divergent selection in speciation, rather than on the geographic context of speciation. In part, this reflects the reclassification of modes of speciation away from process-free descriptions of geographic mode toward descriptions that focus on the evolutionary processes driving genetic divergence (Butlin et al. 2008, Fitzpatrick et al. 2008, Mallet et al. 2009). However, although ecological speciation can occur under any geographic context, it is by no means "geography-free." Geography is important, because it affects the sources of divergent selection and rates of gene flow. Divergent selection is the diversifying process that drives ecological speciation. Gene flow tends to be a homogenizing process that constrains divergence. The interaction of selection and gene flow dictates how speciation proceeds. This chapter focuses on this geographic component of ecological speciation.

I first discuss different definitions of the geography of speciation and then turn to how geography affects the sources of divergent selection. I discuss how geographic contact between populations might constrain or promote ecological speciation, and how speciation can involve multiple geographic modes of divergence. The next part of the chapter focuses on the empirical problem of detecting gene flow, via consideration of comparative geographic, coalescent-based, and genomic approaches. The chapter concludes with a consideration of speciation along continuous environmental gradients (i.e., clines) versus between discrete patches and the unresolved issue of the spatial scale of speciation. These latter considerations explore the question of whether levels of gene flow alone predict speciation, or whether the actual geographic arrangement of populations themselves is also of importance (e.g., a given rate of average gene flow, say 0.1, can arise under a range of different spatial scenarios, such as for example between populations occupying two large versus many small patches of different habitat).

6.1. Geographic views and definitions of speciation

6.1.1. Spatial, demic non-spatial, and spatial population genetic views

Before discussing the geography of speciation, a consideration of the terms normally used to define geographic mode of speciation is required. Along these lines, one might consider three different ways of classifying the geography of speciation: (1) strictly spatial; (2) demic (strictly population genetic); and (3) spatial population genetic.

The terms normally used to describe the geographic arrangement of populations—namely allopatry, parapatry, and sympatry—were initially used in a strict geographic sense, without reference to gene flow (see Mallet et al. 2009 for a review). For example, Mayr (1942, pp. 148–9) wrote "*Two forms or species are* sympatric, *if they occur together, that is if their areas of distribution overlap or coincide. Two forms (or species) are* allopatric, *if they do not occur together, that is if they exclude each other geographically*." However, these strictly geographic definitions did not take into account the dispersal ability of organisms and thus the potential for genetic exchange. This led to the development of the concept of "cruising range." For example, Mayr (1947, p. 269) revised his definition of sympatric speciation to "*the establishment of new populations in different ecological niches within the normal cruising range of individuals of the parental population*." A third term in common use today, parapatry, was invented later, and "*was proposed for situations where ranges are in contact and genetic interchange is geographically possible even without sympatry*" (Smith 1965). The term "parapatry" is often used in its more spatial sense for populations that abut one another.

However, it has been argued that such spatial definitions are inconvenient (i.e., too imprecise) for modeling purposes (Gavrilets 2003). As a result, sympatric speciation is often simplified for modeling purposes to the more tractable assumption that divergence is initiated within a single, randomly mating population separated into two niches (or subpopulations) in which space is ignored. For example: "*we shall call speciation* 'sympatric' *if in its course the probability of mating between two individuals depends on their genotypes only*" (Kondrashov and Mina 1986). Modelers often assume that initial gene exchange rate (m) between diverging demes is maximal, so that mating is random (i.e., $m = 0.5$). Allopatry, in contrast, implies no gene exchange at all ($m = 0$), and all "*intermediate cases when migration between diverging (sub)populations is neither zero nor maximum*" are considered parapatric (Gavrilets 2003, p. 2198). Such purely genetic definitions might be considered "demic" and are useful for modeling purposes. However, they may not address the geographic, spatial issue of speciation in nature (i.e., How common is sympatric speciation?) that was originally posed by Mayr and that remains of interest to evolutionary biologists (Coyne and Orr 2004). For example, under the Gavrilets' formulation, most speciation is, by definition, parapatric (i.e., $m = 0.0001$ and $m = 0.49$ would both be termed parapatric speciation). This led Mallet et al. (2009) to propose "spatial population genetic" definitions of speciation, which combined considerations of gene flow with the geographic arrangement of populations and the dispersal ability of organisms. In essence, Mallet et al. (2009) proposed a more formal derivation of Mayr's concept of cruising range.

The demic and spatial population genetic views are contrasted in Figures 6.1 and 6.2. Verbal and mathematical definitions of allopatry, parapatry, and sympatry under each view are given in Table 6.1. Despite their differences, the demic and

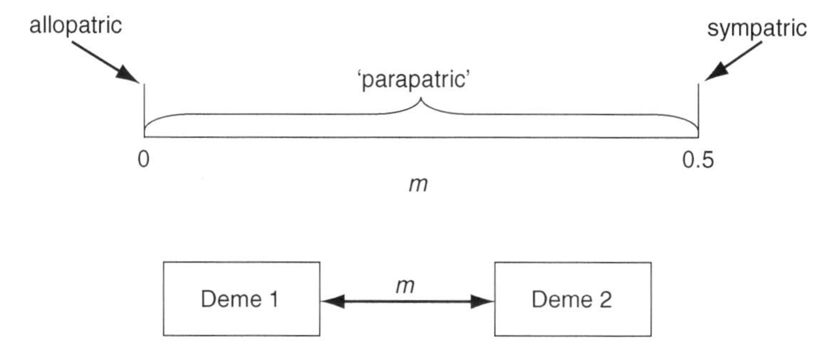

Figure 6.1. Demic view of sympatry and allopatry in Gavrilets' (2003) formulation. Reprinted with permission of John Wiley & Sons Inc.

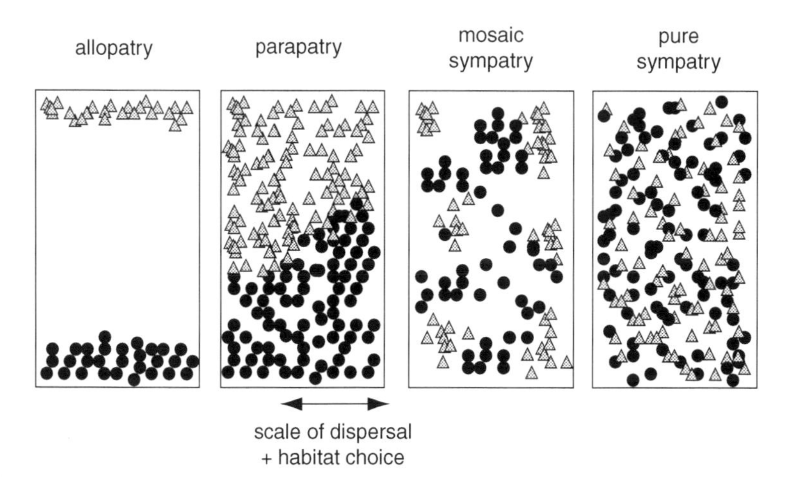

Figure 6.2. The allopatry to sympatry spectrum under the spatial population-genetic definition. Modified from Mallet et al. (2009) with permission of John Wiley & Sons Inc.

spatial population genetic definitions agree on one critical aspect: the terms sympatry, parapatry, and allopatry represent a spectrum ranging from maximum to no gene flow. For the remainder of the chapter, I focus on this issue of degree of gene flow during speciation. When using geographical categories: sympatry refers to scenarios in which reproductively mature individuals are physically capable of encountering one another with moderately high frequency; parapatry to scenarios where groups of populations occupy separate but adjoining regions, such that only a small fraction of reproductive individuals in each encounters the other; and allopatry to scenarios where groups of populations are separated by uninhabited space across which dispersal occurs at very low or zero frequency. Although I focus on the role of space in dictating levels of gene flow, the arguments can readily be modified to consider separation in timing of reproduction (Hendry and Day 2005).

Table 6.1. Different definitions of the sympatry to allopatry spectrum. Modified from Mallet et al. (2009) with permission of John Wiley & Sons Inc.

	Non-spatial, demic		Spatial population genetic	
	Verbal	Mathematical[1]	Verbal[2]	Mathematical[3]
Sympatry	Panmixia	$m = 0.5$	Where individuals are physically capable of encountering one another with moderately high frequency; populations may be sympatric if they are ecologically segregated, as long as a fairly high proportion of each population encounters the other	Distance between populations $< k\sigma_x$ where k is a small number
Parapatry	Non-panmictic	$0 < m < 0.5$	Where groups of populations occupy separate but adjoining regions, such that only a small fraction of individuals in each encounters the other; typically, populations in the contact zone between two forms will be considered sympatric	As above, but only for populations at the abutment zone; most other populations will have very low rates of gene flow
Allopatry	Zero gene flow	$m = 0$	Where groups of populations are separated by uninhabited space across which dispersal and gene flow occur at very low frequency	Distance between regions and populations $>> \sigma_x$

[1] m is the per-generation fraction of individuals exchanged among demes.

[2] Modified from Futuyma and Mayer (1980).

[3] σ_x is the standard deviation in one dimension (x) of dispersal distances between sites of birth and breeding.

6.1.2. Allopatric ecological speciation

During allopatric divergence, ecological speciation may occur owing to differences between disjunct populations in divergent, ecologically based selection. Because selection need not counter the homogenizing effects of gene flow, allopatric speciation may be the easiest form of ecological speciation, and common in nature. Showing ecological speciation in allopatry requires demonstrating a role for ecological divergence in the evolution of reproductive isolation above and beyond that expected from other processes (e.g., genetic drift alone). Case studies in insects and fish have met this requirement (Funk 1998, Nosil et al. 2002, Vines and Schluter 2006, Langerhans et al. 2007). Comparative studies testing the frequency of different geographic modes of ecological speciation are lacking, at least when it comes to studies that pertain to ecological speciation specifically and that rule out alternatives such as mutation-order speciation (Coyne and Orr 2004). Thus, the fundamental question of how common allopatric ecological speciation is, relative to sympatric or parapatric modes, remains unanswered. The brevity of this section should not be misinterpreted as indicative of allopatric speciation being unimportant. Rather, it reflects the fact that allopatric speciation is theoretically uncontroversial.

6.2. Non-allopatric speciation: geographic contact constrains divergence

In contrast to allopatric divergence, when populations are in geographic contact the opportunity for homogenization by gene flow is added. Overall divergence is thus likely to reflect a balance between selection and gene flow. There is ongoing debate about the extent to which divergent selection can counter gene flow to drive speciation (Bush 1969b, a, 1975, Feder et al. 1988, McPheron et al. 1988, Via 2001, Coyne and Orr 2004). Thus, although relatively strong examples of sympatric speciation are accumulating (e.g., Barluenga et al. 2006, Savolainen et al. 2006), these are not without contention (e.g., Schliewen et al. 2006, Stuessy 2006). I turn to exploring these debates from theoretical and empirical perspectives.

6.2.1. The balance between selection and gene flow: theory

Divergence under a balance between selection and gene flow was considered theoretically for Mendelian traits during the 1930s and 1940s (Fisher 1930, Wright 1931, 1940, 1943). For example, Haldane (1930, 1932) considered an island that receives immigrants at rate m from the mainland, where different alleles are favored on the island versus on the mainland. These two populations differ at a single biallelic locus. The frequency (p) of allele A is initially assumed to be close to 1.0 on the island. The frequency (q) of the alternative allele a is assumed to be fixed on the mainland. Haldane showed that when selection (s) acts against maladapted alleles (i.e., selection favors genotypes that are adapted to the island environment, thereby selecting against both immigrants *aa* and hybrids *aA*), the allele conferring adaptation to the island can be maintained, provided $s > m$. The maintenance of the island allele was also possible if selection acted only against hybrids, but required stronger selection relative to migration. Wright (1931, 1940) considered similar scenarios and likewise found that selection can overcome gene flow to cause divergence (see also Bulmer 1972, Slatkin 1982, Vines et al. 2003, Nosil et al. 2005, Mallet 2006, Bolnick and Nosil 2007).

The emergence of cline theory, starting with Haldane (1948) and Fisher (1950), showed how ecological selection and gene flow would balance in a spatial context. This initial cline theory was extended by others (Clarke 1966, Endler 1973, Slatkin 1973, May et al. 1975, Endler 1977), and was developed in detail by Barton and colleagues (Barton 1979, Barton and Hewitt 1985, 1989). Now more than 80 years following the seminal work on the topic, theory concerning selection/gene flow balance has been extended to quantitative traits and applied to a wide range of demographic scenarios (Felsenstein 1976, Slatkin 1985, GarciaRamos and Kirkpatrick 1997, Hendry et al. 2001, Kawecki and Holt 2002, Lenormand 2002). The general conclusion emerging from this theoretical work is that gene flow often acts as a homogenizing process, but that strong selection can overcome moderate gene flow to result in population divergence. Debates persist concerning whether divergence in the face of gene flow can proceed to the extent that strong or complete reproductive isolation evolves (Coyne and Orr 2004, Price 2007, Nosil 2008a).

6.2.2. The balance between selection and gene flow: data

An extremely widespread empirical pattern in nature is an inverse correlation between the degree of adaptive phenotypic divergence between populations and levels of genetic exchange between them. Populations undergoing more gene flow exhibit less adaptive divergence. This pattern has been observed in many organisms, including plants, vertebrates, and invertebrates (Mayr 1963, Riechert 1993, Sandoval 1994b, King and Lawson 1995, Lu and Bernatchez 1999a, Crespi 2000, Riechert and Hall 2000, Hendry et al. 2001, Nosil 2004, Nosil and Crespi 2004, Bolnick et al. 2008, Räsänen and Hendry 2008). Likewise, clines in genotype or phenotype frequency, for example in hybrid zone studies, also often represent divergence under a balance between selection and gene flow (Barton and Hewitt 1985, Mallet et al. 1990, Johannesson et al. 1993). However, the underlying causes of this common pattern have not been well established.

Two general hypotheses exist for the causes of inverse correlations between adaptive divergence and gene flow (Riechert 1993, Hendry et al. 2001, Nosil and Crespi 2004, Räsänen and Hendry 2008). First, gene flow may constrain adaptive divergence. Second, reverse causality is possible, whereby adaptive divergence causes the evolution of barriers to gene flow between populations (i.e., ecological speciation). Experiments manipulating adaptive divergence and gene flow are required to disentangle whether gene flow constrains adaptive divergence, whether

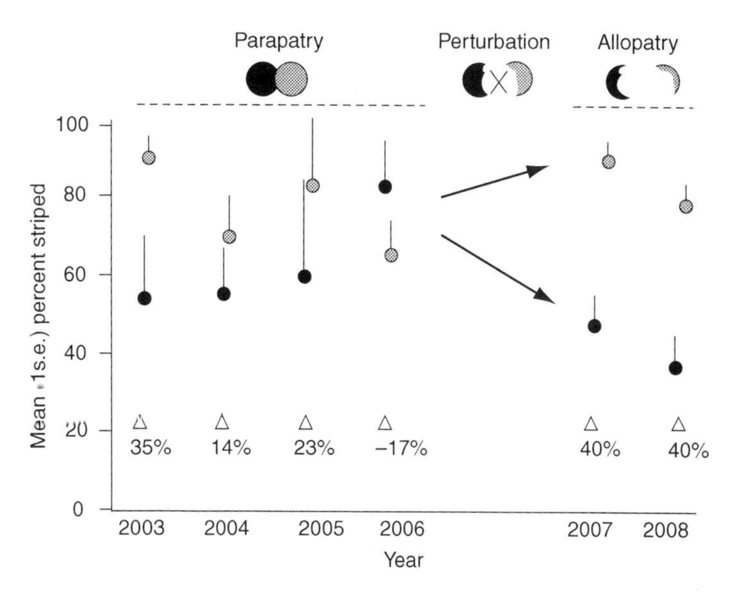

Figure 6.3. Evidence from a natural experiment that gene flow constrains adaptive divergence between two adjacent populations of *Timema cristinae* living on different host plant species. Shown is the frequency (mean ± 1 s.e.) of the striped color-pattern morph on two different host-plant species (black circles = *Ceanothus*, gray circles = *Adenostoma*), from 2003–2008. Δ = percent difference in morph frequency. The site shown is a formerly parapatric site that was made allopatric by road crews, which cut down most plant individuals at the site. At this "perturbed site," Δ increased following the experimental reduction of gene flow (i.e., between 2006 and 2007). In contrast, unperturbed "control" sites did not exhibit such increases in adaptive divergence (not shown). Modified from Nosil (2009) with permission of John Wiley & Sons Inc.

adaptive divergence constrains gene flow, or whether both causal associations occur (Räsänen and Hendry 2008).

One such experiment stems from perturbation of gene flow between *Timema cristinae* walking-stick populations. In this "natural experiment," road crews cut down most host plant individuals in a region where the different hosts used by *T. cristinae* were in contact. This reduced gene flow between hosts and made a parapatric insect population pair become effectively allopatric (see Nosil 2009 for details) (Fig. 6.3). Following this reduction in gene flow at the perturbed site, increased adaptive divergence in cryptic color pattern was rapidly observed in the generations immediately following the perturbation. In contrast, adaptive divergence did not increase at other, nearby, unperturbed ("control") sites during this same time period. These results strongly suggest that gene flow constrained adaptive divergence. However, because the perturbation was not replicated at a second site, unequivocal evidence awaits further work.

The above example concerns adaptive divergence. Is there evidence that gene flow constrains the evolution of reproductive barriers specifically? Nosil (2007)

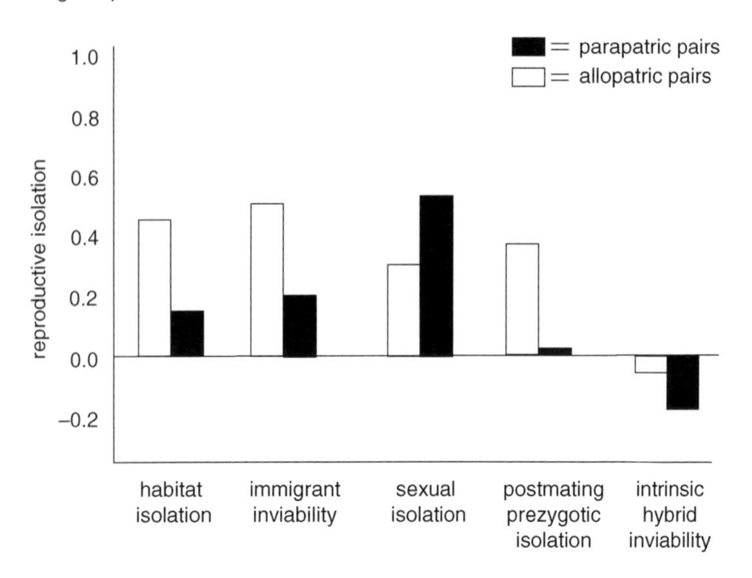

Figure 6.4. The association between geographic arrangement of population pairs (allopatry versus parapatry) and reproductive isolation among populations of *Timema cristinae* feeding on different host species. Multiple population pairs were tested under each geographic scenario, for each form of reproductive isolation (see Nosil 2007 for details and statistical results). In general, parapatric population pairs exhibit significantly weaker reproductive isolation than allopatric pairs, indicating gene flow constrains divergence. The strong exception is the reinforcement of sexual isolation. Modified from Nosil (2007) with permission of the University of Chicago Press.

reviewed a series of studies that quantified the strength of multiple reproductive barriers for allopatric versus parapatric population pairs of the walking-stick ecotypes discussed above (Fig. 6.4). Parapatric pairs are known from experimental and molecular data to undergo more gene flow than allopatric pairs. In general, parapatric population pairs exhibited significantly weaker reproductive isolation than did allopatric pairs, indicating gene flow constrains the evolution of most forms of reproductive isolation. The strong exception was sexual isolation, which was strongest in parapatry owing to reinforcement. However, these experiments were conducted in the lab and did not explicitly address causality. Thus, experiments are needed to establish the causal effects of gene flow on speciation.

6.3. Non-allopatric speciation: geographic contact promotes divergence

As discussed so far, there are situations in which gene flow constrains speciation. However, the role of gene flow in evolution has been debated for more than a century because there are also scenarios under which hybridization in regions of geographic overlap between populations promotes, rather than constrains, divergence. I consider three such scenarios. First, selection against maladaptive hybridization in sympatry or parapatry can drive the evolution of mating discrimination between populations or species. This diversifying process of "reinforcement" requires hybridization between populations. I focus here on reinforcement, but competitive interactions in sympatry or parapatry might have consequences similar to reinforcement, and thus promote non-allopatric speciation (Rosenzweig 1978, Dieckmann and Doebeli 1999, Doebeli and Dieckmann 2003, Kopp and Hermisson 2008, Pennings et al. 2008). Second, the adaptive spread of factors that reduce recombination, such as chromosomal inversions, can be promoted by gene flow. Third, hybridization can act as a source of genetic variation and novelty, thus promoting the adaptive divergence that drives ecological speciation. I consider these three scenarios in turn, using the terms "gene flow" and "hybridization" interchangeably, simply meaning genetic exchange between divergent populations, irrespective of their species status.

6.3.1. The "cascade reinforcement" hypothesis

As discussed in Chapter 3, examples exist of greater sexual isolation between hybridizing populations relative to allopatric ones. This pattern of "reproductive character displacement" is indicative of reinforcement (Rundle and Schluter 1998, Nosil et al. 2003, Pfennig and Rice 2007, Pfennig et al. 2007). Here I consider an extension to the reinforcement hypothesis, which shows how geographic contact might have widespread effects on promoting speciation. Reinforcement is generally invoked to explain levels of sexual isolation between a specific pair of hybridizing species. However, reinforcement might have much more widespread effects, because mating discrimination evolving in response to maladaptive hybridization between species might induce further effects on mate choice within species (see Ortiz-Barrientos et al. 2009 for a review). For example, reinforcement between species has been documented to result in females more readily rejecting males of their own species (Pfennig and Pfennig 2005, Higgie and Blows 2007, Pfennig 2007).

The effects of reinforcement between species on the evolution of premating isolation among populations within species have rarely been explored. An exception concerns different species of frogs from the genus *Litoria*. These species appear to have speciated by reinforcement, but also, because of increases in mating preferences due to reinforcement, they have evolved divergent mate preferences between populations within species (Hoskin et al. 2005). A similar pattern has been reported in another frog (Lemmon et al. 2004, Lemmon 2009) and in *Timema cristinae* walking-stick insects (Nosil et al. 2003, Nosil et al. 2007). In the walking-sticks system, although females are selected to be more discriminating against males from a single adjacent population that is adapted to feeding on a different host-plant species, this selection has indirectly resulted in increased mating discrimination against foreign males from multiple other populations, including even males from other populations that use the same host.

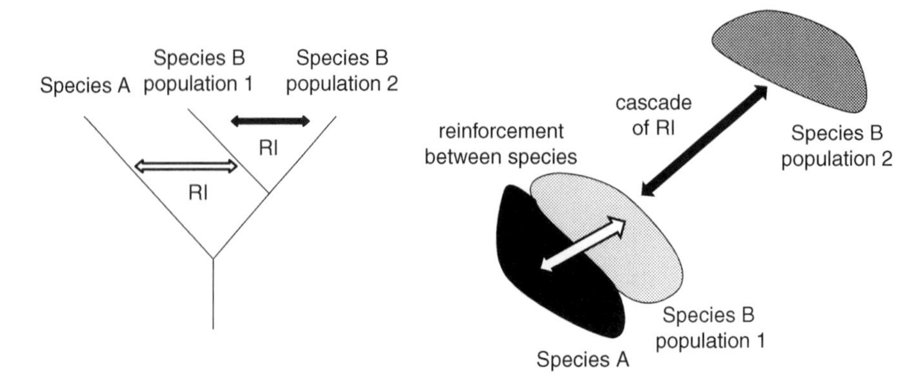

Figure 6.5. A schematic depiction of the "cascade reinforcement" hypothesis. Reinforcement between a specific taxon pair (in this case, between species A and population 1 of species B) cascades to generate reproductive isolation (RI) between allopatric populations within species (in this case, between populations 1 and 2 of species B). See text for details. Modified from Ortiz-Barrientos et al. (2009) with permission of John Wiley & Sons Inc.

Ortiz-Barrientos et al. (2009) proposed that this mechanism by which the effects of reinforcement within a particular taxon pair (e.g., a species pair) cascade to incidentally result in reproductive isolation among other taxon pairs (e.g., among populations within a species in the aforementioned pair) be termed the "cascade reinforcement" hypothesis (Fig. 6.5). Most generally, this hypothesis predicts that the evolution of prezygotic isolation between sympatric species can lead to the evolution of reproductive isolation within species. The hypothesis may often involve sexual selection within species. Such cascade effects of reinforcement on the evolution of reproductive isolation within species may be due to females recognizing and preferring males from their own population based on a "population-specific" trait, instead of an "ecology-specific" or "species-specific" trait (Zouros and Dentremont 1980, Higgie et al. 2000, Hoskin et al. 2005, Higgie and Blows 2007). If such cascade effects are common, then reinforcement could contribute to speciation between ecologically similar pairs of populations and between populations that are geographically separated from one another (Pfennig and Ryan 2006). In all these contexts, geographic contact promotes speciation.

6.3.2. Geographic contact and adaptive spread of chromosomal inversions

Another scenario under which gene flow might promote divergence is by facilitating the spread and fixation of chromosomal inversions that harbor genes involved in ecological speciation. Classic theories proposed that inversions are maladaptive, because they reduce the fitness of hetero-karyotypic individuals. Thus, inversions were assumed to rise to high frequency and fix within populations owing to founder effects and genetic drift in small populations (see Hoffmann and Rieseberg 2008 for a review). Alternatively, recent "genic" theories based on the allelic content of genes within inversions propose that inversions fix because of natural selection in populations exchanging genes (Kirkpatrick and Barton 2006, Feder et al. 2011b). For example, a model of sympatric divergence posits that newly formed inversions can sometimes capture locally adapted alleles at two or more loci in hybridizing populations (Kirkpatrick and Barton 2006). These co-occurring loci confer a fitness advantage to the inversion by keeping well-adapted genotypes intact, thereby causing inverted genomic regions to be favored over collinear ones. In turn, this allows inversions to rise to high frequency and spread via selection. In such a scenario, gene flow can promote speciation because any mechanism that facilitates the spread of inversions might promote speciation. The best potential example concerns *Mimulus guttatus* monkeyflowers (Lowry and Willis 2010). In this example, genes affecting ecologically based reproductive isolation reside within a chromosomal inversion, and experiments with near-isogenic lines (NILs) demonstrate that the inversion is subject to divergent selection, and thus could have spread via the mechanism put forth by Kirkpatrick and Barton (2006).

6.3.3. Hybridization as a source of novelty: the "hybrid swarm" theory of adaptive radiation

In addition to promoting reinforcement and the spread of inversions, hybridization has been argued to promote speciation by providing a source of genetic variation, which allows adaptation to new ecological environments (see Stebbins 1959, Grant and Grant 1992, Arnold and Hodges 1995, Arnold 1997, Seehausen 2004 for reviews). This "hybrid swarm" theory invokes a creative role for hybridization in evolution and can be thought of in two contexts. First, and most generally, hybridization can prevent the exhaustion of genetic variation at loci subject to divergent selection. By maintaining standing genetic variation within species, hybridization can thus facilitate adaptation to new environments (Feder et al. 2003a, Barrett and Schluter 2008). The second context focuses on transgressive segregation, which refers to the observation that hybrid taxa often exhibit novel or extreme characters compared with parental taxa (Rieseberg et al. 1999a, Rieseberg et al. 2003, Seehausen 2004). Transgressive segregation appears common in interspecific hybridization, and Fig. 6.6 depicts an example in *Helianthus* sunflowers (Rieseberg et al. 2003). This process creates new variation for selection to act upon and can facilitate the colonization of environments to which neither parental species can adapt, promoting the formation of new species in such environments. The occurrence of transgressive segregation appears positively related to the genetic distance between species, suggesting that its role in ecological speciation will be maximized when anciently diverged species meet and hybridize (Stelkens and Seehausen 2009a, Stelkens et al. 2009). Finally, although these creative roles of hybridization provide genetic variation, their role in the actual speciation process is indirect, because in these cases hybridization does not directly cause the evolution of reproductive isolation.

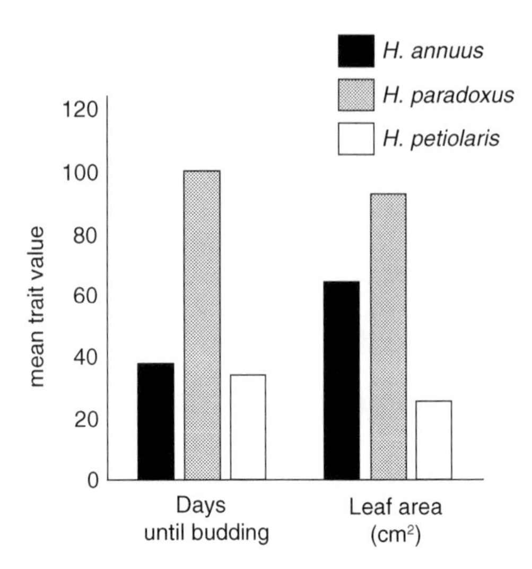

Figure 6.6. An example of transgressive segregation in *Helianthus* sunflowers. *H. paradoxus* is a hybrid species that originated from hybridization between *H. annuus* and *H. petiolaris*. *Helianthus paradoxus* exhibits extreme trait values relative to either parent for a suite of traits, two of which are depicted here. For both traits, differences among species are statistically significant. Data from Rieseberg et al. (2003).

6.4. The balance between constraining and diversifying effects of gene flow

"reinforcement requires some gene flow, but not too much"

Coyne and Orr (2004, p. 371)

As we have seen, gene flow can play both constraining and promoting roles in ecological speciation (summarized in Table 6.2). This section considers the balance between these opposing effects. Numerous theoretical models have demonstrated that high levels of gene flow between diverging populations can erode the effects of reinforcing selection, preventing reproductive character displacement (Sanderson 1989, Servedio and Kirkpatrick 1997, Cain et al. 1999, Servedio and Noor 2003). However, gene flow also generates the opportunity for selection against hybridization to occur in the first place. Thus, gene flow can exert a dual effect during reinforcement (Kirkpatrick 2000).

As noted by the quotation above, a potential prediction is that the effects of reinforcement are maximized when gene flow is intermediate. Nosil et al. (2003) tested this prediction by examining the effects of gene flow on the outcome of reinforcement during ecological speciation in walking-stick insects. The results demonstrate the dual effects of gene flow: the magnitude of female mating discrimination against males from other populations was greatest when gene flow between populations adapted to alternate host plants was intermediate (Fig. 6.7).

Table 6.2. A summary of the mechanisms by which gene flow between populations affects the speciation process.

Effect of gene flow	Effect on speciation
1. Reduces adaptive divergence	Constrains
2. Reduces reproductive isolation that is evolving as a by-product of adaptive divergence	Constrains
3. Allows for reinforcement	Promotes
4. Facilitates the spread of chromosomal inversions	Promotes
5. Allows for the spread of one-allele assortative mating alleles	Promotes
6. Maintains genetic variation in general	Promotes (indirectly)
7. Creates novelty via transgressive segregation	Promotes (indirectly)

In essence, reinforcement was maximized when gene flow was high enough to allow the evolution of reinforcement, but low enough to prevent gene flow from eroding adaptive divergence in mate choice. The generality of this result is unknown, although at least three other studies have documented stronger effects of reinforcement on the species within a hybridizing pair that is less abundant, and

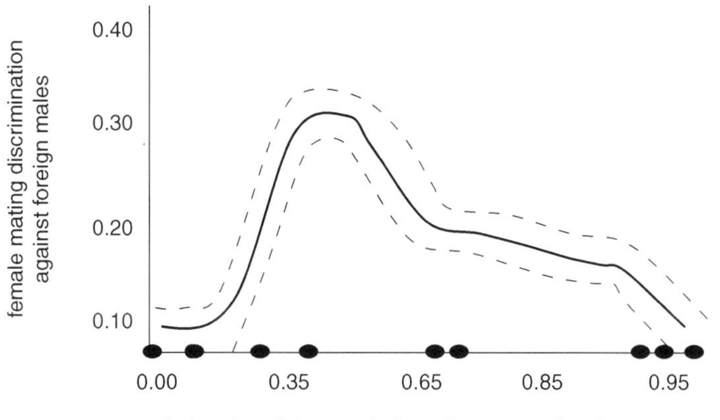

Figure 6.7. Among 12 populations of *Timema cristinae*, female mating discrimination against males from other populations is strongest when gene flow into the population, from an adjacent population adapted to a different host-plant species, is intermediate. The *x*-axis shows the size of the population that is adjacent to a focal study population, relative to the size of the study population itself. This value is positively correlated with the level of gene flow, inferred from molecular data, into the focal study population (relative sizes of populations examined are denoted by circles on the *x*-axis; note that four allopatric populations had value of zero on the *x*-axis). The *y*-axis is mean copulation frequency of females with males from their own population minus mean copulation frequency of females with foreign males. The curve was estimated using the non-parametric cubic spline with standard errors shown from 1000 bootstrap replicates (Schluter 1988). Modified from Nosil et al. (2003) with permission of the Royal Society of London.

thus undergoes more frequent encounters with heterospecifics (Waage 1979, Noor 1995, Peterson et al. 2005). Additionally, a laboratory experimental evolution study has demonstrated that reinforcement can evolve in the face of gene flow, so long as gene flow levels are not too high (Matute 2010). Finally, even after reinforcement occurs, gene flow may be required to spread alleles for mating discrimination outside of the immediate zone of hybridization and across the species range (Coyne and Orr 2004, Ortiz-Barrientos et al. 2009).

In general, further work is required to determine the conditions under which speciation in the face of gene flow can occur, and the extent to which this involves geographic contact having effects that actually promote speciation, versus scenarios in which divergent selection simply overcomes the constraining effects of gene flow.

6.5. Multiple geographic modes of divergence

Discussions of the geography of speciation tend to focus on whether speciation was allopatric, parapatric, or sympatric. However, it could be that multiple geographic modes of divergence characterize any single speciation event (Fig. 6.8). For

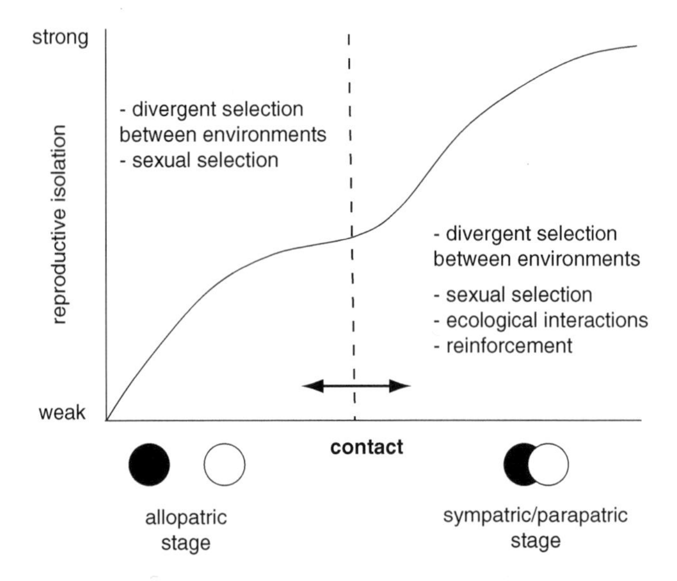

Figure 6.8. A general scenario for speciation under any geographic context. Reproductive isolation between two populations is absent at the beginning of the speciation process (at the left) and evolves to completion (at the right). Populations are initially allopatric, but secondary contact can occur at any time (dashed vertical line), commencing the second stage of the speciation process. The ecological causes of divergent selection by which reproductive isolation may evolve are listed within the panel for each stage. Depicted is an intermediate scenario of "mixed-mode" speciation in which partial reproductive isolation evolves in allopatry, but speciation is completed in sympatry. Modified from Rundle and Nosil (2005) with permission of John Wiley & Sons Inc.

example, speciation could begin in allopatry, and then be completed in sympatry, as might have occurred for present-day sympatric limnetic and benthic threespine sticklebacks (Rundle and Schluter 2004). Recent sequence data from the apple and hawthorn host-races of the apple maggot fly *Rhagoletis pomonella*, a classic case of sympatric speciation, also suggest a more complex geographic scenario (Feder et al. 2003a, Xie et al. 2007). Inversion polymorphisms containing genetic variation affecting ecologically important diapause traits that differ between the host-races trace their origins to allopatric populations in Mexico. Gene flow from the Mexican populations probably introduced this variation into the North American populations. It is unlikely that this introgression was responsible for any immediate reproductive isolation between populations, although it may have provided the genetic variation necessary to facilitate the subsequent host shift. Some key traits that generate partial prezygotic isolation between the host races, such as olfactory preferences for their respective fruits, appear to have evolved recently and in sympatry (Linn et al. 2003). The relative roles of divergence in allopatry and sympatry are not yet fully understood in these examples.

One area in which a mixed geographic mode of speciation has strong implications is for the fixation of chromosomal inversions. As discussed previously, natural selection can promote the spread of inversions in sympatric populations that are undergoing gene flow, provided that inversions capture locally adaptive genotypic combinations within them (Kirkpatrick and Barton 2006). The "purely sympatric" hypothesis of Kirkpatrick and Barton is somewhat restricted, however, to conditions in which migration between hybridizing populations is much lower than selection favoring locally adapted alleles ($m << s$). Lower migration rates are needed to avoid locally favored genes being swamped by gene flow, thus ensuring that a new inversion captures favorable alleles within it, a condition required for the inversion to be favored over collinear regions and spread (Kirkpatrick and Barton 2006). But if migration rates are too low, selection favoring new inversions is weak as well, because well-adapted genotypes are already present at high frequencies in collinear regions. Consequently, in the sympatric model, new inversions will often not establish in populations.

Recent work demonstrated that the adaptive model of inversion fixation is facilitated when divergence involves both an allopatric and a sympatric phase (Feder et al. 2011b). This "mixed mode" model for the spread of inversions is motivated by the observation that inversions are often present at low frequency in allopatric populations (Hoffmann and Rieseberg 2008). Such inversions contain highly locally adapted sets of alleles at loci within them, because allopatric populations are well adapted across the genome. In the mixed mode model, inversions originate in allopatry and are maintained there at a low frequency. Upon subsequent secondary contact and gene flow in sympatry, they have a very strong fitness advantage over collinear regions, because the fit gene combinations within them are not broken up by recombination. This allows the rapid and adaptive fixation of inversions in secondary sympatry. Thus, mixed geographic mode of divergence can promote the spread of inversions, and potentially speciation.

6.6. Two problems with detecting divergence in the face of gene flow

The discussion of gene flow thus far has been quite conceptual. But what of the empirical issue of actually inferring gene flow (Panova et al. 2006)? Demonstrating that speciation occurred in the face of gene flow is hampered by at least two major issues. First, it is difficult to infer confidently that gene flow occurred at any point in the speciation process, because isolating the effects of time since population divergence versus gene flow on levels of molecular genetic differentiation is not a trivial task (Turner and Hahn 2010). For example, weak genetic differentiation between taxa could be observed because of shared ancestral polymorphism due to recent divergence, gene flow, or a combination of these factors (Coyne and Orr 2004, Hey 2006). Second, and even more difficult, is inferring the timing of gene flow during divergence (Strasburg and Rieseberg 2011). Detecting that gene flow occurred at some point in the history of divergence does not necessarily provide information on the timing of gene flow relative to the timing of the evolution of reproductive barriers. For example, low levels of gene flow might occur after strong reproductive isolation evolved during a previous period of strictly allopatric differentiation (Coyne and Orr 2004, Grant et al. 2005). In such a scenario, reproductive isolation may be maintained in the face of gene flow, but it did not originate in the face of it. In such cases, speciation might be thought of as essentially allopatric. Methods are required to determine the timing of gene flow relative to when reproductive isolation evolved. I consider each problem in more detail below, focusing on three approaches: comparative geographic, coalescent-based, and genomic.

6.7. Detecting divergence in the face of gene flow: comparative geographic approaches

One approach for detecting gene flow relies on comparing populations that differ in their geographic arrangement and thus their opportunity for gene flow. The basic premise is that shared ancestral polymorphism affects both allopatric and sympatric/parapatric (sympatric hereafter for simplicity) populations, whereas gene flow affects only sympatric populations (or more strongly affects them). This premise generates the prediction that if shared ancestral polymorphism is responsible for observed levels of (e.g., low) divergence, then allopatric versus sympatric populations should not differ genetically in a predictable manner. In contrast, if gene flow occurs, then allopatric versus sympatric populations should differ in two ways. First, genetic divergence should be consistently greater for comparisons between allopatric populations. Second, linkage disequilibrium should be consistently greater within sympatric populations. The premise for the second prediction is that migration between genetically differentiated populations produces associations between

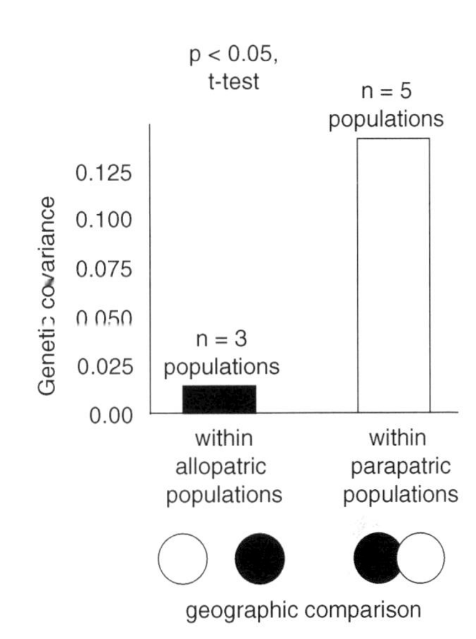

Figure 6.9. Evidence for gene flow between host-associated ecotypes of *Timema cristinae* based upon the comparative geographic approach. a) Genetic divergence (F_{ST}) between populations for mtDNA was greater between allopatric pairs. Similar patterns occur for other molecular markers and for morphological divergence. b) Genetic covariance within populations between host-plant feeding preference and color pattern (a proxy for linkage disequilibrium within populations between loci affecting these traits) was greater within parapatric situations. Data from Nosil et al. (2003, 2006a).

alleles at different loci within populations, even when they are physically unlinked (Kimura 1956, Nei and Li 1973, Kirkpatrick et al. 2002, Nosil et al. 2006a).

Some studies support these predictions (Fig. 6.9). For example, studies of Darwin's finches, walking-stick ecotypes, and cichlid fishes inferred the existence of gene flow via the stronger genetic divergence of allopatric versus sympatric taxon pairs (Nosil et al. 2003, Grant et al. 2005, Mims et al. 2010). Other studies have shown elevated linkage disequilibrium in parapatric versus allopatric populations (Nosil et al. 2006a), or at the center of geographic clines, where gene flow is expected to be greatest (Grahame et al. 2006). This comparative approach has the logistical drawback of requiring the existence of multiple population pairs for study, and ones that differ in their geographic arrangement. When this does not occur, we must turn to other approaches.

6.8. Detecting divergence in the face of gene flow: coalescent approaches

Coalescent-based approaches can estimate gene flow, potentially independent of time since population divergence and effective population sizes (Hey 2006, Wakeley 2008a). The basic premise is that gene flow varies widely across genomic regions. In contrast, genetic drift might act more uniformly across the genome. Along these lines, a history of gene flow is generally indicated if some loci show little divergence and others show strong divergence, such that variation among loci is greater than expected under a model with no gene flow and divergence solely by drift (Wakeley and Hey 1997). Nonetheless, distinguishing between divergence with versus without gene flow is difficult, because variance among loci in the degree to which they share alleles can be large, even under a model without gene flow (Hey 1991, Degnan and Salter 2005, Wakeley 2008a).

6.8.1. Reliably detecting that gene flow occurred at some point in time

An increasingly common coalescent-based approach for isolating time since population divergence independent from gene flow is the "isolation with migration" model (the "IM approach") (Nielsen and Wakeley 2001, Hey and Nielsen 2004, Hey 2006, Pinho and Hey 2010) (Fig. 6.10). For example, a study in salamanders used the IM approach to reject a model of divergence with zero gene flow, thereby supporting ecological speciation in the face of gene flow (Niemiller et al. 2008). Other recent studies employing the IM approach, some of which are depicted in Figure 6.11, report similar conclusions (Won and Hey 2005, Linnen and Farrell 2007, Kotlik et al. 2008, Strasburg and Rieseberg 2008). Notably though, many studies employing IM have reported little or no gene flow between taxa (Counterman and Noor 2006, Dolman and Moritz 2006, see Pinho and Hey 2010 for review). Finally, a number of studies have employed approaches related to IM, which similarly rely on highly discordant patterns in gene genealogies from different loci to infer gene flow (Dopman et al. 2005, Bull et al. 2006, Putnam et al. 2007, Joly et al. 2009, Joyce et al. 2011). Future work might employ approximate Bayesian computational ("ABC") approaches to explicitly test different demographic scenarios.

The power of the IM approach depends on the degree to which the assumptions of the approach are met. One assumption of earlier versions of this approach, from which most published studies stem, is that there are only two populations exchanging genes. Divergence in nature may often occur between multiple populations that are each connected, to varying degrees, by variable levels of migration. Newer versions of IM address this (Hey 2010b, a), and thus future empirical studies may not be plagued by this problem. A second issue is whether the data conform to the mutation models currently incorporated in IM (Hey 2006), a problem that may be

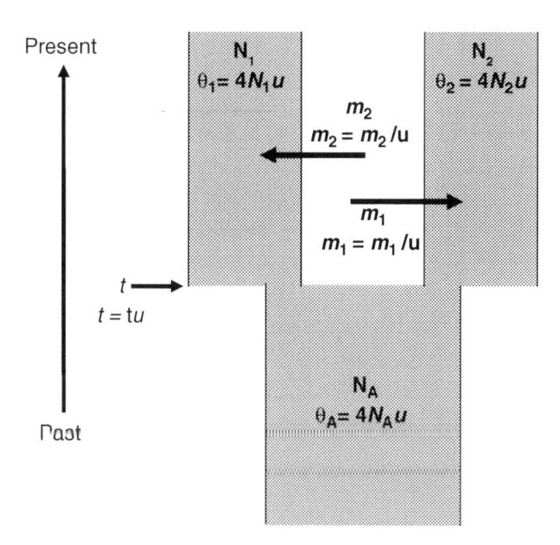

Figure 6.10. The "IM" model, which estimates divergence times (t), effective populations sizes (N), and migration rates (m). Actually empirical estimates of these parameters are scaled by u, the mutation rate. Modified from Hey (2006) with permission of Elsevier Publishing.

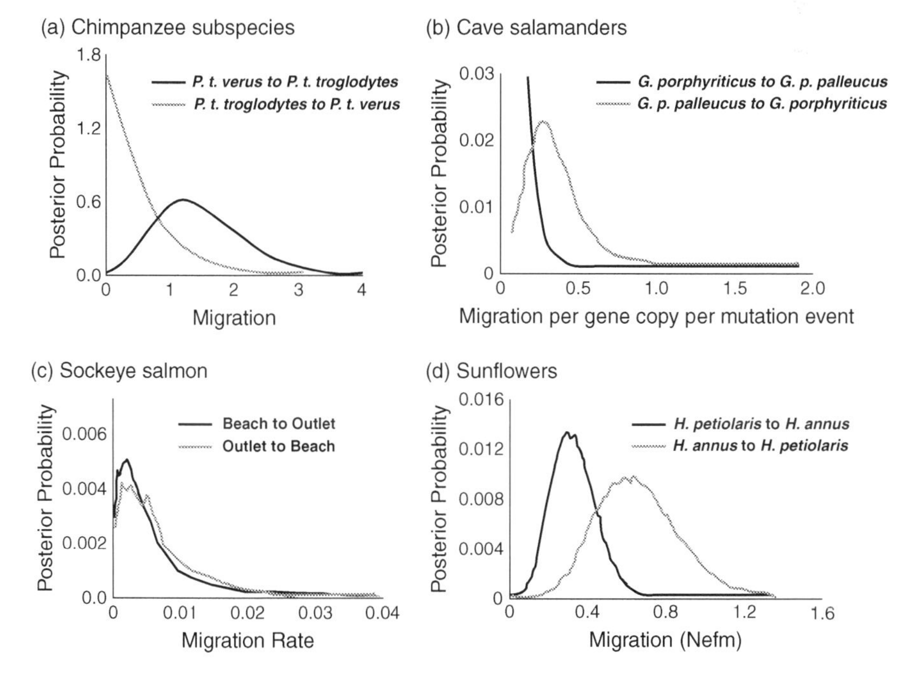

Figure 6.11. Examples of studies employing the "IM" approach to test for gene flow during divergence. The y-axes represent the posterior probability of a given level of migration. In all examples shown here, the probability is maximized at non-zero levels of migration, at least for one direction of migration. The x-axes represent the level of migration (original studies for details). a) *Pan troglodytes* chimpanzee subspecies. Modified from Won and Hey (2005) with permission of Oxford University Press. b) *Gyrinophilous* cave versus surface-dwelling species of salamanders. Modified from Niemiller et al. (2008) with permission of John Wiley & Sons Inc. c) *Oncorhynchus nerka* beach versus outlet forms of sockeye salmon. Modified from Pavey et al. (2010b) with permission of John Wiley & Sons Inc. d) *Helianthus* sunflowers. Modified from Strasburg and Rieseberg (2008) with permission of John Wiley & Sons Inc.

more serious for microsatellites than for DNA sequences. Finally, the assumption of constant population sizes will often be violated in natural populations. Simulation studies by Becquet and Przeworski (2009) and Strasburg and Rieseberg (2010) describe the conditions under which the IM approach works best, as well as its pitfalls. A major remaining issue is that even the most convincing examples of gene flow may represent the maintenance, rather than the origin, of divergence with gene flow. In other words, what was the timing of gene flow in relation to the timing of the evolution of reproductive isolation?

6.8.2. The timing of gene flow in relation to the evolution of reproductive isolation

Perhaps the most difficult aspect of studying speciation with gene flow concerns the timing of gene flow: at what point in the speciation process did gene flow occur? At one extreme, gene flow is ongoing during the entire process. At the other extreme, gene flow ensues upon secondary contact, after a long period of allopatric divergence in which strong reproductive isolation accumulated unimpeded by gene flow. Of course, intermediate scenarios are possible, and in such scenarios it might be even more difficult to determine how much, if any, reproductive isolation evolved in the face of gene flow (Rundle and Nosil 2005, Xie et al. 2007). There are methods for estimating the mean time of migration (Won and Hey 2005), but their power appears highly limited (Strasburg and Rieseberg 2011). Moreover, methods aimed specifically at testing how migration events were distributed throughout time are lacking (Niemiller et al. 2008, Niemiller et al. 2010). Even if migration events can be shown to have occurred throughout time, rather than concentrated into a specific period of secondary contact, it is difficult to rule out the idea that reproductive isolation evolved during periods of allopatry that were interspersed between the periods of gene flow. The degree to which this represents a problem will depend on the duration of periods of allopatry. If such periods are very short, it will be unlikely that at all existing reproductive isolation evolved in allopatry. Further development of approaches for inferring the timing of gene flow is sorely needed (Nosil 2008b, Strasburg and Rieseberg 2008, Muster et al. 2009, Strasburg and Rieseberg 2011). Inferring the timing of gene flow from genetic data will probably remain a difficult task, and thus biogeographical information might be especially useful in this regard.

6.9. Detecting divergence with gene flow: genomic approaches

A number of recent studies have used a more "genomic" approach for inferring gene flow. These approaches are related to coalescent-based ones (and may invoke coalescent-based analyses), but involve examining larger numbers of loci. For example, population genomic methods examining thousands of loci can infer "outlier loci" whose exceptional genetic differentiation statistically exceeds background

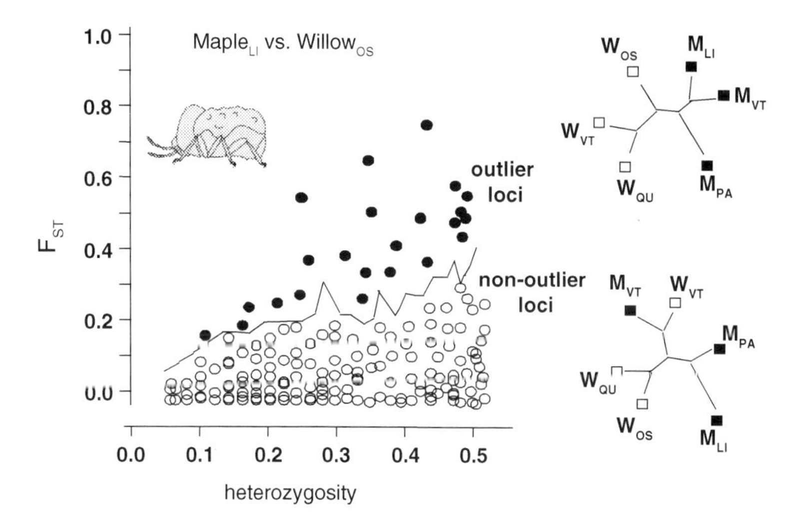

Figure 6.12. Heterogeneous genomic divergence in *Neochlamisus bebbianae* leaf beetle populations. An outlier analysis is shown for a population pair and population trees consider all six study populations. Simulations determine the upper level of genetic divergence expected under neutrality, and loci that exceed this "neutrality threshold" (depicted by solid line, in this case 95% quantile) are "outlier loci" inferred to have evolved under divergent selection. If outliers are highly replicated across population pairs that have diverged ecologically, then population trees from outlier loci (upper right-hand corner) are likely to group populations according to ecology (e.g., box color indicates host plant, subscript indicates population designation), reflecting divergent selection. In contrast, trees from putatively neutral non-outlier loci (lower right-hand corner) may group populations according to geography, reflecting spatial opportunities for gene flow. Modified from Egan et al. (2008) with permission of John Wiley & Sons Inc.

neutral expectations (Lewontin and Krakauer 1973, Bowcock et al. 1991, Beaumont and Nichols 1996, Beaumont 2005, Storz 2005, Vasemagi and Primmer 2005, Foll and Gaggiotti 2008, Stinchcombe and Hoekstra 2008, Excoffier et al. 2009, Nosil et al. 2009a, Butlin 2010). Such outlier loci differentiate between populations more strongly, and introgress less freely, than neutrally evolving regions, and are putatively affected by divergent selection (either directly or via tight physical linkage with such loci). In contrast, putatively neutral loci undergo less-impeded gene flow between taxa, and thus exhibit weak genetic differentiation (Fig. 6.12). It has been proposed that the history of gene flow can be inferred via the proportion and the genomic distribution of outlier loci.

The first set of predictions focuses on the proportion of outlier loci (Kondrashov and Kondrashov 1999, Via 2001). It has been argued that taxa that diverged in the face of gene flow will exhibit strong genetic differentiation at only a few genomic regions: genetic differentiation through most of the genome is homogenized by gene flow and differentiation is maintained only at the few loci that are under selection or contribute to reproductive isolation. The resulting pattern is strong heterogeneity among genomic regions in levels of genetic differentiation and an "L-shaped" distribution of F_{ST} values (i.e., many loci with little or no divergence, and a few loci with strong divergence). Such patterns have been documented in some sympatric taxa (see Nosil et al. 2009a for a review), and sometimes used to argue for divergence with gene flow (Emelianov et al. 2004, Savolainen et al. 2006). A related prediction is that phylogenies built from outlier versus putatively neutral loci will differ, with outlier loci reflecting boundaries between species or ecotypes more strongly than neutral loci (Via 2001, Wu 2001a, Hey 2006). A number of studies have reported such patterns of genealogical discordance (Wilding et al. 2001, Beltran et al. 2002, Campbell and Bernatchez 2004, Dopman et al. 2005, Bonin et al. 2006, Bull et al. 2006, Cano et al. 2006, Geraldes et al. 2006, Putnam et al. 2007, Egan et al. 2008, Nosil et al. 2008, Via and West 2008, Maroja et al. 2009a, White et al. 2009, Ohshima and Yoshizawa 2010).

However, the patterns described above may not necessarily reflect gene flow for at least two reasons. First, similar patterns might be observed under allopatric divergence. Even in the complete absence of gene flow, weak genetic differentiation at most loci could reflect recent origin, with insufficient time for neutral loci to differentiate between populations via genetic drift. Likewise, "outlier" loci are expected even between strictly allopatric populations, with loci subject to divergent selection differentiating more strongly than neutral loci that rely solely on drift to differentiate (Nakazato et al. 2007, Nosil et al. 2008). Although it is true that differential genetic exchange among loci will produce phylogenetic discordance, some such discordance is also expected during recent allopatric divergence. Second, patterns differing from those described above might be observed under divergence with gene flow. Specifically, theoretical studies have shown that when selection acts on more than a few loci, effective migration can be reduced between populations to the extent that widespread divergence occurs across the genome, even during divergence with gene flow (Feder and Nosil 2010). A direct experimental test of this scenario revealed widespread genomic divergence in the classic case of sympatric divergence with gene flow: *Rhagoletis* host races (Michel et al. 2010). Thus, strong divergence in only a few loci within the genome is not necessarily a reliable signature of speciation with gene flow. Further quantification of theoretical expectations is needed.

The second set of predictions involves the genomic distribution of outlier loci. During divergence with gene flow, new mutations might be more likely to diverge between populations if they arise in genomic regions already under divergent selection (i.e., that already exhibit reduced gene flow). Consequently, regions of strong differentiation might accumulate in clusters within the genome, rather than being randomly distributed throughout it (Rieseberg 2001, Navarro and Barton 2003, Gavrilets 2004, Kirkpatrick and Barton 2006, Via 2009, Yeaman and Otto 2011, Yeaman and Whitlock 2011). This contrasts with allopatric divergence where regions of differentiation are not predicted to be highly clustered within the genome, because divergence at all regions (i.e., not only those already exhibiting reduced gene flow) can proceed unimpeded by gene flow. To what degree does strong clustering of differentiation within the genome represents a genetic signature of divergence with gene flow?

Data on this issue are also mixed. A number of empirical studies have reported relatively strong clustering of divergence in taxa putatively undergoing divergence with gene flow (Emelianov et al. 2004, Turner et al. 2005, Via and West 2008). Likewise, low clustering of QTL for reproductive isolation between allopatric taxa has been reported (Brown et al. 2004, Nakazato et al. 2007). However, there are strong exceptions to these trends, with taxa putatively undergoing gene flow exhibiting moderate or weak clustering of differentiated regions (Wilding et al. 2001, Scotti-Saintagne et al. 2004, Achere et al. 2005, Grahame et al. 2006, Harr 2006, Egan et al. 2008, Nosil et al. 2008, Turner et al. 2008, Wood et al. 2008). Additionally, theory indicates that physical linkage of new mutations to selected loci often has little or no effect on the establishment of new mutations, with selection strength on the new mutation being of much greater importance for predicting establishment (Feder et al. 2011a, Yeaman and Otto 2011, Yeaman and Whitlock 2011). In sum, it appears that the properties of genomic clustering of differentiation are currently too poorly understood to be able to make clear predictions about divergence with versus without gene flow. Although genomic approaches hold promise for studying the geography of speciation, further work addressing the issues raised above is required.

6.10. The spatial context of selection: discrete patches versus continuous gradients

Another consideration for the geography of ecological speciation is whether environmental differences change abruptly in space (e.g., discrete patches of different host plant species) or are more continuously spatially distributed along environmental gradients, resulting in geographic clines. Although numerous models and empirical studies of each type exist, there are essentially no studies directly comparing the two in an "all else equal" framework. A notable exception is a simulation study by Cain et al. (1999) examining reinforcement in a clinal hybrid zone versus a mosaic hybrid zone comprised of multiple patches. They concluded that reinforcement occurred under a wider range of conditions in the mosaic hybrid zone. This is likely to be the result of two factors. First, divergent selection between patches acted in the mosaic zone but not in the clinal zone. Specifically, in the mosaic zone one homozygote was most fit in each patch type, such that selection acted against heterozygotes and one (i.e., the foreign) homozygote. In contrast, selection in the clinal zone acted only against heterozygotes. Second, the mosaic zone consisted of multiple contact zones between patches where hybridization occurred and where reinforcing selection could act. This facilitated the origin of mating discrimination and its spread throughout the system. Other such studies are lacking. Thus, it remains unclear what spatial conditions are most conducive for ecological speciation. However, it is likely that spatial scenarios where intermediate environments are not present will be conducive to ecological speciation because regions where intermediate hybrids are more fit than the parental forms will not exist.

Some predictions about spatial variation can be made for environmental gradients specifically (Endler 1977, Doebeli and Dieckmann 2003, Lemmon et al. 2004, Goldberg and Lande 2006). First, when divergence occurs along gradients, competition or reinforcement need not result in the classic pattern of greater divergence of sympatric versus allopatric populations (Fig. 6.13). This is because divergence due to character displacement is overlaid on differentiation occurring as a by-product of adaptive divergence, and the latter can be most pronounced between allopatric populations at the ends of the gradient, which experience the greatest environmental differences but the lowest levels of gene flow. Thus, when testing for character displacement during ecological speciation, attention to the spatial position of populations relative to environmental differences is required.

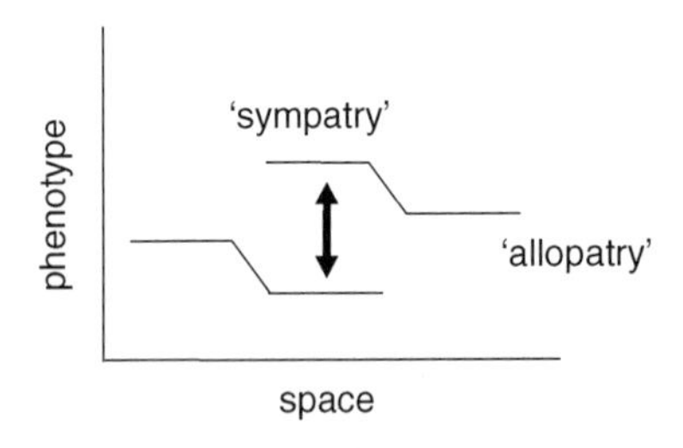

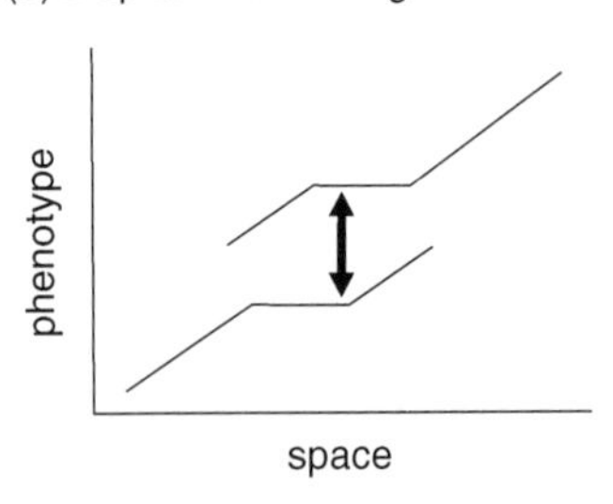

Figure 6.13. Character displacement, here represented by double-headed arrows, in the presence versus absence of environmental gradients. a) Classical character displacement where the optimum phenotype is relatively constant over space. b) The optimum phenotype increases monotonically over space (i.e., divergent selection between the ends of an environmental gradient). Note that the classic pattern of character displacement, greater divergence in sympatry versus allopatry, is only observed when there is no environmental gradient. Modified from Goldberg and Lande (2006) with permission of John Wiley & Sons Inc. (see also Lemmon et al. 2004).

Other predictions concern the steepness (i.e., slope) of environmental gradients. Steep gradients allow for more disruptive selection to emerge due to interactions between physically proximate populations, but also allow for more gene flow because the physical distance between populations is small. In contrast, shallow gradients allow for fewer interactions, but less gene flow. Thus, the gradient steepness most conducive to ecological speciation will depend on how important disruptive selection and gene flow are for driving versus constraining gene flow, respectively (Doebeli and Dieckmann 2003). A study of ecological speciation in lake Victoria cichlids examined five replicate gradients and found that genetic divergence within gradients decreased as gradient steepness increased, supporting the argument for high gene flow along steep gradients limiting divergence (Fig. 6.14) (Seehausen et al. 2008b). However, the same study demonstrated that when data from within and between gradients were combined, divergence appeared strongest at gradients of intermediate steepness. Further work on the effects of the slope of environmental gradients on speciation is needed.

Finally, other areas that could receive attention are how the coarseness of the environment and the shape of geographic ranges affect ecological speciation (Pigot et al. 2010). These factors will influence patterns of selection and gene flow, including possible asymmetries between populations in these parameters. All the factors discussed in this section warrant attention, because they can test whether the actual spatial configuration of populations affects speciation, or whether absolute levels of gene flow alone are the main consideration.

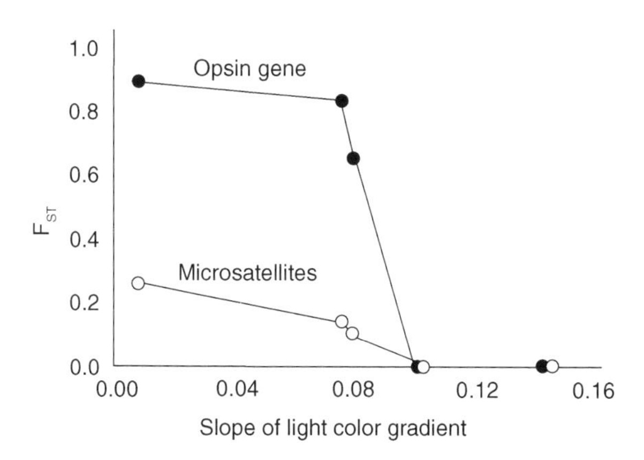

Figure 6.14. Genetic divergence between species of *Pundamilia* cichlid fish at five replicate gradients of water color in lake Victoria as a function of the slope (i.e., steepness) of the color gradient. Shown are F_{ST} values for a divergently selected gene (longwave-sensitive opsin) and for putatively neutral microsatellite markers. Divergence decreases as gradient steepness increases. Values on the y-axis for microsatellites were multiplied by ten for visual clarity. Modified from Seehausen et al. (2008b) with permission of the Nature Publishing Company.

6.11. The spatial scale of speciation

"Species with the greatest geographic ranges have the greatest number of geographic races, and the number of races is roughly proportional to the area occupied"

Blair (1950, p. 261)

A final topic in this chapter is the spatial scale of ecological speciation. The size of a geographical area is assumed to affect speciation rates (MacArthur and Wilson 1967, Endler 1973, Endler 1977, Losos and Schluter 2000, Rosenzweig 2001, Kisel and Barraclough 2010). For example, larger areas might encompass more heterogeneous habitats, leading to stronger and more diverse sources of divergent selection, promoting speciation. Likewise, larger areas can reduce opportunities for gene flow. Despite these obvious predictions, most work on the geography of speciation has focused on patterns of range overlap between populations, rather than on spatial scale per se.

Some empirical studies have shown that spatial scale affects local adaptation, as predicted by theory (Hanski et al. 2011). For example, positive correlations between range size and the number of ecological or geographic races observed have long been noted, as exemplified by the quotation above (Blair 1950). In more modern times, detailed study of color morphs of *Timema cristinae* walking-stick insects reported that host plant used predicted morph frequencies with increasing accuracy as the size of the spatial scale considered increased (Sandoval 1994b). Likewise, local adaptation of an arctiid moth (*Utetheisa ornatrix*) to a plant species whose seeds it predates was more evident at a large versus a small spatial scale (Cogni and Futuyma 2009). Such results support the argument that the adaptive divergence that drives ecological speciation is more likely at larger scales, and might require some minimum area size. However, a meta-analysis in plants found no evidence that spatial scale affected local adaptation (Leimu and Fischer 2008), highlighting the need for further work.

A few insightful studies examined the effects of spatial scale in speciation specifically. For example, Losos and Schluter (2000) examined the probability of in situ speciation of *Anolis* lizards on islands as a function of island size. They reported that indeed speciation was facilitated by larger island size. Kisel and Barraclough (2010) extended this type of analysis to eight disparate groups of plant and animal taxa. The results revealed two major findings. First, within these disparate groups, the probability of speciation tended to increase with island size, as reported in the study of *Anolis*. Second, among groups, the minimum island size required for speciation increased as potential for dispersal and gene flow (estimated from population genetic data) decreased. Figure 6.15 illustrates these findings, which demonstrate that speciation has a spatial scale that depends on levels of gene flow. However, work on ecological speciation specifically is lacking.

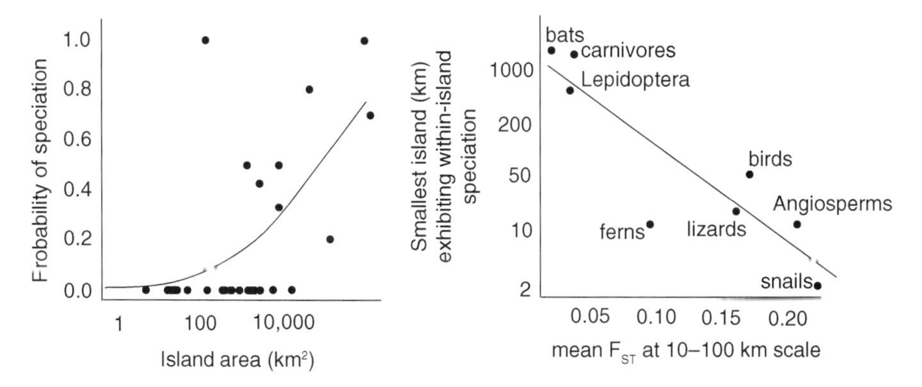

Figure 6.15. Evidence that speciation has a spatial scale which depends on levels of gene flow. a) For lizards, the probability of in situ speciation on an island increases with island size. Similar trends were observed in other groups of organisms. b) Among groups, the minimum island size (measured as the greatest distance between any two points of land within the island or archipelago) where speciation was observed as a function of the opportunity for gene flow (inferred from F_{ST}, with larger values representing lower gene flow). Modified from Kisel and Barraclough (2010) with permission of the University of Chicago Press.

6.12. Geography of ecological speciation: conclusions

"The rapidity with which an isolate is converted into a separate species depends upon . . . the selection pressure to which the isolated population is exposed, and the effectiveness of the isolation" (Mayr 1963, p. 585)

In the absence of gene flow, divergent selection can readily promote speciation. Thus, allopatric ecological speciation is expected to be common. When gene flow does occur, it often constrains divergence, but population differentiation can occur so long as selection is stronger than migration. However, population differentiation does not always lead to the evolution of discontinuities, strong reproductive isolation, and thus speciation per se. The circumstances allowing selection to overcome gene flow to the extent that a discontinuity develops, and how often these circumstances occur, are major remaining questions in speciation research (Nosil 2008a). Related issues concern how often gene flow might promote reinforcement, and whether the effects of reinforcement can cascade to affect reproductive isolation among multiple populations, including those outside of regions of geographic contact. Other roles of gene flow in promoting speciation, such as the creation of genetic novelty, also deserve further attention. Additional work on mixed geographic mode of divergence is warranted, particularly because such modes are conducive to some mechanisms of speciation, such as those involving the fixation of chromosomal inversions. Major challenges in addressing all these questions concern the empirical ability to detect gene flow. In this regard, inferring the timing of gene flow remains especially problematic. Finally, little is known to date concerning how the spatial arrangement and size of geographical areas affect ecological speciation.

7

The genomics of ecological speciation

This chapter focuses on the genomic basis of ecological speciation. It differs from the previous chapter on the genetics of speciation (Chapter 5) by focusing on genome-wide patterns, rather than on specific genes involved in adaptive divergence. Major questions remain concerning how individual "speciation genes" are arrayed within the genomes of diverging taxa, and how this affects ecological speciation. The aim of this chapter is to empirically and theoretically explore this genomic perspective of speciation, leading to a more integrative understanding of ecological speciation. An important and obvious point about the genomics of speciation is that massively parallel sequencing technologies (i.e., "next-generation sequencing" platforms) have removed many time-consuming steps involved in Sanger sequencing and have facilitated sequencing at a fraction of the time and cost previously required (Margulies et al. 2005, Ellegren 2008, Hudson 2008, Vera et al. 2008, Hohenlohe et al. 2010, Wheat 2010). Given these advances, which allow for genome-wide analyses in a wider range of organisms, it is expected that genomic inquiries and questions will come to be an even more dominant part of research on ecological speciation.

The main concept underlying this chapter is that genetic differentiation between populations is expected to be highly variable across the genome, possibly ranging from some regions with little or no differentiation through to fixed differences. I will refer to this pattern of variation among genomic regions in their level of population divergence as "heterogeneous genomic divergence" (following Nosil et al. 2009a). This heterogeneity arises because alleles at loci under divergent selection are pulled apart between populations by selection, and flow less freely between populations than alleles at other (i.e., neutral) loci. This results in stronger genetic divergence between populations at selected loci (Barton 1979, Barton and Hewitt 1989, Harrison 1991, Mallet 1995, Avise 2000, Via 2001, Wu 2001a, Wu 2001b, Ortiz-Barrientos et al. 2002, Gavrilets 2004, Gavrilets and Vose 2005, Mallet 2006, Noor and Feder 2006, Via and West 2008, Nosil et al. 2009a, Thibert-Plante and Hendry 2009). These processes and patterns led to the development of metaphors describing genomic divergence (e.g., genomic islands and genomic continents), which I focus on because they facilitate thinking about genomic divergence.

I initially discuss expectations for genomic divergence at equilibrium (i.e., once a balance has been reached between selection, migration, and recombination). I then treat the transient effects of selective sweeps, both from new mutations and standing genetic variation. The final topic considered is the relatively unexplored role of gene expression in ecological speciation. Research on all these topics is currently undergoing a renaissance, with a recent explosion of studies, many of which will be published between the submission and actual publication of this book. Thus, my goal here is not to be exhaustive, but rather to highlight key conceptual and theoretical points, illustrate these points with empirical examples, and thus provide a glimpse into the future work required.

7.1. Heterogeneous genomic divergence

During population divergence and speciation, genetic differentiation accumulates in some regions, while the homogenizing effects of gene flow or inadequate time for random differentiation by genetic drift precludes divergence in other regions (Wu 2001a, Gavrilets and Vose 2005). Many factors potentially contribute to such heterogeneous genomic divergence, including selection arising from ecological causes (Via 2001) or genetic conflict (Rice 1998, Presgraves 2007a, b), the stochastic effects of genetic drift (Kimura 1968, Ohta 1992), variable mutation rates (Balloux and Lugon-Moulin 2002, Hedrick 2005, Noor and Feder 2006), the genomic distribution and effect size of genes under selection (Orr 2005b), chromosomal structure (Noor et al. 2001, Rieseberg 2001, Ortiz-Barrientos et al. 2002) and the timing of selective sweeps (Barrett and Schluter 2008). I focus here on the contributions of divergent selection.

If divergent selection acts on only a subset of loci, then those loci affected by divergent selection can be identified as those that depart from background levels of genetic differentiation. Specifically, divergent selection will pull apart allele frequencies between populations at loci under selection and those linked to them. This might occur even if the remainder of the genome remains relatively undifferentiated (Fig. 7.1). Thus, loci under selection and those tightly physically linked to them should exhibit greater differentiation than distantly linked or unlinked neutral regions. As noted earlier, divergent selection is thus expected to result in "outlier loci" whose genetic divergence exceeds neutral expectations (Fig. 7.2) (Lewontin and Krakauer 1973, Bowcock et al. 1991, Beaumont and Nichols 1996, Luikart et al. 2003, Beaumont and Balding 2004, Beaumont 2005, Nielsen 2005, Storz 2005, Vasemagi and Primmer 2005, Foll and Gaggiotti 2008, Stinchcombe and Hoekstra 2008). However, it can be difficult to distinguish between the roles of divergent selection versus the other factors listed in the previous paragraph in generating heterogeneous genomic divergence. Along these lines, Box 7.1 provides a brief review of outlier detection methods. More detailed treatments can be found in the references cited in this paragraph.

7.2. The metaphor of genomic islands of divergence

A past review of patterns of heterogeneous genomic divergence in non-model-genetic organisms reached some clear conclusions: (1) population genomic studies regularly detect outlier loci; (2) these putatively selected regions are often dispersed

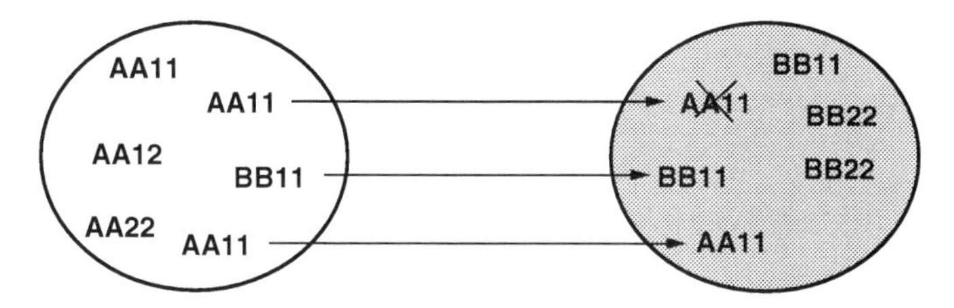

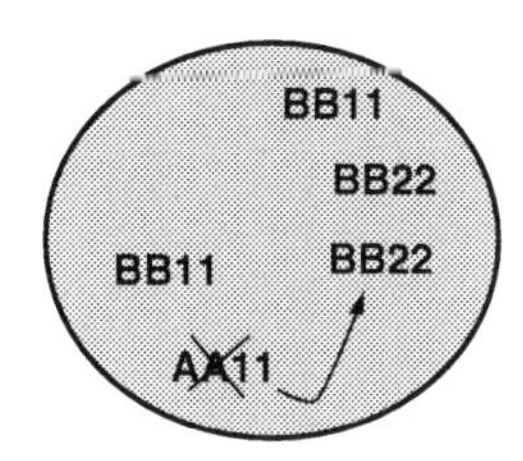

Figure 7.1. A simplified schematic diagram depicting how differential gene flow among gene regions, owing to divergent selection, contributes to heterogeneous genomic divergence. Large circles represent populations in different habitats. Two alleles at two loci are depicted in a diploid organism. Alleles "A" and "B" at locus one are subject to divergent selection, with "A" favored in the unfilled habitat and "B" favored in the gray habitat. Alleles "1" and "2" at locus 2 are neutral. Migration is depicted by straight arrows, selection by X's crossing out genotypes, and recombination by a curved arrow. a) Selection reduces the flow of selected alleles between populations (e.g., immigrant inviability). b) Gene flow occurs more readily at neutral than selected loci because recombination allows neutral alleles to move into a fit genetic background (in the case depicted, allele 1 moves via recombination into a background containing the locally favored B allele). In contrast to neutral alleles, maladaptive alleles at selected loci are continually removed from the population. The end result can be greater genetic divergence between populations at selected versus neutral loci.

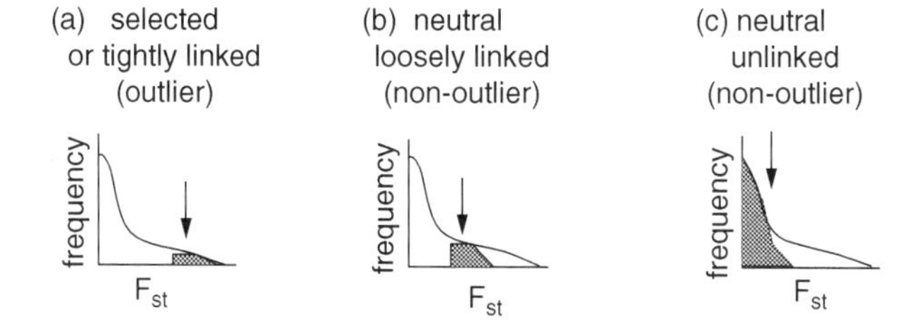

Figure 7.2. Three arbitrary classes of loci embedded within a continuous distribution of F_{ST} values. a) Divergently-selected loci or those very tightly physically linked to them exhibit particularly high F_{ST} and thus outlier status in a genome scan. b) Neutral loci more loosely linked to divergently selected loci (or perhaps weakly selected loci) exhibit F_{ST} values that are somewhat elevated, but insufficiently so to achieve outlier status. c) Neutral loci unlinked to divergently selected loci exhibit low F_{ST}. Modified from Nosil et al. (2009a) with permission of John Wiley & Sons Inc.

Box 7.1. A brief review of outlier detection methods

Population genomic analyses identify outlier loci by contrasting patterns of population divergence among genetic regions. This approach was first proposed by Lewontin and Krakauer (1973), and numerous variations on this original method now exist (Beaumont 2005, Foll and Gaggiotti 2008, Gompert et al. 2010). The initially most commonly employed of these methods, particularly in non-model organisms, was the F_{ST} outlier analysis developed by Beaumont and Nichols (1996). This test contrasts F_{ST} for individual loci with an expected null distribution of F_{ST} based on a neutral model. Under this approach, empirical loci with F_{ST} levels above the upper quantiles (e.g., 95 or 99%) of the simulated (i.e., neutral) distribution are considered candidates for divergent selection. However, many F_{ST} outlier analyses may be biased by departures from the assumed demographic history (Excoffier et al. 2009). An alternative approach is to assume that F_{ST} for individual loci represent independent draws from a common, underlying distribution that characterizes neutral divergence across the genome and that can be estimated from

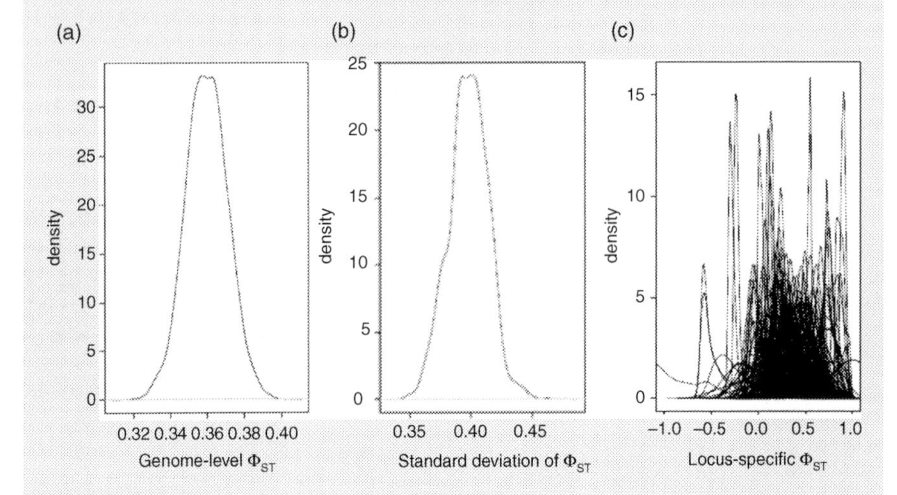

Figure 7.3. An example of outlier detection where statistics of population differentiation (Φ_{ST}) are estimated by Bayesian methods, which provide a probabilistic framework in which to interpret differentiation of individual loci (Gompert et al. 2010). In the example depicted, differentiation among populations of *Lycaeides* butterflies in North America was estimated: a) genome-level estimate of Φ_{ST}; b) average deviation of the genome-level estimate; c) locus-specific Φ_{ST} levels for an arbitrary set of 200 individual loci, which illustrate the high variation among loci in their level of differentiation across the genome. Some of this variation is likely to have been shaped by divergent selection. Modified from Gompert et al (2010) with permission of John Wiley & Sons Inc.

Box 7.1 (*Cont.*)

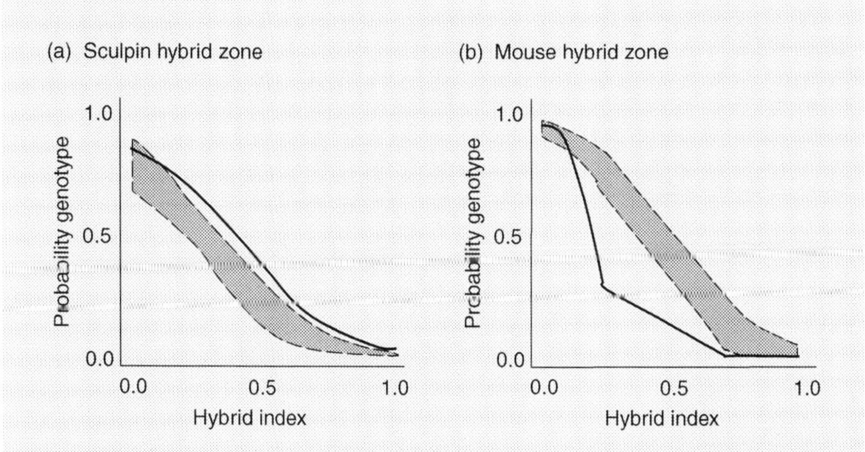

Figure 7.4. Two examples of outlier detection via the "genomic clines" method (Gompert and Buerkle 2009). Note that the term "cline" is used in a different sense than in classical hybrid zone studies. In the genomic cline context, the cline is the probability of various genotypes at individual loci as a function of how admixed individuals are across their genome (i.e., in this case estimated via a hybrid index). This contrasts with classical scenarios in which clines depict genotypes according to spatial position. The gray shaded areas bordered by dashed lines in each plot represent the expected (95% CI) clines for homozygous genotypes. Several loci departed from these background neutral expectations. For illustrative purposes, a single such "outlier" locus is shown here in each example (solid black line). a) Genomic clines between populations of sculpins (*Cottus*) in Germany estimated from 168 microsatellite loci. Hybrid index corresponds to the proportion of alleles with native (versus introduced) ancestry. Modified from Nolte et al. (2009) with permission of John Wiley & Sons Inc. b) Genomic clines in a mouse hybrid zone (*Mus musculus* × *M. domesticus*) in central Europe (Saxony transect) estimated from 41 single nucleotide polymorphisms (SNPs). Hybrid index corresponds to the proportion of marker alleles with *M. musculus* ancestry. Modified from Teeter et al. (2010) with permission of John Wiley & Sons Inc.

multi-locus data (Foll and Gaggiotti 2008, Gompert et al. 2010) (Fig. 7.3). Individual loci that are unlikely to be drawn from the background distribution are deemed outliers. This approach is more robust to different demographic histories. Outlier loci can also be identified via a "genomic clines" framework, where loci exhibiting unusual patterns of introgression are putatively affected by selection (Fig. 7.4) (Gompert and Buerkle 2009, Nolte et al. 2009, Teeter et al. 2010). In all cases, information on the gene function of outlier loci can provide further evidence for or against their involvement in divergent adaptation (Turner et al. 2010).

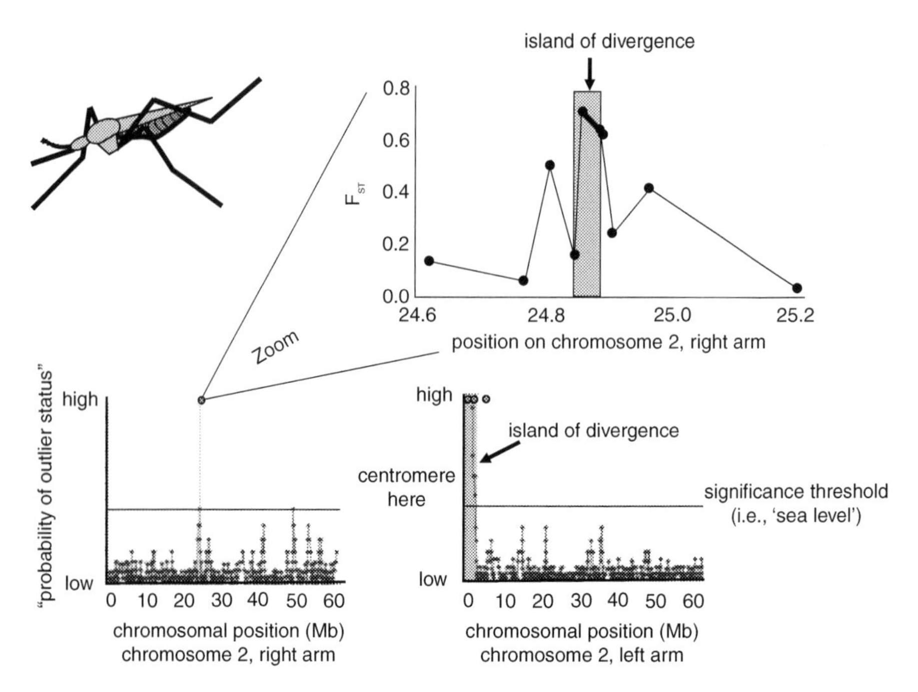

Figure 7.5. An empirical example of "genomic islands of divergence" involving incipient species of *Anopheles gambiae*. The bottom two panels depict patterns of differentiation across chromosome 2. Gray areas were identified as highly differentiated in sliding window analyses, with differentiation further confirmed by sequencing loci within these regions (gray circles). A large island is evident on the left arm, near the centromere. A small island is also evident on the right arm. Modified from Turner et al. (2005) with permission of the Public Library of Science. The top panel treats a subsequent study where portions of all annotated genes within the smaller island were sequenced. Modified from Turner and Hahn (2007) with permission of Oxford University Press.

across the genome; (3) they commonly exhibited replicated divergence across different population pairs; and (4) they are sometimes associated with specific ecological variables (Nosil et al. 2009a). This review further noted that isolation-by-adaptation at putatively neutral loci, caused by selection reducing gene flow such that drift could occur, was relatively common. The overall conclusion was that divergent selection makes diverse contributions to heterogeneous genomic divergence. Nonetheless, much variation was detected in the number, size, and distribution of genomic regions affected by selection. To help understand this variation, one can use the concept of "genomic islands of divergence" (Wu 2001b, Turner et al. 2005, Harr 2006, Turner and Hahn 2007) (Fig. 7.5).

A "genomic island" is any genetic region, be it a single nucleotide or an entire chromosome, that exhibits significantly greater differentiation than expected under neutrality (Fig. 7.5). The metaphor thus draws parallels between genetic differentiation observed along a chromosome and the topography of oceanic islands and the contiguous seafloor to which they are connected. Following this metaphor, sea level represents the threshold above which observed differentiation is significantly greater than expected by neutral evolution alone. Thus, an island is composed of both directly selected and tightly linked (potentially neutral) loci. Factors such as physical proximity between selected and other loci, rates of recombination, and strength of selection then each affect the height and the size of genomic islands. Under this metaphor of genomic islands, the few genes under or physically linked to loci experiencing strong divergent selection can diverge, whereas gene flow will homogenize the remainder of the genome, resulting in isolated genomic islands (but see Noor and Bennett 2010, Turner and Hahn 2010, White et al. 2010).

When it comes to the role of genomic islands in promoting ecological speciation, an important question is how large are the islands expected to be? The larger the genomic island, the larger the proportion of the genome resistant to gene flow, and the greater the possibility that genes within the region contribute to further reproductive isolation. On this issue, empirical evidence is mixed. There are several examples where regions were inferred to be large (Emelianov et al. 2004, Harr 2006, Rogers and Bernatchez 2007a, Via and West 2008). For example, in host races of pea aphids (*Acyrthosiphon pisum*) and lake ecotypes of *Coregonus* whitefish, outlier loci from genome scans reside near quantitative trait loci (QTL) for phenotypic traits more often than expected by chance, yet the average distance between an outlier and the apparently nearest QTL is relatively large (10.6 and 16.2 centimorgans, respectively) (Rogers and Bernatchez 2007a, Via and West 2008) (Fig. 7.6).

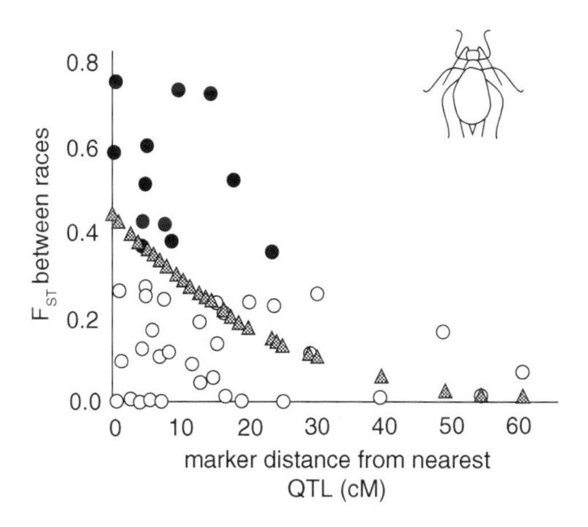

Figure 7.6. An example of a large genomic island of divergence. Genetic differentiation between natural populations of clover- and alfalfa-associated pea aphids as a function of distance away from QTL along a chromosome is shown. Filled circles are outlier loci and open circles are non-outlier loci. Triangles show the predicted values from a logistic regression. Modified from Via and West (2008) with permission of John Wiley & Sons Inc.

Indirect evidence that regions of differentiation are large also stems from the observation that numerous population genomic studies had genomic coverage probably too sparse to detect the actual direct targets of selection, yet they readily detected outliers. This suggests extensive linkage disequilibrium along chromosomes (Payseur et al. 2004, Nosil et al. 2009b), or that outliers do not actually represent loci affected by selection (Noor and Bennett 2010).

However, in contrast to the patterns above, a large number of observations suggest that regions of differentiation are small, including: (1) the tendency for accentuated divergence to be observed only in regions of extensively reduced recombination, such as near centromeres (Turner et al. 2005, Geraldes et al. 2006, Carneiro et al. 2009) or near breakpoints of chromosomal inversions (Machado et al. 2007, Noor et al. 2007, White et al. 2007, Yatabe et al. 2007, Strasburg et al. 2009); (2) a lack of strong genetic divergence at neutral markers physically proximate to sites of divergent selection (Makinen et al. 2008a); and (3) the rapid decay of genetic divergence away from genomic islands (Dopman et al. 2005, Turner et al. 2005, Machado et al. 2007, Noor et al. 2007, Turner and Hahn 2007, Makinen et al. 2008b, Storz and Kelly 2008, Wood et al. 2008, Baxter et al. 2010, Counterman et al. 2010, Scascitelli et al. 2010, Turner et al. 2010, Strasburg et al. 2011). For example, Turner et al. (2005) identified several genomic islands differentiating forms of *Anopheles gambiae* mosquitoes. In a follow-up study, Turner and Hahn (2007) sequenced portions of all annotated genes within one of the islands. As expected, sequence differentiation peaked within the "island," but the fine-scale data allowed more detailed characterization of the nature of the island, showing that differentiation drops off rapidly with distance from the region of maximum differentiation (i.e., the island is very small and steep). A more recent study employing whole-genome sequencing identified many other small regions of differentiation (Lawniczak et al. 2010), but the relative roles of gene flow versus ancestral variation in generating these islands remains unclear (Turner and Hahn 2010, White et al. 2010). Whatever their cause, if islands tend to be small, how is more widespread genomic divergence, which could promote speciation, generated?

7.2.1. Experimental evidence for "genomic continents" of speciation

There are at least two possible mechanisms for generating widespread genomic divergence: substantial divergence hitchhiking around a specific gene, versus genome hitchhiking caused by multifarious selection on numerous gene regions. As reviewed in Chapter 5, current theory predicts that genomic islands created by divergence hitchhiking around a specific, single gene will generally be small (Feder and Nosil 2010, Thibert-Plante and Hendry 2010). In contrast, when multiple loci across the genome are subject to divergent selection, genome-wide reductions in gene flow can allow for widespread divergence across the genome, even for neutral loci unlinked to any selected loci (Feder and Nosil 2010, Feder et al. 2011a). The efficacy of this genome hitchhiking mechanism, as opposed to divergence hitchhiking via physical linkage, warrants further theoretical study (see below for treatment of the "growth" of islands of divergence). Empirically, it raises the question of whether there are examples of widespread genomic divergence generated by multifarious selection acting on numerous genomic regions.

Observational genome scan studies tend to support the island view of genomic divergence (Emelianov et al. 2004, Turner et al. 2005, Nosil et al. 2009a for review). In contrast, evidence for more genomically widespread divergence is rare (but see Johnson and Kliman 2002, Lawniczak et al. 2010, Williams and Oleksiak 2011). However, this may stem from strong limitations in relying on genome scans alone for detecting selection. Genome scans conducted without complimentary selection experiments can predestine an island view because, inevitably, only the most diverged regions will be identified as statistical outliers. Other loci affected by selection, but more weakly, will go unnoticed and be considered part of the mostly "undifferentiated" and neutral genome. In contrast to observational methods, experiments might allow the detection of weak selection acting across numerous regions of the genome. For example, field experiments could be conducted in which experimental populations are created in novel and native (i.e., control) environments (e.g., herbivorous insects could be transplanted to their native host and to a host-plant species used by a close relative). Loci that show consistent allele frequency changes across replicates in the novel, but not in the control, environment are likely to be affected by selection in the novel environment, even if such frequency changes are slight. If properly replicated, experiments could thus detect weak selection and distinguish selection from experimental noise. In short, experimental measurements of selection on the genome are required to determine the fraction of the genome affected by selection.

Michel et al. (2010) conducted an experimental test of the genomic islands hypothesis in apple and hawthorn host races *Rhagoletis pomonella*, a model for sympatric ecological speciation. They reported numerous lines of evidence for widespread divergence and selection throughout the *Rhagoletis* genome, with the majority of loci displaying latitudinal clines, genotype–phenotype associations with adult eclosion time, within-generation responses to selection in a manipulative overwintering experiment, and significant host differences in nature despite levels of gene flow that are too high to allow divergence via genetic drift (Fig. 7.7). The results, coupled with genetic mapping and linkage disequilibrium analyses, provide experimental evidence that divergence was driven by selection on numerous independent genomic regions.

Based on their findings, Michel et al. (2010) proposed that "continents" of multiple differentiated loci, rather than isolated islands of divergence, may characterize even the early stages of speciation (Fig. 7.8). The authors stress, however, that the "island" versus "continent" views of genomic divergence represent ends of a continuum, rather than mutually exclusive hypotheses. For example, continents of divergence can be conceptualized as very large islands with variable topography (e.g., high mountain tops and lowland continental plains all above neutral sea level), depending on variation among genomic regions in selection strength, availability of preexisting genetic variation, linkage relationships, and rates of recombination. Their results illustrate such topography because the divergence observed throughout

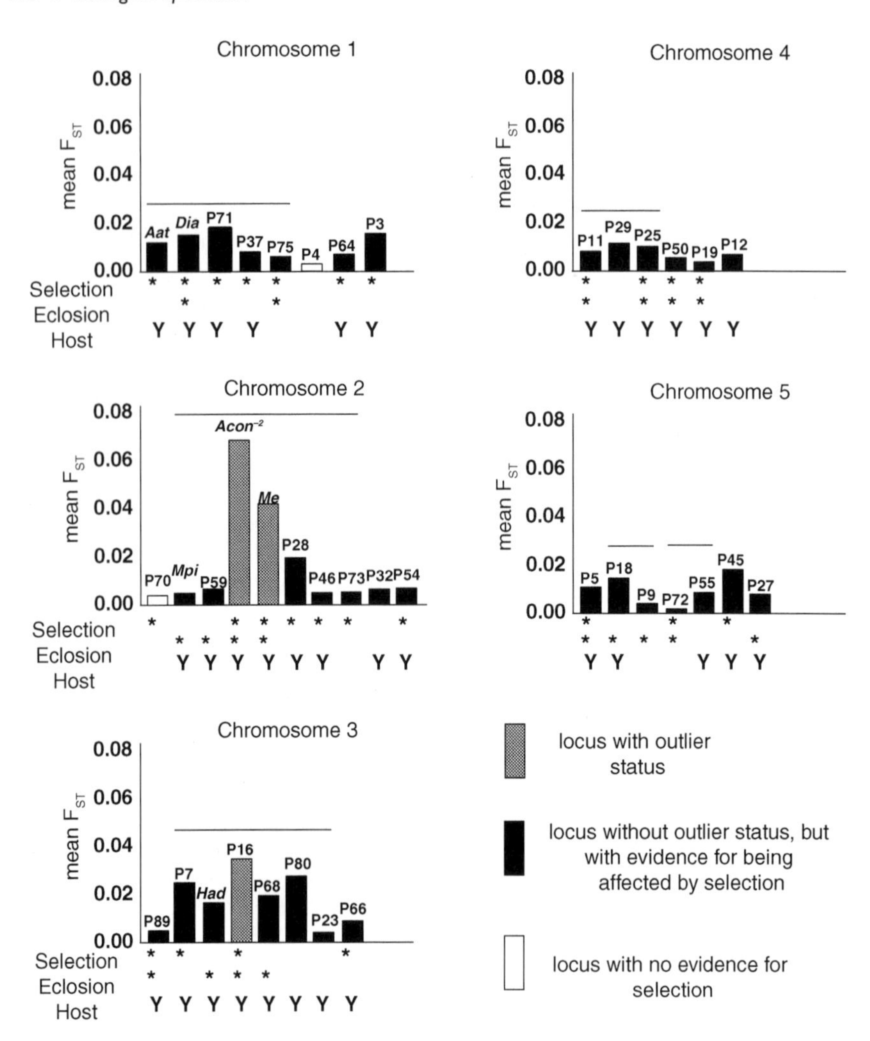

Figure 7.7. Evidence for widespread genomic divergence between apple and hawthorn host races of *Rhagoletis pomonella*. Mean F_{ST} for loci on chromosomes 1–5 is shown. Asterisks below graphs denote loci responding significantly to divergent selection in manipulative overwintering experiment and loci correlated with eclosion time in an association study. Y (yes) and N (no) denote if a locus displayed significant host-related differentiation in nature, despite levels of gene flow too high (i.e., 4%) to allow divergence without effects of divergent selection (i.e., too high for divergence via genetic drift alone). Lines above bars indicate loci in linkage disequilibrium. Modified from Michel et al. (2010) with permission of the National Academy of Sciences USA.

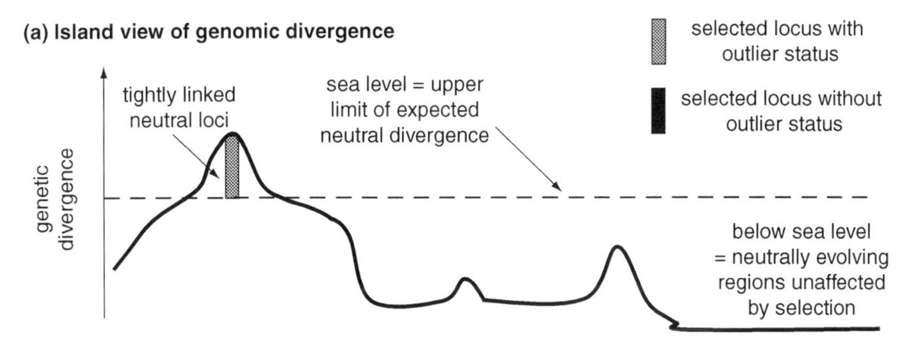

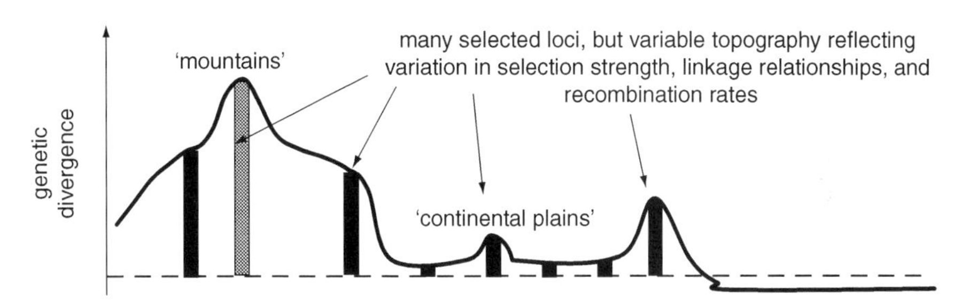

Figure 7.8. Schematic representation of the a) island versus b) continent views of genomic divergence. These views represent ends of a continuum, rather than mutually exclusive hypotheses. For example, "continents" of divergence can be conceptualized as very large islands with variable topography. Outlier status refers to whether a locus would exhibit statistical evidence for unusually high levels of genetic differentiation in an observational genome scan. See text for details. Modified from Michel et al. (2010) with permission of the National Academy of Sciences USA.

the *Rhagoletis* genome was clearly more accentuated in some regions. For example, regions harboring chromosomal inversions had on average twice the genetic divergence of collinear regions. A final point is that standard outlier analyses in this same study were consistent with the genomic island hypothesis: only two independent genomic regions were statistical outliers between the host races. Thus, experimental data and biological information on gene flow in nature was critical for detecting weaker, yet widespread, divergence across the genome.

Further experimental studies of genomic divergence are sorely needed to determine how selection acts across the genome, particularly because patterns of divergence documented in observational studies might often be influenced by the interplay of multiple factors, such as variation in selection strength, rates of gene flow, and recombination rates (e.g., Cutter and Choi 2010, Cutter and Moses 2011). Indeed, the development of a field of "experimental genomics" is especially likely to contribute to our understanding of the genomic basis of ecological speciation. Several cutting-edge experiments surveying genome-wide divergence using next-generation sequencing have now been conducted in the lab using microorganisms (Barrick et al. 2009, Araya et al. 2010, Paterson et al. 2010), but these do not address reproductive isolation per se. A few other experimental genomic studies exist, in *Drosophila* (Burke et al. 2010) and plants (Freeland et al. 2010), and both report widespread selection across numerous gene regions. However, more studies are sorely needed, particularly those focused on ecological speciation.

7.2.2. An integrated view of the size and the number of divergent gene regions

The diverse patterns described so far in this chapter make it clear that regions of divergence in the genome can vary in how large and how numerous they are. Thus, visualization of the spectrum between islands and continents of divergence is best conducted by simultaneous consideration of both these axes of divergence (Via 2009, 2011), coupled with linkage mapping or genomic analyses testing whether different divergent regions are contiguous along chromosomes or not.

The range of possibilities along these dimensions is depicted in Figure 7.9. For example, one might observe a small region of divergence surrounding a single

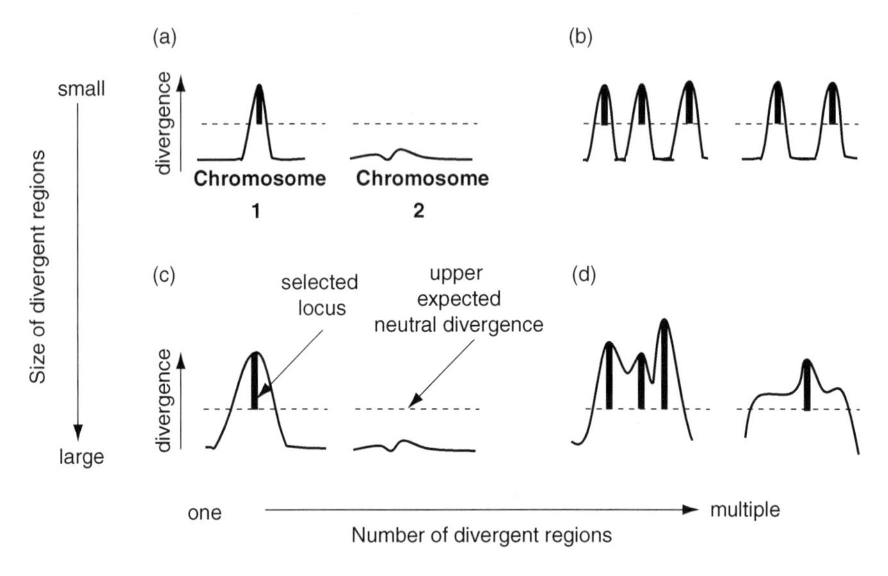

Figure 7.9. Simultaneous consideration of the size (y-axis) and the number (x-axis) of divergent regions within the genome. For simplicity, two classes of size and of number of divergent regions are depicted, but these represent opposing ends of a continuum of possibilities. Two chromosomes are depicted in four scenarios: a) one small island; b) many, small islands; c) one large island, for example formed by divergence hitchhiking; d) genomic continents, for example formed by selection on many loci within chromosomes, perhaps facilitated by genome hitchhiking (e.g., note that the right-hand chromosome contains only a single selected region, but widespread divergence could nonetheless occur via genome hitchhiking).

selected site, representing a single, small genomic island. If selection affects multiple, but not numerous, different regions dispersed across the genome, one might observe many small islands of divergence on different chromosomes. If divergence hitchhiking around one or a few loci generates large regions of divergence, on might observe large regions of divergence surrounding a single selected site, representing a large genomic island. In contrast, if widespread selection acts across the genome, one might observe large regions of divergence comprised of many different selected loci, representing genomic continents.

These depicted scenarios illustrate how careful consideration of both the number and the size of divergent genomic regions are required for accurate characterization of genomic divergence. However, clear metrics for quantifying such divergence are lacking, and future work developing such metrics is sorely needed. Once divergence is well characterized, one can begin to explore its consequences for speciation.

7.2.3. The growth of genomic regions of divergence

Previous sections focused on the size and number of genomic islands under equilibrium conditions. The theory and data presented did not encapsulate the fixation probabilities of new mutations that arise within regions of existing divergence, versus outside of them, to consider how regions of divergence in the genome might grow in size. I now outline three, non-mutually-exclusive, models for the growth of genomic islands: an allopatric model; an ecological model; and a structural model (Nosil et al. 2009a).

Genetic divergence and reproductive isolation during allopatric differentiation is unimpeded by gene flow and generally increases with time (Coyne and Orr 2004). Genomic divergence might still be heterogeneous during allopatric differentiation, but with genomic islands not representing regions of reduced gene flow, but rather regions that have differentiated more strongly than possible under genetic drift alone (Noor and Bennett 2010, Turner and Hahn 2010, Nachman and Payseur 2011). Compared with models of divergence with gene flow (see below), regions of differentiation might not be as highly clustered within the genome during allopatric divergence, because divergence at all regions can proceed unimpeded by gene flow. Thus, the "allopatric model" predicts many small to modest-sized islands, whose number and height are a positive function of time and selection strength, and which are distributed throughout the genome. There is some support for these predictions. For example, widely distributed regions of divergence were reported in allopatric populations of ferns (Nakazato et al. 2007) and fruit flies (Brown et al. 2004).

The growth of islands in the face of gene flow could differ from their growth in allopatric conditions. Genes under divergent selection, and those tightly linked to them, will experience reduced introgression relative to neutral, unlinked loci. The chance that a new mutation (whether adaptive or neutral) will persist and increase in frequency has been argued to be highest in regions of already reduced introgression, because reduced gene flow in these regions should facilitate divergence (Gavrilets

2004, Via and West 2008, Via 2009) (Fig. 7.10). Therefore, under an "ecological model" with gene flow, differentiated loci are expected to accumulate in genomic regions that already harbor genes under divergent selection, leading to increases in the number of genes within an island and thus in island size. This model predicts that genes affecting local adaptation will form clusters within the genome rather than being more evenly distributed across it. QTL studies demonstrating that different adaptive traits map to similar genomic regions are consistent with this prediction, although pleiotropy could readily explain such patterns (Hawthorne and Via 2001, Kronforst et al. 2006, Rogers and Bernatchez 2007a, Albert et al. 2008, Shaw and Lesnick 2009).

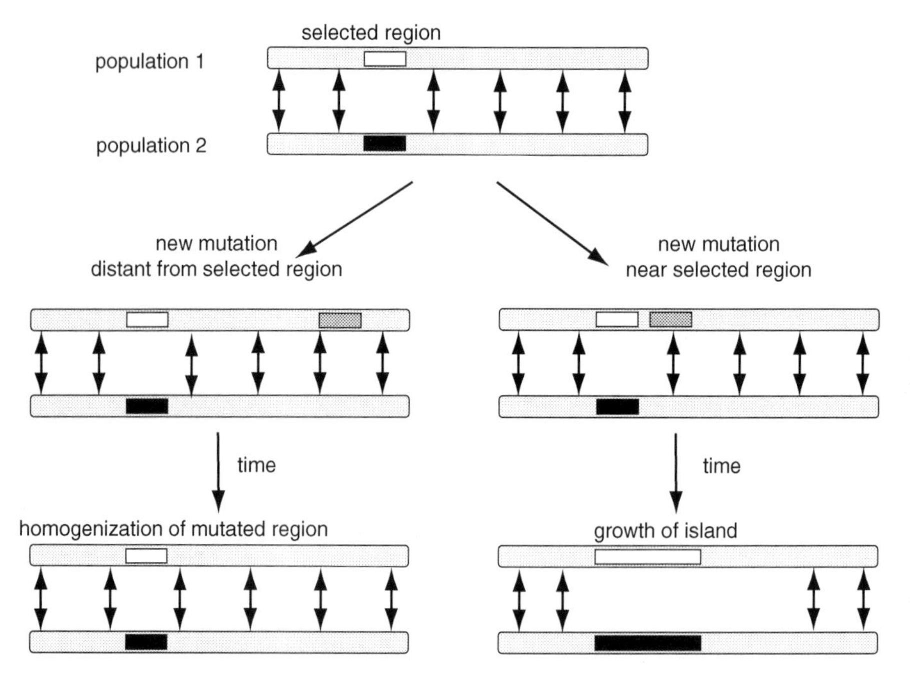

Figure 7.10. Conceptual model for the growth of genomic islands of divergence. Long horizontal bars represent chromosomes. White and black boxes within chromosomes represent regions of the genome that are affected by divergent selection and thus are undergoing reduced gene flow between populations. Two-headed arrows connect undifferentiated regions where genetic exchange between populations is high. A new mutation (gray box within chromosome) arising near a genomic region under divergent selection might have a higher likelihood of differentiating between populations than a new mutation arising in higher gene flow regions (but see text and Figure 7.11 for strong caveats associated with this model). Likewise, arising in a region of reduced recombination might also facilitate the establishment of a new mutation. Modified from Nosil et al. (2009a) with permission of John Wiley & Sons Inc.

A number of issues will affect the generality of the ecological model. A major one is the extent to which physical linkage to a selected site actually increases the probability that a new mutation establishes (Yeaman and Otto 2011, Yeaman and Whitlock 2011). A recent study used simulation and analytical models to explore this issue (Feder et al. 2011a). Specifically, it quantified the probability that a new mutation establishes when: (1) it is the first and only mutation to arise in an undifferentiated genome; (2) it arises in physical linkage to a locus already diverged via selection; and (3) it arises unlinked to any selected loci, but within a genome that has some diverged loci. This allowed the partitioning of how various processes aid the establishment of new mutations. For example, the effect of arising in physical linkage to a diverged region represents the effect of divergence hitchhiking. In contrast, the effect of arising unlinked to selected regions but in an already diverged genome represents the effects of genome-wide reductions in gene flow caused by selection (i.e., genome hitchhiking). The results showed that by far the most important predictor of establishment for new mutations was the strength of selection acting directly on a new mutation, with both forms of hitchhiking having smaller effects in comparison. In essence, so long as selection was stronger than migration, new mutations could establish. Nonetheless, divergence and genomic hitchhiking could help mutation establishment somewhat under some conditions (Fig. 7.11).

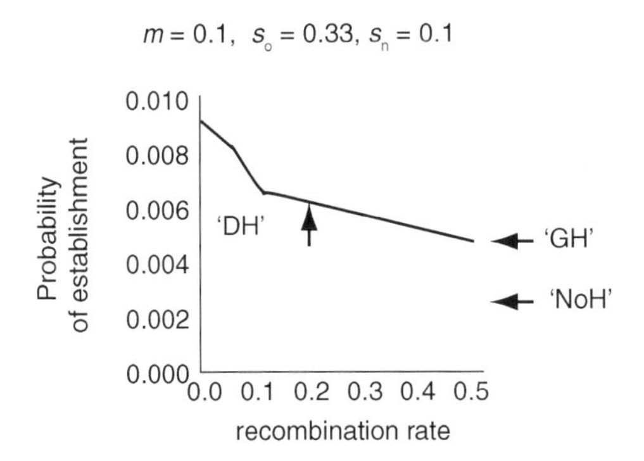

Figure 7.11. Simulation results demonstrating how the role of different processes in the probability of establishment of a new beneficial mutation can be partitioned. "NoH" = the probability of establishment of a new mutation when it is the first and only mutation to arise in an undifferentiated genome. "DH" = the probability of mutation establishment when a new mutation arises in physical linkage (i.e., $r > 0.0$ but < 0.5) to a locus already diverged via selection (this represents the effects of divergence hitchhiking). "GH" = the probability of mutation establishment when a new mutation arises unlinked to any selected loci (i.e., $r = 0.5$), but within a genome that had a single locus diverged before the origin of the new mutation (this represents the effects of genome hitchhiking). m = migration rate, s_o = selection strength on the originally diverged mutation, s_n = selection strength on the new mutation. In general, the strength of selection acting directly on a new mutation was a more important predictor of mutation establishment than either DH or GH (not shown). Modified from Feder et al. (2011a).

Thus, although arguments about hitchhiking and the growth of islands are verbally compelling, their explanatory power deserves further attention.

Another question concerns the capacity of natural selection to favor the evolution of tighter linkage among loci, for example to keep beneficial genotypic combinations together (Kimura 1956, Butlin 2005). This process could proceed via the evolution of modifier loci that suppress recombination (Kouyos et al. 2006) and facilitate the growth of genomic islands. Support for the evolution of tighter linkage is provided by the evolution of "supergenes": that is, groups of neighboring genes on a chromosome that are inherited together (Sheppard 1953, Turner 1967a, b, Charlesworth and Charlesworth 1975, Sinervo and Svensson 2002, Joron et al. 2006).

Finally, under a "structural model," the origin and growth of sizeable islands is facilitated by structural factors within the genome that reduce recombination, such as chromosomal inversions (Noor et al. 2001, Rieseberg 2001, Ortiz-Barrientos et al. 2002, Butlin 2005). New mutations arising within regions of reduced recombination might be particularly likely to diverge between populations, so long as they arise in environments and genetic backgrounds in which they are favored (Kirkpatrick et al. 2002, Ortiz-Barrientos et al. 2002, Feder et al. 2011a).

7.3. Selective sweeps and adaptation from standing variation versus new mutations

Two final issues regarding genomic divergence are: (1) the transient effects of selective sweeps; and (2) the consequences of genetic divergence building up from new mutations versus standing genetic variation (see Barrett and Schluter 2008 for review). Compared with equilibrium levels of divergence, divergence in the genome could be much different during the very initial stages of adaptive divergence to a new habitat, where strong selection on some loci rapidly sweeps linked neutral markers to high frequency. In this case, genetic divergence could initially be at a high maximum over extended areas of chromosomes, and then decay over time via recombination (Nielsen 2005, Barrett and Schluter 2008). Moreover, even when populations are old, new mutations may occasionally sweep through them, generating new hitchhiking events, and temporarily elevating divergence across much of the genome. The implications of the transient effects of such selective sweeps for ecological speciation require more attention.

If the sweeps stem from new mutations, then divergence is expected to be elevated across particularly large regions of the chromosome, especially if selection is strong and the sweep occurs rapidly. However, adaptation, particularly initial adaptation to a new habitat, might often stem from preexisting rather than new mutational variation (Feder et al. 2003a, Barrett and Schluter 2008). In such cases, recombination has already been acting for some time to break up associations among genomic regions. Thus, sweeps from standing variation can reduce the magnitude of neutral differentiation surrounding selected sites relative to that observed for new mutations, resulting in "soft sweeps" (Orr and Betancourt 2001, Przeworski 2003,

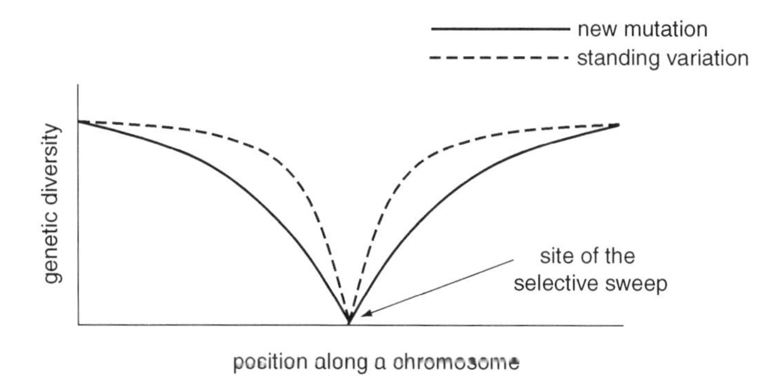

Figure 7.12. Adaptation from new mutations versus standing genetic variation. Genetic diversity within a population is decreased for larger regions of a chromosome when selective sweeps occur from new mutations versus standing genetic variation. Note that these effects of sweeps are transient, and will decay as recombination occurs over time. Also note that reduced diversity within populations generally translates to greater divergence between populations. Modified from Barrett and Schluter (2008) with permission of Elsevier.

Hermisson and Pennings 2005) (Fig. 7.12). Adaptation from standing variation also differs from that involving new mutations in that it might allow for more rapid evolution, for example because there is no waiting time for beneficial new mutations and because preexisting beneficial alleles can start at a frequency that is much higher than $1/2N_e$. However, in very large populations the distinction between adaptation from standing variation versus new mutations becomes blurred, because multiple new mutations can arise readily and contribute to adaptation (Hermisson and Pennings 2005, Pennings and Hermisson 2006b, a, Barton 2010).

What is known empirically about the role of standing variation versus new mutations in ecological speciation? Not much yet, but some key studies provide initial insight. For example, as predicted by theory, regions of divergence in two species of *Heliconius* butterflies were larger in the species likely to have undergone a more recent sweep (Baxter et al. 2010, Counterman et al. 2010). Additionally, some examples of ecological speciation clearly involve an element of divergence from standing variation. For example, genetic variation contributing to important diapause timing differences between sympatric apple and hawthorn races of *Rhagoletis* flies in the United States has a much older (i.e., preexisting) origin in Mexican populations (Feder et al. 2003a). Other examples of adaptation from standing variation exist, such as armor evolution in sticklebacks (Colosimo et al. 2005) and coloration in beach mice (Hoekstra et al. 2006). Nonetheless, examples of adaptation from new mutations also exist, as documented in deer mice (Linnen et al. 2009) and in the classic example of industrial melanism in the peppered moth (van't Hof et al. 2011). Finally, Karasov et al. (2010) report that rapid adaptation to insecticides in *Drosophila melanogaster* involved multiple new mutations but was apparently not mutation-limited because large population sizes allowed for numerous mutations to arise, precluding a need for standing variation. All these examples concern adaptation, and thus the implications for reproductive isolation specifically deserve future attention. In general, further studies examining the effects of sweeps, both from new mutations and standing variation, will probably lead to a more integrated understanding of ecological speciation.

7.4. Gene expression and ecological speciation

The final topic considered in this chapter is gene expression, which is shaped by both genetic and environmental components, and thus can be considered a "molecular phenotype" (Ranz and Machado 2006). For example, the transcription rate of a gene can vary among genotypes, among environmental conditions, or both (Schadt et al. 2003, Gibson and Weir 2005, Whitehead and Crawford 2006, Roelofs et al. 2009). The role of gene expression in ecological speciation is relatively unexplored (Pavey et al. 2010a). To what extent does ecological speciation involve genetic changes in coding regions versus divergence of regulatory regions (Hoekstra and Coyne 2007, Stern and Orgogozo 2008, Chan et al. 2009)? What can gene expression tell us about ecological speciation?

Gene expression might provide novel insights into speciation because it facilitates the ability to uncover phenotypes that would not readily be visible via traditional approaches. Our understanding of evolution has often been limited by our ability to define relevant phenotypes (Nevins and Potti 2007). For example, initial progress in understanding ecological speciation has necessarily focused on easily measured morphological, and to some extent, behavioral traits. In essence, gene expression might allow us to circumvent these limits by uncovering hidden phenotypes potentially of ecological relevance: phenotypes that are perhaps difficult or counterintuitive to measure. This could be especially crucial given that most genome annotations to date currently stem mostly from model genetic organisms, and consequently lack ecological relevance (Aubin-Horth and Renn 2009). Identifying ecologically relevant expressed genes will thus probably increase the efficacy of genomics to address questions related to ecological speciation (Carroll and Potts 2006, Landry and Aubin-Horth 2007, Pena-Castillo and Hughes 2007).

Furthermore, gene expression warrants consideration because two events need to occur during the process of ecological speciation (following Schluter 1998), and gene expression might strongly affect each of them. First, ecological speciation requires that newly founded populations persist in the colonized environments. Ernst Mayr (1947, 1963) especially espoused this "persistence view" of the role of ecology in speciation (see Schluter 1998, Levin 2004 for review). Second, populations in different environments need to evolve genetically based reproductive isolation. Gene expression might therefore promote ecological speciation in two ways: (1) indirectly by promoting population persistence; or (2) more directly by affecting adaptive genetic divergence in traits causing reproductive isolation

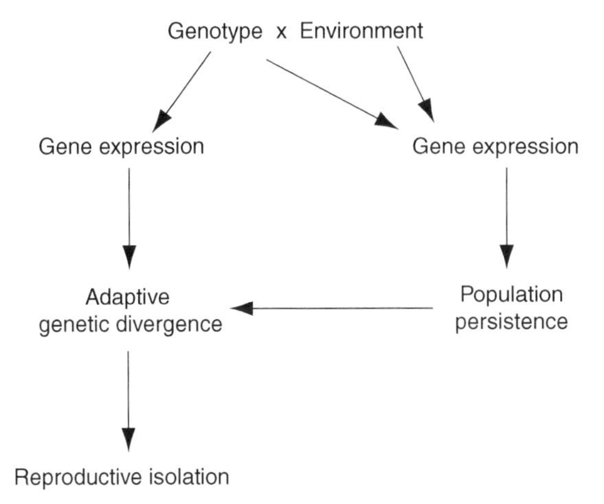

Figure 7.13. Conceptual diagram of the different ways the genetic and environmental components of gene expression might contribute to ecological speciation. Both components might contribute to population persistence, which is required for eventual speciation. The genetic components of gene expression could contribute to the adaptive genetic divergence that drives the evolution of reproductive isolation during ecological speciation. Modified from Pavey et al. (2010a) with permission of John Wiley & Sons Inc.

(Price et al. 2003) (Fig. 7.13). Concerning the first aspect, there are many factors other than just gene expression that could promote population persistence, but gene expression might be a powerful one, as it allows for rapid phenotypic change without a long waiting time for new mutations and substitutions in coding regions. I focus here on the second aspect, because this is where gene expression might be causally involved in generating reproductive isolation. I refer readers to Pavey et al. (2010a) for a more thorough treatment of gene expression in ecological speciation.

7.4.1. Gene expression and adaptive genetic divergence

Gene expression might affect the adaptive genetic divergence that drives ecological speciation. While many studies have discovered gene expression differences between diverging populations, these differences need not represent heritable genetic divergence. Studies documenting heritable expression differences between adaptively divergent populations are emerging (e.g., Lexer et al. 2004, Colosimo et al. 2005, Derome and Bernatchez 2006, Roff 2007, Steiner et al. 2007, Whiteley et al. 2008, Linnen et al. 2009). A first class of studies quantified genetic differences using common garden experiments that remove or reduce environmental variation (Lai et al. 2008). For example, St-Cyr et al. (2008) quantified variation in gene expression for almost 4000 genes in species pairs of lake whitefish (*Coregonus clupeaformis*) under common garden conditions. They found that 14% of genes exhibited heritable differences in transcription. The genes differentially expressed between species pairs in the common environment were similar to those that differed in expression in the wild (Derome and Bernatchez 2006). This suggests a strong genetic component to differential transcription between species.

A second class of studies uses QTL mapping to examine the genetic basis of expression divergence. Traditional, or "phenotypic," quantitative trait loci (pQTL) uncover associations between genetic regions and traditional phenotypic traits such as morphology. Expression QTL (eQTL) map transcript abundance in the same manner as pQTL map "traditional" traits. eQTL is emerging as a useful technique for localizing genomic regions contributing to gene expression divergence (Gilad et al. 2008). Although eQTL studies are still in their infancy, a general pattern that has been observed is the existence of genomic regions associated with the expression level of many transcripts: so-called eQTL "hotspots" (Gibson and Weir 2005, Whiteley et al. 2008). These hotspots have revealed several things about the genetics of ecological speciation.

First, they show that both types of QTL, eQTL and pQTL, can map to the same genomic region, as observed in dwarf and normal lake whitefish species pairs (Rogers and Bernatchez 2007b, Derome et al. 2008, Whiteley et al. 2008). Second, eQTL studies indicate that genomic regions involved in ecological speciation can be non-randomly distributed across the genome. For example, in the same lake whitefish species pairs noted above, 50% of 249 eQTL identified in the brain were associated with only 12 hotspots distributed over eight linkage groups (Whiteley et al. 2008). A similar pattern was observed in muscle, with 41% of eQTL mapping to six hotspots across four linkage groups (Derome et al. 2008). These findings hint at the existence of localized "genomic islands" of expression divergence. Third, eQTL have been associated with genomic regions harboring outlier loci between natural populations. For example, in the whitefish species pairs, ten loci were identified as outliers, and three of these also corresponded to eQTL hotspots (Bernatchez et al. 2010).

7.4.2. Gene expression and reproductive isolation

Even after adaptive genetic components of gene expression are identified, a major question remains: are these components associated with reproductive isolation? With the exception of studies demonstrating that reductions in hybrid fitness can be due to gene expression (Landry et al. 2007, for review see Ortiz-Barrientos et al. 2007, Renaut et al. 2009), studies of gene expression have generally not made links to genetically based reproductive isolation. This is usually because reproductive isolation itself was not considered or underlying mutations have not been identified. I consider below two examples that illustrate both the promise and the difficulties associated with linking gene expression directly to ecological speciation.

The first example concerns Darwin's finches, in which divergence in beak morphology arose from divergent selection stemming from competition and use of seeds

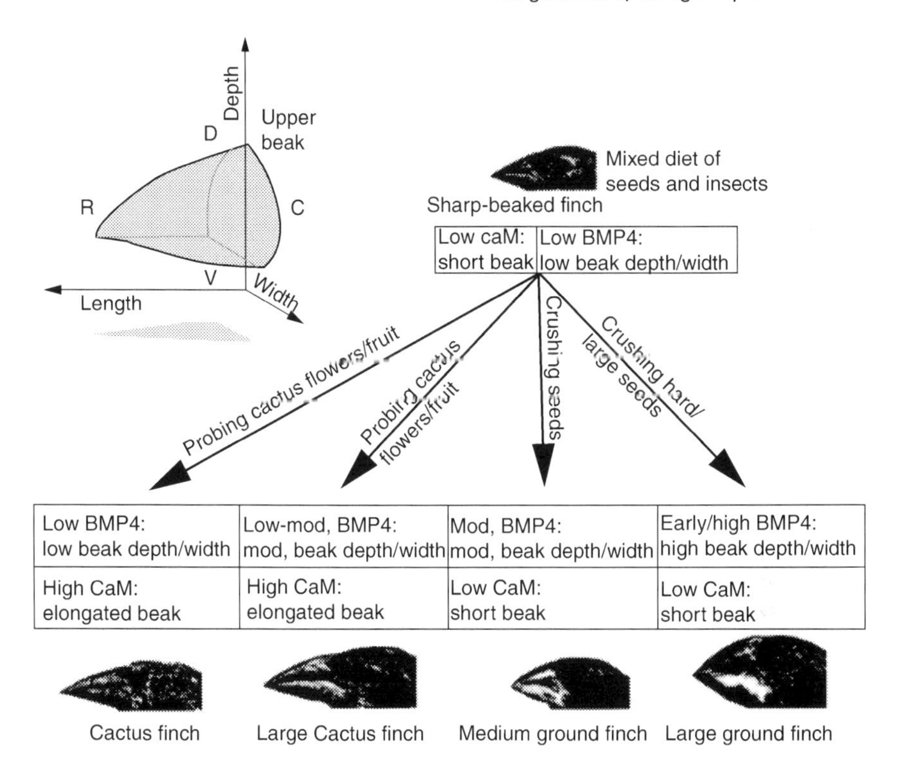

Figure 7.14. An example of the effects of gene expression on phenotypic traits of likely importance for ecological speciation. Specifically, evidence that bone morphogenetic protein 4 (*Bmp4*) and calmodulin (*CaM*) regulated growth along different axes of bill morphology in *Geospiza* Darwin's finches, facilitating the evolution of distinct beak morphologies. A beak of the sharp-beaked finch reflects a basal morphology for *Geospiza*. Abbreviations: C, caudal; D, dorsal; R, rostral; V, ventral. Modified from Abzhanov et al. (2006) with permission of the Nature Publishing Group.

of differing size and hardness (Schluter and Grant 1984b, Grant 1986, Grant and Grant 2006). Beak morphology also probably contributes to reproductive isolation via song divergence and selection against immigrants and hybrids (Podos 2001, Grant and Grant 2008b, Hendry et al. 2009b). Among species, higher levels of the bone morphogenetic protein 4 (*Bmp4*) expression are correlated with deeper beak shapes. Also, manipulation of the expression level of *Bmp4* in chicken embryos altered beak development in the predicted direction (Abzhanov et al. 2004). These results demonstrate that gene expression variation from *Bmp4* affects morphology in Darwin's finches (Fig. 7.14). Similar results occur for another gene, calmodulin (*CaM*) (Abzhanov et al. 2006). However, owing to a lack of common garden or mapping studies, there is as of yet no evidence that heritable differences in beak morphology are affected by *Bmp4* or *CaM*. In other words, the mutations underlying beak size differences in Darwin's finches have not been identified. Thus, although there is good evidence that regulatory changes underlie morphological divergence among species of Darwin's finches, the ultimate link between gene expression and reproductive isolation is yet to be made.

The second example concerns ecological speciation in the threespine stickleback complex. Ancestral marine and most derived freshwater sticklebacks have a robust pelvic apparatus, but at least 24 independent freshwater populations exhibit a greatly reduced or completely absent pelvic structure (Reimchen 1980, Bell 1987, Gow et al. 2008, Chan et al. 2009). Repeated parallel evolution is itself an indication that divergent selection drove evolution, with experimental evidence pointing to predation and differences in ion concentration as the mechanisms of selection (Reimchen 1980, Bell et al. 1993, Reimchen and Nosil 2002, Vamosi 2002). Recent studies have examined the genetic basis of pelvic reduction. Chan et al. (2009) reported that expression of the gene *Pitx1*, which is implicated in pelvic reduction, is driven by a small (501 bp) tissue-specific enhancer (*Pel*). Remarkably, small deletions functionally inactivated *Pel* in nine of 13 tested pelvic-reduced populations. These regions exhibiting recurrent deletions, rather than the *Pitx1* gene itself, appear to have been subject to positive selection (Chan et al. 2009). These results demonstrate that genetically based expression divergence contributed to adaptive divergence in pelvic morphology. However, direct links to reproductive isolation remain to be established. The ability to conduct manipulative experiments in semi-natural ponds (e.g., Schluter 1994, Barrett et al. 2008) indicates that linking gene expression at *Pitx1* to reproductive isolation, for example reduced fitness of immigrants and hybrids, is a distinct possibility.

7.4.3. Gene expression and ecological speciation: conclusions

Gene expression might facilitate population persistence and adaptive genetic divergence. There are clear examples of gene expression having effects on phenotypic traits and adaptive genetic divergence, but links to the evolution of reproductive isolation itself remain indirect. Gene expression during adaptive divergence often seems to involve complex genetic architectures controlled by gene networks, regulatory regions, and "eQTL hotspots" (Wentzell et al. 2007). Nonetheless, approaches for isolating the functional mutations contributing to adaptive divergence are beginning to prove successful. Future work on gene expression are likely to uncover novel, previously "hidden," phenotypes involved in ecological speciation and inform the types of genetic changes involved.

7.5. The genomics of ecological speciation: conclusions and future directions

The study of the genomics of speciation is still in its infancy. Nonetheless, much has already been learned, including widespread documentation of the heterogeneous nature of genomic divergence. This chapter was not meant to be exhaustive, but rather to offer a glimpse of what the future might hold, especially because many topics have only begun to be explored. For example, we have empirical examples of both large and small genomic regions of differentiation, leading to a debate about whether genomic divergence tends to represent "islands" or "continents," and the extent to which these arise via divergence versus genome hitchhiking. Population genomic studies, by virtue of focusing on extremely differentiated outlier loci, are predestined to support an island view. Further experimental studies are required to determine how often divergence is localized versus widespread across the genome. Theory predicts that continents of divergence can occur, but that this probably requires selection acting on numerous loci throughout the genome. If selection acts on only a few isolated loci, small islands of divergence may be the norm. Major questions also remain about how genomic regions of divergence grow in size during the speciation process. To what extent can reduced gene flow surrounding selected sites facilitate the divergence of new mutations that arise in such regions? Future work should also focus on how answers to these outstanding questions might be modified when selective sweeps are recent, or stem from standing genetic variation. Finally, clear causal links between gene expression and reproductive isolation during ecological speciation have rarely been established. Thus much work remains to be done, but advances in sequencing technologies and computational methods will allow these questions to be pursued with increased vigor and sophistication.

8

The speciation continuum: what factors affect how far speciation proceeds?

"... these forms may still be only... varieties; but we have only to suppose the steps of modification to be more numerous or greater in amount, to convert these forms into species... thus species are multiplied" (Darwin 1859, p. 120)

This chapter focuses on the degree to which ecological speciation unfolds once it is initiated. The central concept underlying the chapter is that speciation is often an extended and continuous process. Thus divergent selection, as well as various other genetic, time-based, and geographic factors, can affect the point in the speciation continuum that is achieved by a population pair. I begin the chapter by discussing evidence for the continuum of speciation in more detail. I note that examples of partial reproductive isolation abound, and explore the extent to which they represent a stable outcome versus an intermediate state that is bound to progress to complete reproductive isolation (or eventual extinction). I then turn to explanations for variation in how far speciation proceeds, starting with non-selective hypotheses, and then turning to hypotheses based upon the specific nature of divergent selection itself, namely how strong selection is and how many traits or genes it acts on. I conclude by discussing the importance of temporal stability in selection and outcomes alternative to speciation that might occur when divergent selection acts.

8.1. The speciation continuum

"if we are able to discover several steps in the [speciation] process... not in one and the same species, but in different species... and trace an uninterrupted series from the first origin of the... variety to the full development of the... species"

(Walsh 1864, p. 411)

It has been argued that speciation is often a continuum (Mallet et al. 2007, Hendry 2009, Hendry et al. 2009a, Nosil et al. 2009b, Peccoud et al. 2009, Gourbiere and Mallet 2010, Merrill et al. 2011). For example, Darwin's view of gradual species formation proposed a continuum of divergence from populations, to well-marked varieties, to species (Darwin 1859). Early evolutionary entomologists (Walsh 1864, Walsh 1867) and botanists (Clausen 1951, Lowry unpublished) also adopted this view, describing how "host races" and "ecotypes" were at intermediate stages in the speciation process. A number of recent studies are consistent with these views (Fig. 8.1) (Seehausen et al. 2008b, Berner et al. 2009, Peccoud et al. 2009, Merrill et al. 2011). For example, a study of butterfly hybridization showed that the number of hybrids observed between races and species decreased gradually with genetic distance, with no obvious qualitative break in the frequency of hybrids across the "species boundary" (Mallet et al. 2007) (Fig. 8.2). However, although different points in the speciation process are highly evident in nature, most examples tend

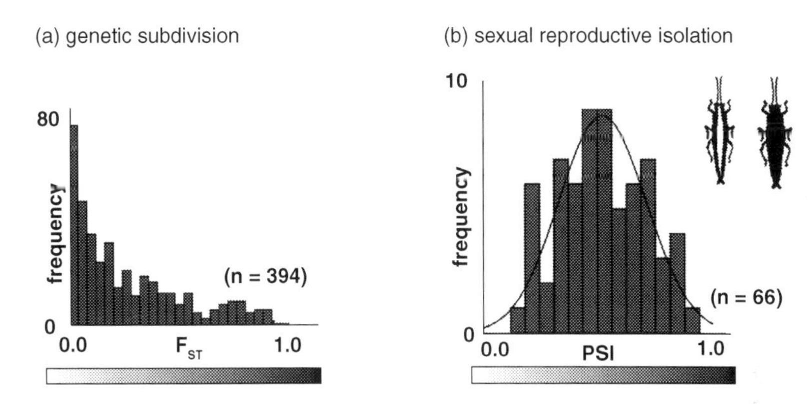

Figure 8.1. Examples of the quantitative nature of divergence during the speciation process. a) Levels of genetic subdivision (F_{ST}) between 394 pairs of animal populations in putatively neutral markers can vary from absent to complete. Data from Morjan and Rieseberg (2004). b) Levels of sexual isolation between 66 pairs of *Timema cristinae* walking-stick insect populations vary from nearly absent to nearly complete (PSI is an index of reproductive isolation, where zero is random mating and one is complete reproductive isolation). Data from Nosil et al. (2003).

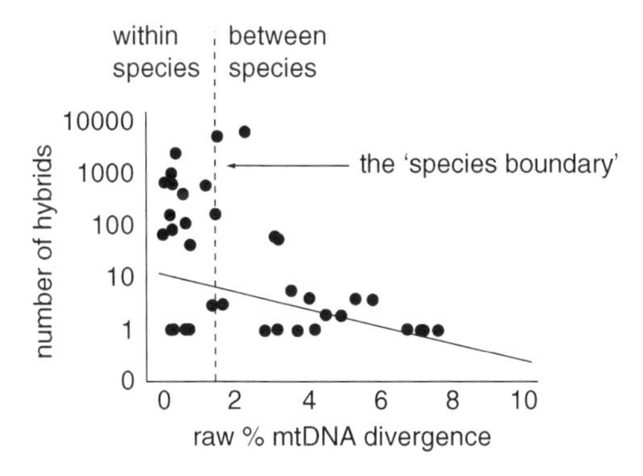

Figure 8.2. Evidence for the continuous nature of the speciation process based upon the number of hybrids observed between populations and species of *Heliconius* butterflies. The number of observed hybrids decreases gradually with genetic distance, with no obvious qualitative break in the frequency of hybrids across the "species boundary." For illustrative purposes, numerous points of value zero for number of hybrids detected, which were distributed across the *x*-axis, were deleted. Modified from Mallet et al. (2007) with permission of BioMed Central.

to stem from very different study systems that have each achieved a different degree of divergence. Thus, evidence that gradual transitions occur within a single study system, from populations, to ecotypes, to species, is still rare. Is speciation really a continuum, and if so, how does one reconstruct the continuum?

Several approaches to addressing these questions exist. Most directly, one could use experimental evolution across generations to study the speciation continuum by examining reproductive isolation build-up over time. However, owing to the time-scales involved, the speciation continuum in many instances can only be studied indirectly, for example by examining multiple taxon pairs at different points in the continuum, and reconstructing retroactively how divergence may have unfolded. At best, one would compare different populations within a single species pair that vary from weak to very strong or complete reproductive isolation, as was done for *Pundamilia* cichlids (Seehausen 2008, Seehausen et al. 2008b, Seehausen and Magalhaes 2010). Another example is the study by Peccoud et al. (2009) that used populations of *Acyrthosiphon pisum* pea aphids on numerous host-plant species to document the gradual evolution of less-differentiated forms into distinct species. Such approaches allow strong inferences about how transitions along the speciation continuum unfold from beginning to end within a single taxon pair. In other instances, one could compare multiple pairs of "ecotypes" within species to multiple recently diverged species pairs (or compare different independently derived species pairs that vary in their degree of reproductive isolation). Here, one cannot as directly infer how a single taxon pair underwent transitions along the speciation continuum, but one can nonetheless test which factors reliably predict differences between ecotypes and species (Langerhans et al. 2007, Berner et al. 2009, Nosil and Harmon 2009, Nosil et al. 2009b, Merrill et al. 2011).

These approaches are best applied using closely related lineages that nonetheless differ in their degree of reproductive isolation. Unlike experimental evolution, such approaches do not provide a direct "time-slice" of how speciation unfolds, and will have difficulties ruling out contingencies that might cause even closely related lineages to differ (Seehausen 2008). However, it may often be the only approach feasible for natural populations, and recent applications provide strong evidence for the speciation continuum (Langerhans et al. 2007, Nosil and Sandoval 2008, Seehausen 2008, Seehausen et al. 2008b, Berner et al. 2009, Merrill et al. 2011) (Fig. 8.3). These case studies of single systems, coupled with the examples documented in the first chapter of the book, provide important confirmation for the Darwinian view of gradual species formation.

If we accept that speciation is often a continuum, a highly interesting problem emerges: What factors predict how far speciation proceeds along this continuum? Divergent selection promotes speciation, but often to varying degrees. It is likely that divergent selection can drive divergence along the full length of the speciation continuum, causing speciation to unfold from beginning to end. Again, perhaps the best examples stem from *Pundamilia* cichlids and *Acyrthosiphon* pea aphids (Seehausen 2008, Seehausen et al. 2008b, Peccoud et al. 2009, Seehausen and Magalhaes 2010).

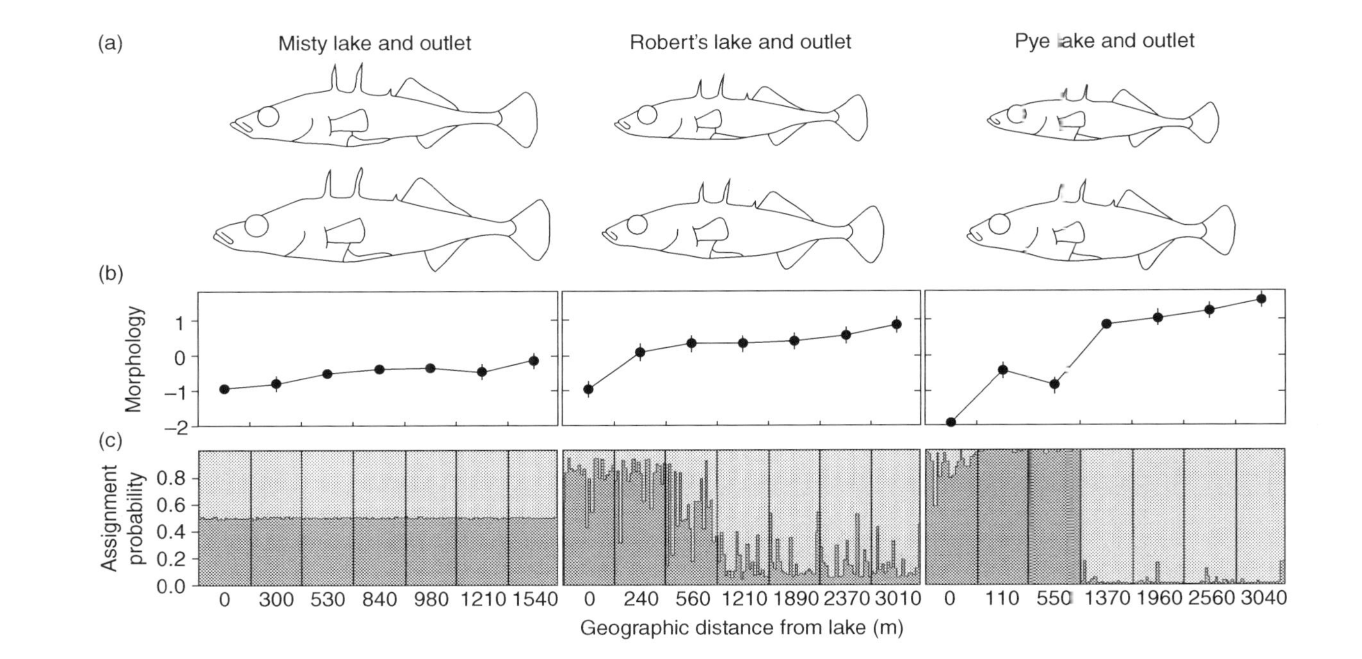

Figure 8.3. An example of the continuous nature of the speciation process, stemming from morphological and genetic clustering between different pairs of lake versus stream forms of stickleback fishes. a) Average body shape for fish from the lake site (top) and the last stream site (bottom) in each system. The drawings are based on geometric–morphometric consensus shapes. b) Clinal shifts in morphology, here expressed as the size-standardized principal component of body shape and gill raker number. c) Genetic clustering visualized by individual assignment probabilities to two predefined populations (dark gray, light gray), based on microsatellite data and the software Structure (Pritchard et al. 2000). More solid coloration reflects speciation having proceeded further. Modified from Hendry et al. (2009a) with permission of John Wiley & Sons Inc.

However, in other cases, only intermediate points in the speciation continuum have been achieved, or ecological divergence fails to generate particular forms of reproductive isolation (Raeymaekers et al. 2010). For example, ecological divergence is accompanied by only moderate reproductive isolation and weak neutral genetic differentiation in host-plant ecotypes of *Timema* walking-stick insects (Nosil 2007), *Rhagoletis* flies (Feder et al. 1994), *Ostrinia nubilalis* corn borers (Dopman et al. 2005, Dopman et al. 2010), and other host-associated insects (Drès and Mallet 2002, Funk et al. 2002). Similarly, partial reproductive isolation has been documented between intertidal shore ecotypes of *Littorina snails* (Sadedin et al. 2009), head and body forms of human head lice (Light et al. 2008), and parasitic indigobirds (Balakrishnan et al. 2009). A particularly compelling example stems from a population of freshwater sticklebacks that is subject to strong disruptive selection and mates assortatively by diet: these are the conditions most theoretically conductive to speciation, yet this population remained a phenotypically unimodal, single species (Snowberg and Bolnick 2008). In other instances, divergent selection appears to not even initiate speciation (Crispo et al. 2006, Rueffler et al. 2006, Svensson et al. 2009), or distinct species collapse back into a single population: i.e., "speciation reversal" (Taylor et al. 2006, Seehausen et al. 2008a). The main question addressed in this chapter is: What factors explain this variability in how far ecological speciation proceeds?

8.2. The stability of partial reproductive isolation

"Once genetic changes underlying speciation start, they go to completion very rapidly. . . the quantitative theory predicts . . . short duration of intermediate stages in speciation" Gavrilets (2003, p. 2213).

Before tackling the causes of variability in how far speciation proceeds, I will explore whether cases of partial reproductive isolation tend to be transitory stages that will proceed to complete reproductive isolation or instead represent equilibrium levels of divergence, for example a stable outcome maintained at a balance between selection and gene flow. In the latter case, further divergence is not inevitable, or even expected, and partial reproductive isolation is not necessarily a precursor to full speciation (McAllister et al. 2008).

8.2.1. The "stuck partway" view of speciation

Certain mathematical models have shown that partial reproductive isolation can evolve and be maintained as a "stable outcome" or equilibrium state (Matessi et al. 2001, Kirkpatrick and Ravigné 2002, Gavrilets 2003, 2004, Bolnick and Fitzpatrick 2007, Otto et al. 2008, Sadedin et al. 2009, Thibert-Plante and Hendry 2009, 2010). For example, Matessi et al. (2001) studied the evolution of assortative mating in a model that included a selected trait. In this model, they found that even in the absence of costs to mating, reproductive isolation increased from absent to complete only when selection was strong. In fact, they report that only weak or moderate reproductive isolation, corresponding to a continuously stable equilibrium, tends to evolve when selection is below a certain threshold. Likewise, a critical threshold of linkage disequilibrium between selected and mating loci needs to be attained in some models before the speciation process ensues (Dieckmann and Doebeli 1999, Doebeli and Dieckmann 2003). However, some models predict that the strength of selection required to initiate speciation is not much different from that required to complete it (Fry 2003). In general, if costs to being choosy are included in theoretical models, only weak or partial reproductive isolation often evolves (Gavrilets 2004, Bolnick and Fitzpatrick 2007).

Empirically, examples of partial reproductive isolation between relatively old population or species pairs support the argument that rapid transitions from partial to full reproductive isolation need not occur, and thus that partial isolation could be relatively stable (see Table 8.1 for examples). For example, Price (2007) reviewed hybrid zones between bird species and documented 12 hybrid zones between species that diverged more than a million years ago. In these zones, hybrids are often quite common, indicating that reproductive isolation is far from complete (mean percent hybrids in the zone = 46%). If partial reproductive isolation is truly a stable outcome, it indicates that populations can get stuck at intermediate stages of speciation. Essentially, populations diverge until reaching equilibrium levels of differentiation, and then remain "stuck partway" on the path to complete reproductive isolation (Berlocher and Feder 2002), at least until environmental conditions change (e.g., selection gets stronger, or gene flow lower), or genetic constraints on speciation are overcome (e.g., via the emergence of new mutations).

8.2.2. The "feedback loop" view of speciation

The "stuck partway" view of speciation strongly challenges an alternative view, where speciation is driven by ongoing feedback loops between selection and gene flow (Gourbiere and Mallet 2010). Associations between adaptive divergence and gene flow can arise because adaptive divergence reduces gene flow, gene flow itself reduces adaptive divergence, or both causal associations operate. These two processes could generate a positive feedback loop whereby low gene flow allows adaptive divergence, which in turn further reduces gene flow, which feeds back to increase adaptive divergence, which reduces gene flow, and so on (Hendry et al. 2001, Nosil and Crespi 2004, Seehausen 2008, Gourbiere and Mallet 2010). Such a feedback loop will allow increasingly weakly selected differences to diverge as gene flow gets lower, and new advantageous mutations could be protected by already reduced gene flow. In principle, the feedback could continue until levels of gene flow became infinitesimally reduced, completing speciation.

Under this view, divergence itself facilitates divergence. Relatedly, the first few incompatibilities to arise between populations should "snowball" into the

Table 8.1. Examples of partial reproductive isolation between taxon pairs that diverged a million or more years ago.

Taxa	Inferred age	Evidence for partial isolation	References
Timema cristinae host ecotypes on *Adenostoma* versus *Ceanothus*	Roughly one to two millions years, based upon levels of mitochondrial and nuclear DNA sequence divergence among allopatric population pairs	Molecular studies demonstrate gene flow between parapatric populations; experiments indicate RI is far from complete	(Nosil 2007)
Heliconius butterflies (melpomene/ cydno and silvaniform clades)	Divergence occurred millions of generations ago, perhaps as many as 30 million, based upon mitochondrial DNA	Historical introgression inferred from sequence variation in nuclear genes and AFLPs identified individuals of mixed ancestry, supporting ongoing introgression	(Kronforst 2008)
Fish in the Centrarchid family	Up to 20 million years, based upon molecular data	Hybrids documented in the wild between species that diverged millions of years ago	(Bolnick and Near 2005, Near et al. 2005)
Helianthus annuus and *H. petiolaris* sunflowers	One million years, based upon anonymous nuclear genes	Introgression inferred from microsatellite loci and nuclear genes	(Strasburg and Rieseberg 2008, Kane et al. 2009)
Haplochromine cichlids in Lake Malawi	Five million years between *Astatotilapia* species, based on mitochondrial DNA	Introgression inferred from AFLPs	(Joyce et al. 2011)
Mammals, such as forest and savannah elephants, wolves, and coyotes	One to two million years in both the case of elephants and of wolves/coyotes, based on mitochondrial DNA	Hybrids observed	(Seehausen et al. 2008a)
Numerous bird species pairs in hybrid zones	Among 12 hybrid zones where taxa diverged at least one million years ago; mean age of divergence was 2.7 million years ago (minimum = 1.1, maximum = 4.5)	Among the 12 hybrid zones, mean percentage of hybrids occurring in the zone was 46% (minimum = 3, maximum = 100), indicating that RI is far from complete	(Price 2007)

Abbreviations: AFLP, amplified fragment length polymorphism; RI, reproductive isolation.

accumulation of many incompatibilities and eventual speciation (Orr 1995). Another argument against the "stuck partway" view is that it is difficult to imagine a general mechanism to completely stabilize partial reproductive isolation. In the absence of such a stabilizing mechanism, why would reproductive isolation remain intermediate (Gourbiere and Mallet 2010)? Rather, it would seem that partial isolation is a kind of knife edge, with no mechanism to keep it as such, and the trend will generally be for greater reproductive isolation through time (or divergence will collapse).

This feedback view seems logical, but if such feedback loops exist, it seems that once speciation is initiated, it should inevitably go through to completion, as predicted by some models (Gavrilets 2003, 2004). Contrary to this prediction, we see population pairs that diverged from one another long ago, but exhibit only partial reproductive isolation. Under a feedback view, such pairs should have passed rapidly through intermediate stages of speciation once the process was initiated. This raises an interesting question: Is the feedback view inconsistent with the existence of partial reproductive isolation in nature?

8.2.3. Reconciliation of views: time-dependence of stability and the "speciation slowdown"

How might the "stuck partway" versus "feedback loop" views of speciation be reconciled? A simple answer emerges if increased divergence takes extended periods of time, particularly for the latter stages of the speciation process. In this case, whether partial reproductive isolation is defined as "stable" or not depends on the timescale examined: it could be stable in the short term, but inevitably unstable over the long term. Thus further divergence will be expected, eventually, but sometimes not for quite some time. In this case, existing levels of partial reproductive isolation tend to represent a "quasi-stasis" or "pseudo-equilibrium." In essence, partial reproductive isolation will never be completely stable, but will appear so for certain (shorter) time periods. This situation is particularly likely if: (1) speciation involves just a few, rare genes of major effect, such that the waiting time between substitutions causing increased reproductive isolation is long (Bolnick and Near 2005); and (2) the rate of the evolution of reproductive isolation slows down over time, such that the latter stages of speciation take extended periods of time, as reported by Goubière and Mallet (2010).

On the second point above, why might speciation slow down? The reasons are numerous and reviewed elsewhere (Gourbiere and Mallet 2010). I touch upon a few scenarios for illustrative purposes. One example has to do with reinforcement. After gene flow is reduced to a low level, there will be limited reinforcing selection to strengthen premating isolation even further (Nosil and Yukilevich 2008). In fact, once gene flow is very low, there may even be selection to avoid inbreeding that encourages occasional outcrossing to other populations (Ingvarsson and Whitlock 2000, Ebert et al. 2002). Inbreeding avoidance might sometimes counteract processes driving divergence and thus stabilize intermediate points in the speciation process. However, this inbreeding mechanism does not apply to allopatric taxa that have no opportunity to outcross, might not apply during the initial stages of speciation where gene flow is still high, and thus is not likely to be a universally applicable stabilizing mechanism. Finally, the temporal stability of divergent selection could be of key importance, with periods of weaker, or even reversed, selection weakening or counteracting divergence (Reimchen 1995, Grant and Grant 2002, Reimchen and Nosil 2002, 2004, Seehausen et al. 2008a, Siepielski et al. 2009), and disrupting or even reversing feedback loops.

In sum, further work is needed to reconcile different theoretical models and empirical observations supporting the "stuck partway" versus "feedback loop" views (Coyne and Orr 2004, Nosil et al. 2005, Nosil et al. 2009b). Ideally, we would like to know just how truly "continuous" the speciation process is. Does speciation have distinct stages (Mendelson et al. 2004), and does it tend to halt and recommence, or even reverse, predictably? Only by further quantification of the speciation continuum will answers to these questions emerge.

8.3. Non-selective explanations for how far speciation proceeds

"without an extrinsic reduction of gene flow, the ecological variability cannot become a primary source of discontinuity" (Mayr 1947, p. 281)

As exemplified by Mayr's quotation, it was long considered that although ecological divergence could initiate the evolution of reproductive isolation, stronger divergence (i.e., the evolution of discontinuities) required geographic isolation. Much work on speciation has thus focused on geographic factors, and it is now established that speciation can be strongly promoted by geographic barriers to gene flow (Coyne and Orr 2004). The role of time in explaining how far speciation proceeds has also been established, as exemplified by documented positive relationships between genetic distance and levels of reproductive isolation in multiple taxa (Coyne and Orr 2004). Likewise, some forms of sexual selection, such as unidirectional mating preferences (e.g., a general preference for larger males), might constrain population divergence, thereby affecting how far speciation proceeds (Schwartz and Hendry 2006, Takahashi et al. 2010). Finally, as reviewed in Chapter 5, a number of genetic factors, such as pleiotropy and one-allele assortative mating mechanisms, can also promote speciation. These hypotheses have increased our understanding of the factors driving and constraining the speciation process and provide explicit, albeit not mutually exclusive, alternatives to the selective factors discussed below (Nosil et al. 2009b).

8.4. Ecological explanations for how far speciation proceeds

I begin my exploration of how the nature of divergent selection affects the degree of speciation by considering how speciation is promoted by different kinds of ecological shifts. I then describe two hypotheses for how such ecological shifts so strongly promote speciation (Hutchinson 1957, 1959, Rice and Hostert 1993, Dambroski and Feder 2007, Price 2007, Nosil and Sandoval 2008, Seehausen 2008, Seehausen et al. 2008b, Nosil and Harmon 2009, Nosil et al. 2009b). Under the "stronger selection" hypothesis, how far speciation proceeds is positively related to the strength of selection on a single trait, with very strong selection on one or a few traits driving speciation to its latter stages. Under a "multifarious selection" hypothesis, how far speciation proceeds is positively related to the number of genetically independent traits subject to selection, with selection on many traits required for speciation to proceed far. These two hypotheses can be visualized in terms of the metaphor of an adaptive landscape: Does speciation proceed further by increased divergence between adaptive peaks in a single dimension, or via the generation of peaks that are separated in multiple dimensions (Johnson and Kliman 2002, Gavrilets 2004, Nosil and Harmon 2009)?

To keep the hypotheses ecologically rooted, and because selection has thus far been measured mostly at the phenotypic level, I first focus on phenotypic traits. However, the hypotheses can be applied to the genetic level, for example by considering selection on few versus many genes. In fact, the hypotheses are most precisely tested at the genetic level, by considering per-locus (rather than per-trait) selection coefficients, albeit coupled with knowledge of the traits involved. I thus conclude by outlining a framework for studying the degree of speciation using genomic data, and argue that some of the limitations of phenotypic measurements of selection can be overcome with genomic data.

It is crucial to understand that more overall or "total" selection will generally promote speciation, no matter how it is distributed among traits or genes. Thus, I stress that the ideal circumstances for speciation are likely to be strong *and* multifarious selection, such that total selection summed across all traits/genes is very high. However, the total amount of divergent selection to which a pair of populations is exposed is finite. The question I thus focus on is how the speciation process is affected by a finite amount of selection distributed across a few, versus many, traits or genes.

8.4.1. Dimensionality of ecological shifts

The nature of ecological shifts can affect how far speciation proceeds (Nosil et al. 2009b). Under one scenario, slight shifts along a single niche dimension initiate speciation, but more extreme shifts along that same dimension are required to complete speciation (Gavrilets 2004, Funk et al. 2006, Price 2007). An example comes from ecological and phylogenetic studies of galling Australian thrips (Thysanoptera), which suggest that extreme shifts in host-plant use promote speciation more strongly than smaller shifts (Crespi et al. 2004). In addition to how extreme a shift in one dimension is, the number of niche dimensions differing between taxa might affect speciation, with speciation proceeding further when divergence occurs in many niche dimensions (Rice and Hostert 1993, Gavrilets 2004, Price 2007, Seehausen 2008, Seehausen et al. 2008b). These ideas have seen few tests because most speciation studies: (1) consider only two categories of ecological divergence (Funk et al. 2002, Rundle and Nosil 2005), i.e., presence or absence of ecological divergence, precluding a test of how reproductive isolation varies with the quantitative degree of divergence along a niche dimension; and (2) have not statistically isolated independent niche dimensions (Schluter 1996a, 2000b, Nosil 2007), potentially confounding the magnitude of an ecological shift in a single direction with the dimensionality of the shift.

How do the types of ecological shifts referred to above affect divergent selection? This question is difficult to answer because one-to-one mapping is not expected between the nature of an ecological shift (i.e., how extreme or multidimensional it is) and the nature of divergent selection (i.e. its strength and how many traits it acts upon). Thus both types of ecological shift noted above might cause stronger selection on a given single trait, selection on a greater number of traits, or both. For example, an extreme ecological shift along a single niche dimension might cause stronger selection on a trait that was previously under weaker selection, or it might result in more (i.e., new) traits being subject to selection. This means that actual selection estimates, rather than environmental data on niche divergence, are required to distinguish the stronger versus multifarious selection hypotheses.

8.5. Multifarious versus stronger selection: theory

The probability of speciation under stronger versus multifarious selection can vary according to a number of factors, including the total strength of divergent selection, per-trait selection coefficients, and the nature of correlations between selected traits and those causing reproductive isolation (Rueffler et al. 2006).

8.5.1. Contributions to total selection

Two arguments concerning total selection strength suggest that multifarious selection can be important for causing speciation to proceed further. First, multifarious selection can be required to generate increased total strength of divergent selection in natural populations, because the strength of selection on any single trait is dictated by the ecological setting, and thus can be low and never increase (Endler 1986, Schluter 2000b, Kingsolver et al. 2001). Second, even if divergent selection on one trait is strong, extreme divergence in that trait can be constrained by a lack of suitable genetic variation (Bush 1969b, Futuyma et al. 1995, Gavrilets and Vose 2005) or functional constraints (Lande 1979, Arnold 1992). In such a scenario, multifarious selection on many traits can be required to generate an overall degree of trait divergence that is large enough to complete speciation. Empirical studies of selection strength and levels of genetic variation are required to test these ideas.

8.5.2. Per-trait selection coefficients and correlated evolutionary response

There are reasons to suspect crucial differences in how genetic divergence occurs under the stronger versus multifarious selection hypotheses, even when the total strength of divergent selection is held constant. First, the hypotheses differ in the expected magnitude of per-trait selection coefficients, and thus in the ability to counter gene flow. Specifically, for a given total strength of selection, per-trait selection coefficients will increase as the number of traits under selection decreases. Divergence in a given trait is a function of its selection coefficient and rates of gene flow (Endler 1973, Endler 1977, Gavrilets 2004, Mallet 2006, Nosil et al. 2009b). The implication is that strong selection on a few traits will sometimes be most effective at causing adaptive divergence of specific traits in the face of gene flow.

Second, divergent selection on a trait can cause divergence in other correlated traits, referred to as a "correlated evolutionary response." As the number of divergent traits subject to selection increases, the number of correlated traits that also diverge will probably increase (Rice and Hostert 1993, Johnson and Porter 2000). Thus, in a "sampling model," multifarious selection samples divergence across traits more widely than selection on few traits, perhaps making multifarious selection more likely to incidentally cause differentiation in traits of key importance for speciation. At the genetic level, as more and more loci diverge, this should lead to a "snowball" of the accumulation of genetic incompatibilities (Orr 1995, Orr and Turelli 2001). The counterargument here is that weak multifarious selection might cause little divergence at all. In such a case, strong selection on a few traits might actually cause the greater correlated response, even if this response is restricted to few traits. Another argument in favor of multifarious selection is that divergence in many traits might allow the combined effects of many types of reproductive isolation to be strong, even if the effect of single barriers on their own is weak (Matsubayashi and Katakura 2009). Thus, controlling for a fixed total strength of selection, some predictions emerge.

1) Strong selection on one or a few traits is better at causing adaptive divergence in the face of gene flow than is multifarious selection. However, because selection on a single trait often causes little correlated response, it will often result in single trait polymorphism rather than speciation.
2) Multifarious selection will sometimes be too weak to strongly overcome gene flow, precluding divergence in the selected traits and any correlated response. However, when multifarious selection does cause widespread divergence, it may be more effective at incidentally causing reproductive isolation.

8.5.3. Differential importance at different stages of the speciation process

A final point is that the importance of the two hypotheses will probably vary among stages of the speciation process. For example, strong selection on one or a few traits may initiate speciation, thereby causing some reduction in gene flow and the evolution of a genetic polymorphism, which in turn allows divergence in others traits that are under (weaker) multifarious selection. In such a scenario, single trait polymorphisms may be converted to speciation (McKinnon and Pierotti 2010). Because most past work has focused on the early stages of ecological speciation, future studies that examine multiple stages are required to avoid a bias toward understanding only a part of the speciation process. Additionally, future studies testing the stronger versus multifarious selection hypotheses should consider multiple reproductive barriers, as different barriers might be differentially affected by each hypothesis. For example, if premating isolation evolves early in the speciation process (Jiggins and Mallet 2000, Mendelson 2003, Coyne and Orr 2004, Lowry et al. 2008a), it might often be driven by strong selection on a few traits that can readily overcome gene flow. In contrast, forms of reproductive isolation that involve epistatic interactions between multiple loci might evolve via multifarious divergence in many traits (Orr 1995).

8.6. Multifarious versus stronger selection: phenotypic tests

A number of phenotypic tests of the two hypotheses are possible. The general idea is to measure selection on multiple traits for taxon pairs that vary in how far speciation has proceeded. One can then test the extent to which increased degree of speciation is associated with stronger selection on a single trait, selection on more traits, or both. Figure 8.4 depicts schematically such an approach for three taxon pairs. The approach depicted could be extended to a multiple regression framework (i.e., the ERG test discussed in Chapter 2), potentially incorporating multiple variables and numerous taxon pairs (Funk et al. 2002, Funk et al. 2006). Nosil et al. (2009b) review further tests, for example methods where the dimensionality of phenotypic divergence serves as a proxy for the dimensionality of selection.

8.6.1. Phenotypic support for the stronger selection hypothesis

Support for the stronger selection hypothesis stems from the study of Funk et al. (2006), which reported that divergence in one phenotypic trait (body size) was positively correlated with reproductive isolation, independent from time. Assuming that greater divergence in size arises via stronger divergent selection on size, the results support the stronger selection hypothesis. Similar results stem from positive relationships between body size divergence and levels of premating isolation in stickleback fishes (Nagel and Schluter 1998, McKinnon et al. 2004), body size divergence and levels of intrinsic postzygotic isolation in *Centrarchid* fishes (Bolnick et al. 2006), body shape divergence and premating isolation between *Gambusia* fish ecotypes (Langerhans et al. 2007), and the magnitude of color-pattern shifts in relation to levels of premating isolation in *Heliconius* butterflies (Jiggins et al. 2004). A final example concerns *Pundamilia* cichlids, in which reproductive

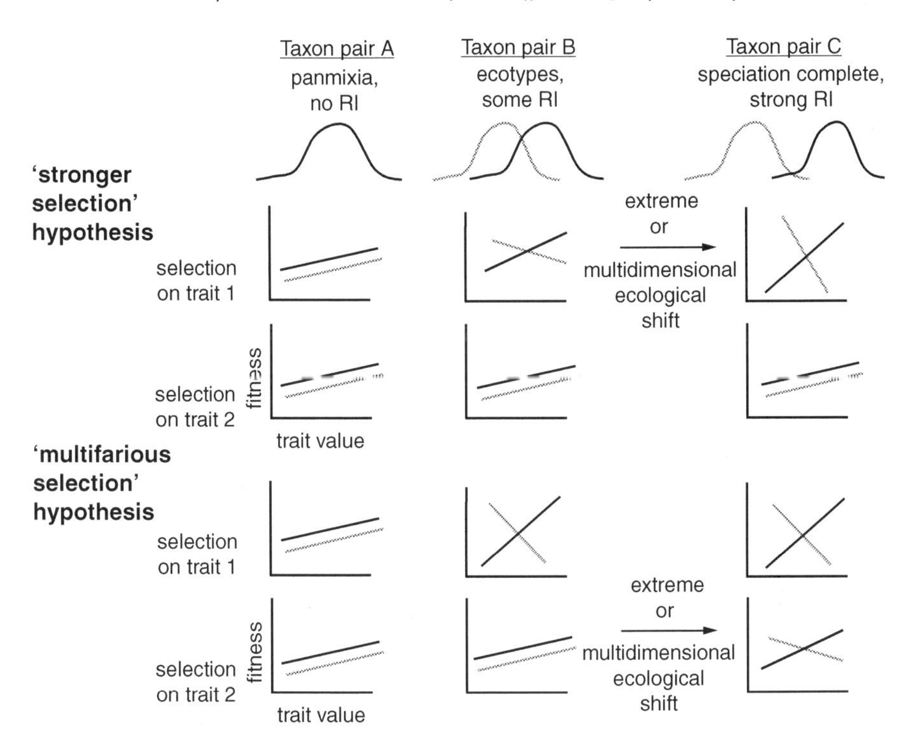

Figure 8.4. Predictions of the "stronger selection" and "multifarious selection" hypotheses at the phenotypic level. The three taxon pairs depicted vary in how far speciation has proceeded (RI, reproductive isolation). Selection might act on two phenotypic traits (e.g., morphology and physiology) and is measured in each of two populations. Graphs represent fitness functions, where x-axes represent trait values, y-axes represent fitness, and one fitness function is shown for each population. Crossing fitness functions are indicative of divergent selection, with steeper lines indicating stronger divergent selection. Arrows label the critical change predicting the evolution of strong RI (i.e., speciation having proceeded far) under each hypothesis. Note that both extreme shifts along one niche dimension and multidimensional niche shifts can cause stronger selection on a given trait, selection on a greater number of traits, or both. Modified from Nosil et al. (2009b) with permission of Elsevier.

isolation is positively related to the degree of divergence in opsin genes (Seehausen et al. 2008b).

8.6.2. Phenotypic support for the multifarious selection hypothesis

"Laboratory experiments collectively indicate that multifarious . . . divergent selection can readily lead to complete reproductive isolation, but that single-factor. . . divergent selection will typically lead to only incomplete reproductive isolation" (Rice and Hostert 1993, p. 1647).

The multifarious selection hypothesis most clearly traces its roots to a review of experimental evolution studies in *Drosophila* (Rice and Hostert 1993). Although the quotation above still holds today, it is crucial to note that the acid test has not been conducted: some experiments selected on one trait, others on multiple traits, and no single experiment has actually manipulated the number of traits under divergent selection to test the effects of multifarious selection on the evolution of reproductive isolation (Nosil and Harmon 2009). Moreover, despite being intuitive, there are almost no tests of this hypothesis in nature, perhaps owing to the difficulty of generating the required selection estimates. In general, multi-trait divergence between ecotypes or species provides circumstantial evidence that speciation involves multifarious selection (Reimchen 1983, Singer and McBride 2010), but experiments measuring selection are required for definitive evidence.

A few key systems in which selection has been measured in multiple taxon pairs provide some preliminary information. In herbivorous insects, divergent selection between populations on different host plants might act on many different types of traits, for example on cryptic coloration used to evade visual predation or on physiology used to detoxify plant chemicals. Selection was estimated on both these traits in three taxon pairs of *Timema* walking-stick insects (Fig. 8.5) (Nosil and Sandoval 2008). The results revealed that strong divergent selection on cryptic coloration is associated with host ecotype formation and intermediate levels of reproductive isolation. In contrast, stronger reproductive isolation between a species pair was associated with divergent selection on both cryptic coloration and physiology, rather than on cryptic coloration alone. The results are consistent with the multifarious selection hypothesis, but further replication is required for a robust test. Another potential example comes from *Rhagoletis* flies, in which diapause life history traits create a strong ecological barrier to gene flow. Different diapause traits, such as initial diapause depth, timing of diapause termination, and post-diapause development rate, are genetically uncoupled and are each subject to divergent selection, such that the barrier to gene flow is created by multifarious selection (Dambroski and Feder 2007). The need for further work is evidenced by the fact that neither of these studies, despite providing the best tests available to date, explicitly isolated an effect of multifarious selection independent from selection strength on individual traits.

8.6.3. Problems with phenotypic tests

Tests of the stronger and multifarious selection hypotheses conducted at the phenotypic level suffer from several difficulties, and potentially severe ones. One major problem is that of unmeasured traits. It will be impossible to be certain that one has measured all the phenotypic traits that are subject to divergent selection, because some traits will be very difficult to measure empirically, and others we might never think of measuring. Selection on unmeasured traits might create a bias against supporting the multifarious selection hypothesis. Nonetheless, considering major

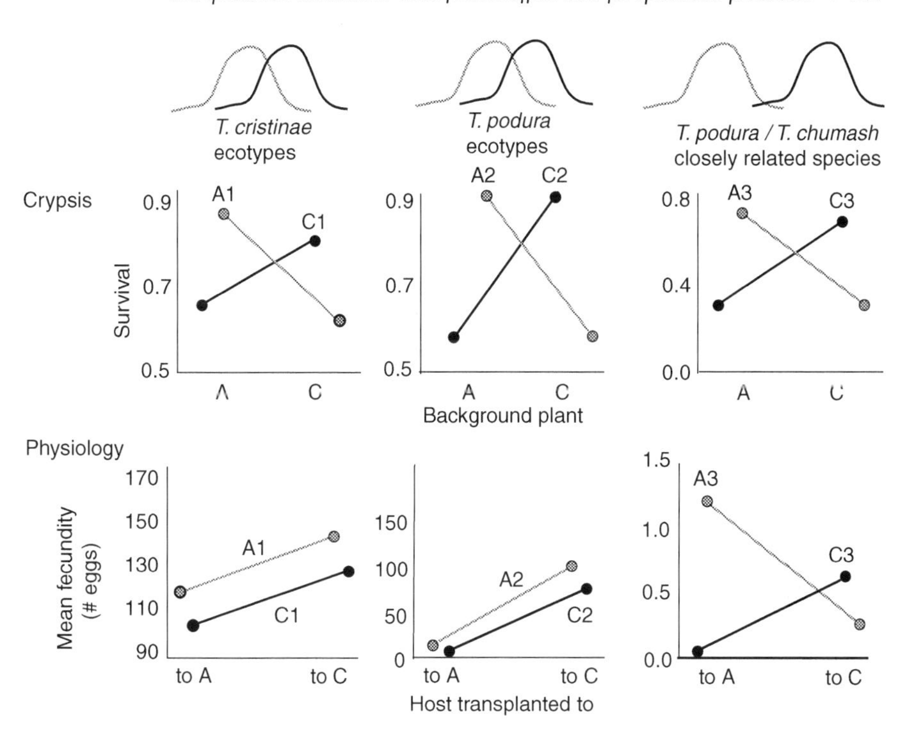

Figure 8.5. Evidence consistent with the multifarious selection hypothesis. In the graphs, crossing fitness functions are indicative of divergent selection. Shown here is the nature of selection on crypsis (i.e., survival in the face of visual predation) and physiology (i.e., fecundity in the absence of visual predation) for three taxon pairs of *Timema* transplanted to their native and an alternative host. A1 and C1 refer to ecotypes of *T. cristinae* (A, *Adenostoma* and C, *Ceanothus* hereafter). A2 and C2 refer to ecotypes of *T. podura*. A3 and C3 refer to the species pair *T. podura* and *T. chumash*, respectively. The ecotype pairs exhibit weaker divergence in morphology, host preference, and mitochondrial DNA than the species pair (denoted here by greater overlap of trait distributions for the ecotype pairs). The ecotype pairs are also subject to divergent selection on fewer traits than the species pair. Modified from Nosil and Sandoval (2008) with permission of the Public Library of Science.

"classes" of traits such as morphology, physiology, and behavior, provides a starting point, and might still be informative (Fig. 8.5). Another problem is that is can be difficult to determine whether some aspect of the phenotype represents a "single" trait or "multiple" traits (or even a single or multiple "class" of traits), particularly when correlations among traits are considered. Finally, even single traits might vary in the number of underlying genes, such that selection on one phenotypic trait could represent possibilities ranging from strong selection on one or a few genes through to multifarious selection on many loci. These issues are not insurmountable. For example, the number of genes affecting traits might be quantified. However, these difficulties do indicate that perhaps the hypotheses are most effectively and precisely tested at the genetic level (but see Johnson and Kliman 2002).

8.7. Multifarious versus stronger selection: genomic tests

As divergence along the speciation continuum increases, the strength of selection on individual loci, and the number of selected loci, can both increase (Fig. 8.6). Thus, the "stronger" and "multifarious" selection hypotheses might be tested at the genetic level by quantifying selection across the genome (note here that I refer to the total number of genes under selection across *all* traits, not the number of genes affecting a *single* trait). If genomic coverage is broad, then most or all of the major targets of selection might be identified. Thus, genomic tests of the hypotheses can potentially overcome the problem of unmeasured traits that plagues phenotypic tests. In addition, estimates of selection on the genome can be coupled with linkage mapping to quantify the genomic distribution of selected regions. This will allow tests of whether divergent regions are contiguous along chromosomes ("genomic continents") or represent many small islands of differentiation on different chromosomes. Genomic data could thus test the extent to which genetic divergence during speciation involves physical linkage to a few selected sites versus genome-wide reductions

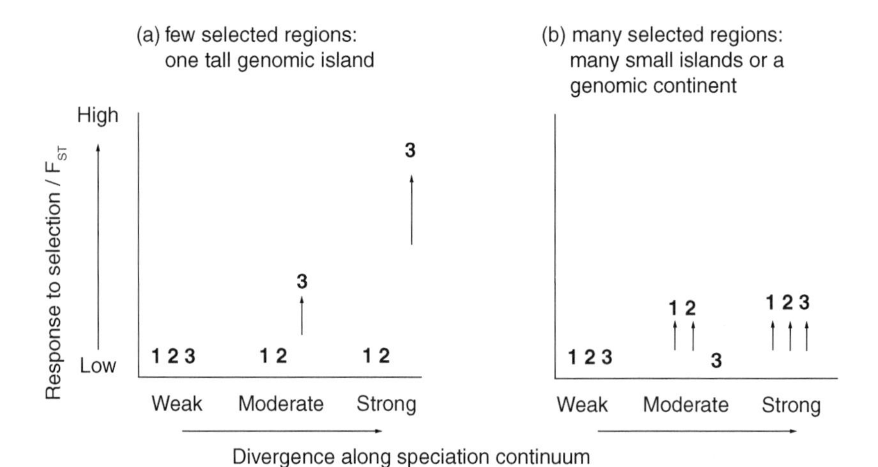

Figure 8.6. Patterns expected if few versus many genomic regions diverge as the speciation continuum unfolds. Ideally, selection would be measured directly in experiments, but in some cases F_{ST} might act as a proxy for selection, with higher values possibly indicating greater net effects of selection. Numbers represent three different loci within the genome. a) Divergence along the speciation continuum is associated with stronger differentiation of one locus ("gene 3"). b) Divergence along the speciation continuum is associated with divergence in a larger number of loci. Note that in b) divergence of each individual locus is weaker than in a), such that the "total" selection strength summed across loci is similar between the two scenarios.

in gene flow caused by multifarious selection on numerous loci. Thus, genomic tests of the hypotheses will probably inform not only our understanding of speciation, but of genome evolution more broadly.

8.7.1. Genomic support for the stronger and multifarious selection hypotheses

There are no direct genetic tests of the stronger selection hypothesis, but some indirect evidence suggests strong selection on a few loci can drive speciation. For example, examples exist of strong reproductive isolation caused by divergence in one or a few genes (Bradshaw and Schemske 2003, Coyne and Orr 2004, Schluter and Conte 2009, Nosil and Schluter 2011). Some genome-scan studies report that speciation is far from complete even in cases where selection is acting on numerous gene regions, suggesting stronger selection on a few regions might be required for speciation (Turner et al. 2008, Nosil et al. 2009a). However, none of these lines of evidence directly demonstrates that stronger selection causes speciation to proceed further. Additionally, as discussed in the chapter on genomics of ecological speciation, observational genome scans alone are biased toward detecting only a few regions under strong selection. Although observational studies do sometimes report widespread genomic divergence (Lawniczak et al. 2010, Williams and Oleksiak 2011), definitive tests distinguishing between the stronger and multifarious selection hypotheses will probably need to experimentally measure selection on the genome directly. One such experimental test reported that speciation between host races of *Rhagoletis* flies involves selection acting on many gene regions (Michel et al. 2010). However, this example did not examine multiple stages of the speciation process, and thus multifarious selection on the genome has yet to be linked to speciation having proceeded further.

8.7.2. Hypothetical experimental genomic tests

Here I outline experimental approaches that could be taken to explicitly test the stronger versus multifarious selection hypotheses at the genomic level. Unlike observational approaches alone, experiments such as those proposed here can detect weaker selection acting on many regions across the genome, and thus could truly distinguish how many gene regions are affected by selection, and how strongly. I discuss three possible experiments, but many variations on these themes are possible. First, individuals from a genetically variable natural population could be transplanted into new areas, with some individuals moved into a novel environment and others moved into their native environment (e.g., herbivorous insects could be transplanted to their native host or a host-plant species used by a close relative). This could be done in a paired block fashion, representing treatment and control replicates, respectively. One could then directly quantify how selection differentially builds up divergence across various regions of the genome according to which regions are more affected by selection.

Second, individuals from paired, currently allopatric populations that are genetically differentiated and divergently adapted could be moved into zones of parapatry or sympatry. This would represent a secondary contact situation and would ideally be conducted with pairs of populations that would hybridize, at least to some extent, upon transplantation. One could then directly quantify how gene flow differentially breaks down (rather than how selection builds up) divergence across the genome, again according to how many regions are affected by selection, and how strongly so. In this regard, one might be able to use instances of the breakdown of species barriers and the collapse of species pairs (i.e., "speciation reversal") as natural experiments of this sort (Seehausen et al. 2008a).

Third, mark–recapture methods could be used to measure the relationship not only between phenotype and fitness, but also between "genometype" and fitness. This could be done if material that would allow genome-wide sequencing could be extracted from the same individuals used in the recapture experiment (e.g., prior to release, take fin or tail clips from fish or frogs, keep older molts of insects, etc.). This approach could yield estimates of selection on the genome within a generation, and thus could be applied to study systems with generation times that are too long to allow for quantification of evolution across generations. All three types of powerful experiments might be conducted using various types of hybrid crosses or recombinant inbred lines.

Ideally, experiments such as those proposed above would be conducted using multiple closely related taxa that span the speciation continuum. This would allow reconstruction of how genomic divergence changes as speciation unfolds, and could test how the relative importance of strong versus multifarious selection might change as speciation proceeds across its various stages. Experiments such as these are yet to be conducted in nature, and are likely to be time-consuming and difficult to implement (Johnson and Kliman 2002), or ethically questionable owing to conservation issues. Moreover, their results will probably be highly dependent on the experimental design and the characteristics of the study organisms used (e.g., starting allele frequencies, etc.). Nonetheless, such experiments could be conducted in some systems in nature (Bolnick 2004, Losos et al. 2004, Nosil and Crespi 2006b, Calsbeek and Smith 2008), or in experimental semi-natural mesocosms (Barrett et al. 2008). Such experiments could collectively address both the mechanisms and the genomic basis of speciation, and thus represent one of the most promising avenues for further research on ecological speciation.

8.8. Other factors affecting the speciation continuum

Two factors other than those discussed so far might affect how far speciation proceeds. I focus here on these, citing past reviews that offer more in-depth discussion.

8.8.1. Temporal stability of divergent selection

The process of ecological speciation requires divergent selection to act for time periods that are long enough to allow the evolution of substantial reproductive isolation. This raises two questions. First, how long does the evolution of such reproductive isolation take? The answer is unclear because we still know little about the rate at which ecological speciation proceeds (Hendry et al. 2007). The process might be initiated quickly—as quickly as adaptive divergence itself—but further movement along the speciation continuum may require longer time periods (Gavrilets and Vose 2007, Gavrilets et al. 2007, Sadedin et al. 2009). Second, how temporally consistent and stable is divergent selection? This second question can only be addressed using long term field studies that measure selection across time periods. Long-term studies of Darwin's finches on the Galápagos Islands (Grant and Grant 2002) and stickleback fish on the Haida Gwaii archipelago in Canada (Reimchen 1995, Reimchen and Nosil 2002, 2004) clearly indicate that selection can be highly variable across time. These studies demonstrate that selection can vary seasonally, yearly, and over longer timescales. A recent review indicates that temporal fluctuations in selection are a widespread phenomenon, and thus that selection estimates based upon single points in time can be highly misleading as to the true nature of selection (Siepielski et al. 2009). The temporal stability of selection may be a major determinant of how far speciation proceeds, and further work on how it affects the rate and degree of speciation is strongly warranted.

8.8.2. Alternative outcomes to speciation: sexual dimorphism and phenotypic plasticity

A final topic concerning how far speciation proceeds concerns outcomes alternative to speciation when divergent selection acts (Rueffler et al. 2006). Consider two possibilities, each of which might limit the degree to which speciation proceeds. First, divergent selection acting on a population might result in divergence between the sexes, rather than between populations or ecotypes (Butler et al. 2007). Ecologically based sexual dimorphism often occurs in nature (Selander 1966, Reimchen and Nosil 2004), and a theoretical model has explored the factors determining whether sexual dimorphism or speciation arise (Bolnick and Doebeli 2003). In general, when genetic correlations between the sexes are weak, as sometimes occurs in nature (Bonduriansky and Rowe 2005), it can be easier for sexual dimorphism than for speciation to occur.

A second alternative to ecological speciation is the evolution of phenotypic plasticity (West-Eberhard 2005, Pfennig et al. 2010). Rather than generating genetic divergence and speciation, variable and divergent environments might result in the evolution of generalized and highly plastic populations in which individuals can express a range of phenotypes according to the environment they find themselves in. Recent theoretical work examined the competition between genetic divergence versus evolution of plasticity. This work found that when environments were unstable the evolution of increased phenotypic plasticity was the more likely outcome of the two (Svanback et al. 2009). In general, more work on the role of phenotypic plasticity in speciation is needed. For example, phenotypic plasticity might promote speciation by facilitating the colonization of new environments, but constrain speciation by limiting the need for genetic divergence and adaptation (Price et al. 2003, Pavey et al. 2010a, Thibert-Plante and Hendry 2011). The balance between these positive and negative effects of plasticity on speciation warrants further work.

8.9. The speciation continuum: conclusions and future directions

Ecological speciation often proceeds to varying degrees. Major questions remain about both the general nature of the speciation continuum, and the factors that determine which point in the continuum is achieved. Some models and data indicate speciation often gets stalled at an intermediate point. However, feedback loops between adaptation and gene flow, coupled with a lack of a stabilizing mechanism to keep taxon pairs at an intermediate point in the speciation process, support the argument that speciation should eventually proceed further (or be reversed). An obvious resolution to these divergent views is that speciation often halts for some period of time, but not indefinitely. Another possibility is that further divergence is occurring, but has slowed down to the point that it is not empirically detectable (Gourbiere and Mallet 2010). Further quantification of taxa lying at different points in the speciation continuum is required to test these scenarios and to quantify for how long populations get stuck at intermediate points in the speciation process. Only with such data will we be able to determine if there are predictable gaps in the speciation continuum.

Numerous hypotheses have been put forth concerning the factors determining which point in the speciation process is achieved. This chapter focused on two, the stronger and multifarious selection hypotheses, which invoke the nature of selection itself. Some circumstantial evidence in support of each hypothesis exists, but explicit tests represent a pressing avenue for further research. In particular, experimental tests of selection on the genome are required, as they might most readily detect all the targets of selection—strong or weak, obvious or not—acting on populations.

It seems likely that the importance of the hypotheses will shift across different points in the speciation continuum. For example, speciation may often be initiated via divergence in the few specific gene regions directly subject to strong divergent selection. This period may then transition in a second phase, where gene flow is reduced in localized regions of the genome surrounding selected sites, and divergence hitchhiking may then act to facilitate differentiation of regions physically linked to those under selection. As further loci diverge, and perhaps new mutations come to differentiate populations, effective gene flow then gets further reduced across the genome. As this proceeds, further loci diverge and a transition to widespread genomic divergence, facilitated by genome hitchhiking, occurs. The end result then is strong reproductive isolation and widespread genetic divergence. Questions remain about the conditions under which each of these phases occurs, and their durations (Feder et al. 2011a).

Although the idea that selection acting on more dimensions will promote speciation is intuitive, there are some strong counterarguments against it. For example, selection spread over many gene regions may make it difficult to overcome gene flow. Additionally, divergence in many dimensions could result in "holey" rather than rugged adaptive landscapes, where diversification is driven by "drift" along ridges of high fitness, rather than divergent selection per se (Gavrilets 2003, 2004). Future studies reconstructing adaptive landscapes using selection estimates on a range of genotypes and phenotypes are required to determine whether multidimensional divergent selection truly promotes speciation. Finally, much new insight into the continuous and potentially fluid nature of the speciation process could stem from better integrating studies of "speciation reversal" with those of how far speciation proceeds (Seehausen et al. 2008a).

9

Conclusions and future directions

This final chapter of the book has three explicit goals. First, to summarize what we already know about ecological speciation, focusing on findings with strong support. Second, for the topics covered in this book, to outline the key missing components in our understanding, thereby outlining clear avenues for further research. Third, I touch upon aspects of ecological speciation that were not thoroughly treated in this book.

9.1. What we know about ecological speciation

Several aspects of ecological speciation are now well established. Foremost, we know the process occurs in nature. We also now understand aspects of each of the three main components of ecological speciation: (1) a source of divergent selection; (2) a form of reproductive isolation; and (3) a genetic mechanism to link the two. For example, it is relatively clear that the sources of divergent selection can be numerous, including differences between environments, interactions between populations, and ecologically based divergent sexual selection. The review of different forms of reproductive isolation revealed at least some evidence for most forms of reproductive isolation evolving as a result of divergent selection, including forms such as sexual isolation and intrinsic postmating isolation, which are not "inherently" ecological. It is also clear that multiple forms of reproductive isolation can act simultaneously during ecological speciation.

The genetic basis of ecological speciation remains poorly understood. Nonetheless, a few things are known. Pleiotropy is expected to be a very effective mechanism for linking selection to reproductive isolation, and examples of pleiotropic effects of genes involved in divergent adaptation on reproductive isolation exist. There are also putative examples of ecological speciation driven by linkage disequilibrium between selected genes and those causing reproductive isolation. In such cases, the underlying genes sometimes lie in genomic regions of low recombination, although it is unclear if this is required to drive speciation. We now know of genes affecting ecological characters subject to divergent selection, but none of these have yet been demonstrated to unequivocally affect extrinsic, ecologically based reproductive isolation. Thus, although we have very strong candidates, "ecological speciation genes" are yet to be unequivocally demonstrated. Also on the topic of genetics, heritable gene expression divergence has been implicated in ecological speciation, but underlying mutations have rarely yet been discovered.

Some aspects other than those concerning the three main components of ecological speciation are also well established. For example, there are examples of the process of ecological speciation occurring in sympatry, parapatry, and allopatry. Thus, we know that ecological speciation can occur, at least to some extent, under any geographic arrangement of populations. Nonetheless, there is evidence that divergence occurs most easily when rates of gene flow are low. It is also known that single instances of ecological speciation can occur under multiple geographic modes, with some divergence occurring in allopatry and some in the face of gene flow. Little is known about the broader genomic basis of ecological speciation. There are examples of both islands and continents of divergence, but further data, particularly those stemming from experiments, are required to determine which are more common, or expected. Finally, it is now known that speciation proceeds to highly varying degrees. Determining the factors affecting which degree is achieved is a major avenue for further work.

9.2. Future work: 25 unresolved issues in ecological speciation

Table 9.1 outlines 25 specific unresolved issues or "questions" in ecological speciation, and how each might be addressed. These each represent promising avenues for future work.

9.3. Competing hypotheses deserving further work

In addition to the specific unresolved questions discuss in Table 9.1, there are a number of general hypotheses that require further testing (Table 9.2). In cases where there are alternative hypotheses, the alternatives are not mutually exclusive, and thus could be operating simultaneously.

9.4. Issues warranting further work that were not covered in detail

9.4.1. Tempo and rate of ecological speciation

This book focused on the mechanisms, rather than the rate, of speciation. Thus, there was little discussion of the tempo and mode of ecological speciation. To what extent does ecological speciation proceed gradually, versus in more punctuated bursts of adaptive divergence? How long is the "waiting time" for the commencement of a new ecological speciation event once an existing instance of speciation is more or less complete? What is the average duration of ecological speciation and at what rate does reproductive isolation evolve? How do gene flow and genetic architecture affect the rate of speciation? Some predictions might be made from what we already know. For example, ecological speciation can be initiated as quickly as adaptive

Table 9.1. Twenty-five major, yet unresolved, questions in ecological speciation.

Unresolved issue/question	Context	Some possibilities for addressing the issue
1. Diverse tests for ES have been conducted, but caveats associated with their weaknesses require future attention	tests of ES[2]	Ensure criteria outlined for each type of test are met, and caveats are considered
2. Causal associations between adaptive divergence and gene flow have not been explicitly tested in nature	tests of ES[2]	Conduct manipulative experiments in nature to test the extent to which adaptive divergence constrains gene flow, gene flow constrains adaptive divergence, or both causal associations act
3. What are the relative importances of different sources of selection during ES?	sources of selection[3]	Estimate the strength of divergent selection stemming from different sources and test how likely each is to generate RI
4. How do different sources of divergent selection, such as predation and competition, interact?	sources of selection[3]	Examine study systems in which multiple sources of selection act, and test whether and how they interact
5. Which forms of RI are most common during ES?	forms of RI[4]	Conduct tests of ES for different forms of RI
6. What are the relative importances of different individual forms of RI to total RI at any single point in time?	forms of RI[4]	Measure multiple forms of RI in single study systems that are at a particular point in ES
7. Across time, in what order do different forms of RI tend to evolve?	forms of RI[4]	Measure multiple forms of RI across taxon pairs spanning different points in the ES process
8. How often do genes under selection have pleiotropic effects on RI?	genetic mechanisms[5]	Find genes underlying adaptive divergence, and then test if they have effects on RI
9. How effective is physical linkage in allowing linkage disequilibrium to drive ES?	genetic mechanisms[5]	For examples of RI evolving as a result of linkage disequilibrium, examine how tight physical linkage of genes causing RI is to selected loci
10. How strongly do factors that reduce recombination promote ES?	genetic mechanisms[5]	Examine genetic divergence in regions of reduced recombination versus other genomic regions and test for ES in systems in which underlying genes lie, versus do not lie, in regions of low recombination
11. Are genetic constraints important for ES?	genetic mechanisms[5]	Measure genetic variation in single traits and correlations among traits (e.g., the G-matrix) and test if they bias the direction or rate of adaptive divergence and associated evolution of RI

Unresolved issue/question	Context	Some possibilities for addressing the issue
12. How many loci underlie ES, and do their numbers have relevance for the process?	genetic mechanisms[5]	Quantify the number of loci involved in ES and test whether the degree of RI varies according to the number of loci identified
13. Which genes are involved in ES, and what are their effect sizes?	genetic mechanisms[5]	Test for genes meeting the four criteria for an "ecological speciation gene"
14. What are the relative importances of the constraining versus diversifying roles of gene flow for ES?	geography of ES[6]	Compare adaptive divergence and RI in population pairs exhibiting variable levels of gene flow during divergence
15. How common and important are mixed geographic modes of ES?	geography of ES[6]	Attempt the difficult, but critical, task of distinguishing between purely allopatric, purely sympatric, and mixed modes of divergence
16. How can gene flow be more reliably detected during ES?	geography of ES[6]	Use integrative approaches to test for gene flow, perhaps coupling observational and experimental approaches
17. How much of the genome diverges during ES?	genomics of ES[7]	Requires experimental measurements of selection on the genome such that weak selection on many loci can be detected (should it occur)
18. Does genomic divergence during ES tend to represent genomic islands or continents?	genomics of ES[7]	Use linkage mapping analyses or whole genome sequences to test whether regions affected by selection are contiguous along chromosomes or more isolated from one another, and to quantify the decay of genetic divergence away from selected sites
19. How do genomic regions of divergence grow during ES?	genomics of ES[7]	Test which factors determine the establishment of new mutations, particularly distinguishing between the roles of divergence and genome hitchhiking
20. Do the transient effects of selective sweeps affect ES?	genomics of ES[7]	Compare patterns of genomic divergence between populations that have, versus have not, reached equilibrium levels of divergence
21. How often does ES stem from standing variation versus new mutations?	genomics of ES[7]	In studies of ecological speciation, test whether divergence stems from new mutations or not
22. What is the role of gene expression in causing RI during ES?	genomics of ES[7]	Find functional changes and mutations underlying adaptive, genetically based gene expression divergence and determine how they contribute to RI

Table 9.1. *Cont.*

Unresolved issue/question	Context	Some possibilities for addressing the issue
23. What is the role of phenotypic plasticity in ES?	genomics of ES[7]	Test whether phenotypic plasticity facilitates speciation only indirectly, by allowing population persistence in new environments, or by playing a more direct role in adaptive divergence
24. How often is RI a stable outcome, or at least a pseudo-stable outcome?	ES continuum[8]	Quantify the ES continuum and test whether any "gaps" in it exist; also determine whether examples of partial RI represent recently or more anciently diverged taxa
25. What are the roles of stronger versus multifarious selection in making ES proceed further?	ES continuum[8]	Test whether proceeding further in ES is associated with stronger selection on one or a few key traits/genes, selection distributed across a greater number of traits/genes, or both.

Superscripts refer to relevant chapter numbers. Abbreviations: ES, ecological speciation; RI, reproductive isolation.

divergence itself, but probably takes much longer to complete (Hendry et al. 2007). Another prediction is that divergence from standing genetic variation might occur more quickly than divergence from new mutations. However, most questions about the rate of ecological speciation remain unknown or unexplored. Future work on these issues might use comparative approaches to examine how reproductive isolation builds up with time since population divergence and with ecological differences (Funk et al. 2006). Comparative studies of ecological speciation testing how quickly and regularly branching events occur in phylogenetic trees are also warranted (Kozak et al. 2005, Alfaro et al. 2009). Reviews of the tempo and mode of speciation in general can be found in Coyne and Orr (2004) and Gavrilets (2003, 2004).

9.4.2. How common is ecological speciation?

"no marked ecological change has accompanied speciation in the [Peromyscus] leucopus group" (Blair 1950, p. 269).

Table 9.2. Specific unresolved hypotheses concerning ecological speciation.

Hypotheses	Context	Predictions	Notes about existing and further required data
1. Rugged versus holey adaptive landscapes[3]	Causes of phenotypic and genetic divergence among populations: true divergent selection versus "drift" along ridges of high fitness	Rugged landscapes predict strongly reduced fitness of intermediate forms in multidimensional space, whereas holey landscapes do not	Fitness of intermediate forms needs to be measured more often, and adaptive landscapes reconstructed; number of dimensions measured needs to be considered
2. Fitness trade-offs versus information-processing hypotheses[4]	The evolution of habitat isolation: active selection against switching between habitats versus selection for increased efficiency of information processing	Fitness trade-offs hypothesis predicts character displacement of habitat preference, whereas the information-processing hypothesis does not	Further measurements of habitat preference in allopatric populations, and of individual efficiency in relation to individual specificity, are required
3. Interference hypothesis[4]	Different reproductive barriers can interfere with each other's evolution: e.g., sexual versus habitat isolation during reinforcement	Evolution of strong habitat preferences can interfere with the evolution of sexual isolation via reinforcement, by reducing reinforcing selection on mating preferences	Multiple reproductive barriers need to be studied within single systems, in the context of how they might facilitate or interfere with one another's evolution
4. Divergence versus genome hitchhiking hypotheses[5]	The role of physical linkage versus genome-wide reductions in gene flow in facilitating genetic divergence	Both forms of hitchhiking facilitate divergence, but under different conditions; selection on many loci is effective at generating widespread genomic divergence	The decay of genetic divergence away from selected sites needs to be quantified, and compared among cases where the overall number of selected loci in the genome varies
5. Cascade reinforcement hypothesis[6]	Effects of reinforcement within a specific zone of hybridization can cascade to effect pre-mating isolation in other geographic regions	Reinforcement between species in zones of sympatry can result in mating discrimination between sympatric and allopatric conspecific populations	Measurements of mating discrimination both within and between species, in zones of both sympatry and allopatry, are required

Table 9.2. *Cont.*

Hypotheses	Context	Predictions	Notes about existing and further required data
6. Genomic island versus genomic continent hypotheses[7]	The number, size, and genomic distribution of divergent genomic regions can vary	The very initial stages of speciation might comprise islands, but subsequent selection on many loci readily generates genomic continents	Experimental measurements of selection on the genome are required to distinguish these hypotheses, as observational approaches are biased to detecting only a few islands of divergence
7. "Stuck partway" versus "feedback loop" views of speciation[8]	How far speciation proceeds: Is partial reproductive isolation a stable outcome, or is further divergence via feedback loops more likely?	Partial reproductive isolation might persist over long time periods, but there is no stabilizing mechanism to keep it as such, thus further divergence (or extinction) should eventually occur	Quantification of the speciation continuum is required to determine whether predictable gaps in the continuum exist and, if so, how long they persist
8. Strong versus multifarious selection hypotheses[8]	How far speciation proceeds: does speciation proceed further via stronger selection on a few key traits/genes or via selection on a greater number of loci?	Selection on a few loci is better at overcoming gene flow, but selection on many loci might be more likely to cause widespread genomic divergence and have incidental effects on multiple forms of reproductive isolation	Determine the number of divergent traits and genes for taxon pairs at different points in the speciation continuum; again experimental approaches are desirable
9. Sampling model[5,9]	How far speciation proceeds: Does sampling divergence across more of the genome generate stronger reproductive isolation?	Divergence in more genomic regions is more likely to cause reproductive isolation than divergence in fewer regions	Quantify whether the degree of reproductive isolation increases as the number of divergent genomic regions increases

Superscripts refer to relevant chapter numbers.

This book focused on how and why ecological speciation occurs. A critical question remains: How common is ecological speciation? Some insight into this question stems from the comparative study by Funk et al. (2006), which demonstrated that ecological divergence promotes the evolution of reproductive isolation, at least to some extent, in a wide range of organisms. Importantly, the datasets examined by Funk et al. (2006) were unlikely to harbor an ascertainment bias towards supporting ecological speciation because they were not chosen *a priori* to test the ecological speciation hypothesis (i.e., the datasets were originally collected to examine the relationship between reproductive isolation and time, without reference to ecological differences) (see Funk and Nosil 2007 for a review). Another observation is that forms of reproductive isolation that are certainly involved in ecological speciation (e.g., immigrant inviability) are common in nature, indicating that the process of ecological speciation itself is often occurring (Schluter 2009) (Fig. 9.1).

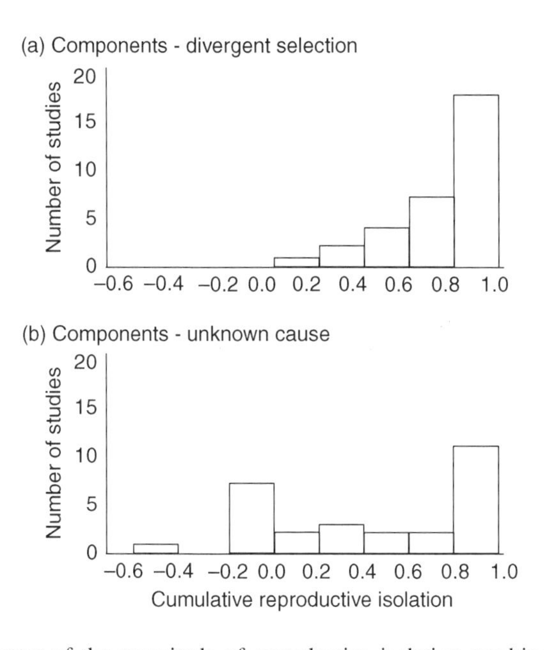

Figure 9.1. Estimates of the magnitude of reproductive isolation resulting from divergent selection (top), compared with other components of reproductive isolation currently lacking identifiable causes (bottom), following Schluter (2009). Components of reproductive isolation associated with divergent selection include those attributable to active divergent selection on traits (immigrant inviability and extrinsic postzygotic isolation) and to trait-based assortative mating (habitat preference, floral isolation, and breeding time). The components lacking a known cause include intrinsic hybrid inviability, sexual selection against hybrids, pollen competition, and reduced hybrid fecundity. These unattributed components might have evolved as a result of divergent selection, or other processes. A negative value indicates that hybrids had higher fitness than the parental species for at least one component of postzygotic isolation. Original data were from Nosil et al. (2005) and Lowry et al. (2008a).

Despite this evidence that ecological speciation is common, it is well known that factors other than ecology can promote speciation, as exemplified by numerous ecologically similar sister-species pairs of birds (Price 2007). Likewise, it is also known that divergence in obvious ecological factors does not always strongly promote the evolution of reproductive isolation (Coyne and Orr 2004, Presgraves 2007a, b, McCormack et al. 2010, Raeymaekers et al. 2010). Advances in addressing questions concerning the frequency of ecological speciation will probably stem from further broad comparative and phylogenetic analyses (see Funk and Nosil 2007 for a review, Nyman et al. 2010), particularly ones that avoid ascertainment biases. The outcomes of such analyses will depend on how far ecological speciation proceeds once adaptive divergence causes the evolution of at least some reproductive isolation. For example, if adaptive divergence often initiates, but rarely completes, speciation, then: (1) adaptive divergence and the degree of (partial) reproductive isolation will generally be associated; but (2) branching (i.e., speciation) events in a phylogeny will not necessarily be associated with ecological shifts. In this case, one might predict evidence for ecological speciation being common in studies that examine reproductive isolation directly, but less so in analyses relying on branching events in phylogenies to infer speciation.

9.4.3. Joint action and interactions between speciation models

Models of speciation alternative to ecological speciation were introduced in Chapter 1. However, the different models are not mutually exclusive. Thus, in principle, multiple speciation models might be acting simultaneously and the different models might even interact (Schluter 2009). In practice, the joint action of multiple models, and their potential interaction, has received little empirical study. Cytonuclear interactions are often thought to evolve via genomic conflict, but there are examples in which the cytoplasm is divergently adapted to different environments (Sambatti et al. 2008). Does ecological speciation interact with speciation scenarios involving various types of conflict? Likewise, how do Fisherian runaway processes interact with divergent selection? Another consideration is the joint action of divergent selection and random genetic drift. Ecological speciation requires ecological divergence. However, ecological shifts into new environments might often be accompanied by reductions in population size (i.e., a population bottleneck associated with colonization of a new environment). Thus, if we see speciation associated with ecological shifts (e.g., Mayr 1947, Schluter 1998, Winkler and Mitter 2008, Nyman et al. 2010), is this causally related to divergent selection or to founder effects? Do the processes interact? For example, do reductions in population size during the colonization of new environments affect the efficacy of selection (Barton 2010)? Studies addressing these possibilities are almost completely lacking (but see Rundle 2003). Future work examining multiple models of speciation in concert will almost certainly lead to a more holistic understanding of species formation.

9.4.4. Consequences of ecological speciation

A final consideration that was not treated in this book, but certainly warrants future work, is the consequences (rather than causes) of ecological speciation. How might ecological speciation affect broader patterns of biological diversity (Butlin et al. 2009)? To what extent does ecological speciation affect the organization and the richness of ecological communities, both in time and in space? These questions form the focus of the emerging field of "eco-evolutionary dynamics," which explores not only how ecological processes affect evolution, but in turn, how evolution might affect ecological processes (Bailey et al. 2009, Pelletier et al. 2009). For example, levels of adaptation and gene flow within a focal species might have effects on the communities in which the focal species is found. This could occur by the focal species affecting competitive interactions, or via it attracting predators or parasites to the community. Interactions between species within the community could be relatively direct, or mediated more indirectly by population density. For detailed treatment of such dynamics, I refer readers to recent reviews on these topics (Bailey et al. 2009, Pelletier et al. 2009). The field of eco-evolutionary dynamics holds great promise for understanding the consequences of ecological speciation.

9.5. Final conclusion

The study of ecological speciation has come a long way. Mechanisms have been clarified, specific predictions have been recognized, and much data has been collected. Most importantly, numerous case studies of ecological speciation have emerged. Nevertheless, a detailed understanding of the process still eludes us, even in the best-studied model systems. The reason is that ecological speciation is complex and can encompass many different scenarios. Divergent selection can have various ecological causes, numerous forms of reproductive isolation can result, and there are different genetic mechanisms than can link them. And all of this can occur under different geographic contexts, have complex underlying genomic bases, and unfold over extended time periods. It will be no small task to evaluate all of these possibilities to develop a general understanding of how speciation proceeds from beginning to end. Understanding the influence of all these factors will require ecological and experimental studies that also integrate molecular, population, and quantitative genetics, and that consider the phylogenetic history of the system (e.g., Bernatchez et al. 1999, Bernatchez et al. 2010).

For many topics, it is the classic ecological processes that have received the least attention. For example, we know only a little about the role of competitors and predators in the evolution of reproductive isolation, and even less concerning other possibilities such as parasites, mutualists, or facilitators. Factors other than ecology also warrant attention. For example, population structure is common in nature and is known to affect many evolutionary processes. However, its effect on ecological speciation has received little attention. The influence of shared ancestry is also not known. Closely related populations may share biases in their standing genetic variation and in their production of new variation (Schluter et al. 2004). How such biases affect adaptive divergence and the evolution of reproductive isolation has not been deeply considered.

Like many rapidly growing fields, much of the evidence for ecological speciation is indirect, relying on observational and comparative studies. In some taxa, a detailed understanding of ecological speciation should permit at least some stages of the process to be recreated experimentally in replicate field or lab populations, providing some of the strongest evidence possible. The study of ecological speciation is yet to enter a truly experimental phase, where manipulations are commonly employed to isolate causal associations between the factors driving and constraining speciation. Nevertheless, I close by noting that the future holds promise, particularly as rapidly developing genomic methodologies continue to be further integrated with classical ecological approaches. Much progress has been made, and where gaps in our knowledge exist, it is often clear what needs to be done and the tools to do so are generally available.

References

Abrahamson, W. G, and C. P. Blair, editors. 2008. Sequential radiation through host-race formation: herbivore diversity leads to diversity in natural enemies. University of California Press, USA, Los Angeles.

Abrams, P. A. 2000. The evolution of predator-prey interactions: Theory and evidence. Annual Review of Ecology and Systematics **31**:79–105.

Abrams, P. A., C. Rueffler, and G. Kim. 2008. Determinants of the strength of disruptive and/or divergent selection arising from resource competition. Evolution **62**:1571–86.

Abzhanov, A., W. P. Kuo, C. Hartmann, B. R. Grant, P. R. Grant, and C. J. Tabin. 2006. The calmodulin pathway and evolution of elongated beak morphology in Darwin's finches. Nature **442**:563–7.

Abzhanov, A., M. Protas, B. R. Grant, P. R. Grant, and C. J. Tabin. 2004. *Bmp4* and morphological variation of beaks in Darwin's finches. Science **305**:1462–5.

Achere, V., J. M. Favre, G. Besnard, and S. Jeandroz. 2005. Genomic organization of molecular differentiation in Norway spruce (Picea abies). Molecular Ecology **14**:3191–201.

Agrawal, A, A. J. L. Feder, and P. Nosil. 2011. Ecological divergence and the evolution of intrinsic postmating isolation with gene flow. International Journal of Ecology **2011**:1–15.

Agrawal, A. F. and J. R. Stinchcombe. 2009. How much do genetic covariances alter the rate of adaptation? Proceedings of the Royal Society B-Biological Sciences **276**:1183–91.

Albert, A. Y. K., S. Sawaya, T. H. Vines, A. K. Knecht, C. T. Miller, B. R. Summers, S. Balabhadra, D. M. Kingsley, and D. Schluter. 2008. The genetics of adaptive shape shift in stickleback: Pleiotropy and effect size. Evolution **62**:76–85.

Albert, A. Y. K. and D. Schluter. 2004. Reproductive character displacement of male stickleback mate preference: Reinforcement or direct selection? Evolution **58**:1099–107.

Alfaro, M. E., F. Santini, C. Brock, H. Alamillo, A. Dornburg, D. L. Rabosky, G. Carnevale, and L. J. Harmon. 2009. Nine exceptional radiations plus high turnover explain species diversity in jawed vertebrates. Proceedings of the National Academy of Sciences of the United States of America **106**:13410–14.

Araya, C. L., C. Payen, M. J. Dunham, and S. Fields. 2010. Whole-genome sequencing of a laboratory-evolved yeast strain. BMC Genomics **11**:88.

Arbuthnott, D. 2009. The genetic architecture of insect courtship behavior and premating isolation. Heredity **103**:15–22.

Arbuthnott, D. and B. J. Crespi. 2009. Courtship and mate discrimination within and between species of *Timema* walking-sticks. Animal Behaviour **78**:53–9.

Arbuthnott, D., M. G. Elliot, M. A. McPeek, and B. J. Crespi. 2010. Divergent patterns of diversification in courtship and genitalic characters of *Timema* walking-sticks. Journal of Evolutionary Biology **23**:1399–11.

Arnold, M. L. 1997. Natural hybridization and evolution. Oxford University Press, Oxford.

Arnold, M. L. and S. A. Hodges. 1995. Are natural hybrids fit or unfit relative to their parents. Trends in Ecology & Evolution **10**:67–71.

Arnold, S. J. 1992. Constraints on phenotypic evolution. American Naturalist **140**:S85–S107.

Arnqvist, G. and L. Rowe. 2005. Sexual Conflict. Princeton University Press, Princeton, NJ.

Aubin-Horth, N. and S. C. P. Renn. 2009. Genomic reaction norms: using integrative biology to understand molecular mechanisms of phenotypic plasticity. Molecular Ecology **18**:3763–80.

Avise, J. C. 2000. Phylogeography: the history and formation of species. Harvard University Press, Cambridge, MA.

Badyaev, A. V., R. L. Young, K. P. Oh, and C. Addison. 2008. Evolution on a local scale: Developmental, functional, and genetic bases of divergence in bill form and associated changes in song structure between adjacent habitats. Evolution **62**:1951–64.

Bailey, J. K., A. P. Hendry, M. T. Kinnison, D. M. Post, E. P. Palkovacs, F. Pelletier, L. J. Harmon, and J. A. Schweitzer. 2009. From genes to ecosystems: an emerging synthesis of eco-evolutionary dynamics. New Phytologist **184**:746–9.

Balakrishnan, C. N., K. M. Sefc, and M. D. Sorenson. 2009. Incomplete reproductive isolation following host shift in brood parasitic indigobirds. Proceedings of the Royal Society B-Biological Sciences **276**:219–28.

Balloux, F. and N. Lugon-Moulin. 2002. The estimation of population differentiation with microsatellite markers. Molecular Ecology **11**:155–65.

Barluenga, M., K. N. Stolting, W. Salzburger, M. Muschick, and A. Meyer. 2006. Sympatric speciation in Nicaraguan crater lake cichlid fish. Nature **439**:719–23.

Barrett, R. D. H., S. M. Rogers, and D. Schluter. 2008. Natural selection on a major armor gene in threespine stickleback. Science **322**:255–7.

Barrett, R. D. H. and D. Schluter. 2008. Adaptation from standing genetic variation. Trends in Ecology & Evolution **23**:38–44.

Barrick, J. E., D. S. Yu, S. H. Yoon, H. Jeong, T. K. Oh, D. Schneider, R. E. Lenski, and J. F. Kim. 2009. Genome evolution and adaptation in a long-term experiment with Escherichia coli. Nature **461**:1243–U1274.

Barton, N. 2010. Understanding adaptation in large populations. Plos Genetics **6:** e1000987.

Barton, N. and B. O. Bengtsson. 1986. The barrier to genetic exchange between hybridizing populations. Heredity **57**:357–76.

Barton, N. H. 1979. Dynamics of hybrid zones. Heredity **43**:341–59.

Barton, N. H. 1995. Linkage and the limits to natural selection. Genetics **140**:821–41.

Barton, N. H. 2000. Genetic hitchhiking. Philosophical Transactions of the Royal Society of London Series B-Biological Sciences **355**:1553–62.

Barton, N. H. 2001. The role of hybridization in evolution. Molecular Ecology **10**:551–68.

Barton, N. H. and G. M. Hewitt. 1985. Analysis of hybrid zones. Annual Review of Ecology and Systematics **16**:113–48.

Barton, N. H. and G. M. Hewitt. 1989. Adaptation, speciation and hybrid zones. Nature **341**:497–503.

Basolo, A. L. 1995. Phylogenetic evidence for the role of a preexisting bias in sexual selection. Proceedings of the Royal Society of London Series B-Biological Sciences **259**:307–11.

Basolo, A. L. and J. A. Endler. 1995. Sensory biases and the evolution of sensory systems. Trends in Ecology & Evolution **10**:489–489.

Baxter, S. W., N. Nadeau, L. Maroja, P. Wilkinson, B. A. Counterman, A. Dawson, M. Beltrán, S. Perez-Espona, N. Chamberlain, L. Ferguson, R. Clark, C. Davidson, R. Glithero, J. Mallet, W. O. McMillan, M. Kronforst, M. Joron, R. ffrench-Constant, and C. D. Jiggins. 2010. Genomic hotspots for adaptation: the population genetics of Müllerian mimicry in the *Heliconius melpomene* clade. Plos Genetics **6**:e1000794.

Beaumont, M. A. 2005. Adaptation and speciation: what can F-st tell us? Trends in Ecology & Evolution **20**:435–40.

Beaumont, M. A. and D. J. Balding. 2004. Identifying adaptive genetic divergence among populations from genome scans. Molecular Ecology **13**:969–80.

Beaumont, M. A. and R. A. Nichols. 1996. Evaluating loci for use in the genetic analysis of population structure. Proceedings of the Royal Society of London Series B-Biological Sciences **263**:1619–26.

Becquet, C. and M. Przeworski. 2009. Learning about modes of speciation by computational approaches. Evolution **63**:2547–62.

Bell, M. A. 1987. Interacting evolutionary constraints in pelvic reduction of threespine sticklebacks, *Gasterosteus aculeatus* (Pisces, Gasterosteidae). Biological Journal of the Linnean Society **31**:347–82.

Bell, M. A., G. Orti, J. A. Walker, and J. P. Koenings. 1993. Evolution of pelvic reduction in threespine stickleback fish—a test of competing hypotheses. Evolution **47**:906–14.

Beltran, M., C. D. Jiggins, V. Bull, M. Linares, J. Mallet, W. O. McMillan, and E. Bermingham. 2002. Phylogenetic discordance at the species boundary: Comparative gene genealogies among rapidly radiating *Heliconius* butterflies. Molecular Biology and Evolution **19**:2176–90.

Bengtsson, B. O., editor. 1985. The flow of genes through a genetic barrier. Cambridge University Press, Cambridge.

Benkman, C. W. 2003. Divergent selection drives the adaptive radiation of crossbills. Evolution **57**:1176–81.

Benkman, C. W., W. C. Holimon, and J. W. Smith. 2001. The influence of a competitor on the geographic mosaic of coevolution between crossbills and lodgepole pine. Evolution **55**:282–94.

Berlocher, S. H. and J. L. Feder. 2002. Sympatric speciation in phytophagous insects: Moving beyond controversy? Annual Review of Entomology **47**:773–815.

Bernatchez, L., A. Chouinard, and G. Q. Lu. 1999. Integrating molecular genetics and ecology in studies of adaptive radiation: whitefish, *Coregonus* sp., as a case study. Biological Journal of the Linnean Society **68**:173–94.

Bernatchez, L., S. Renaut, A. R. Whiteley, D. Campbell, N. Derôme, J. Jeukens, L. Landry, G. Lu, A. W. Nolte, K. Østbye, S. M. Rogers, and J. St-Cyr. 2010. On the origin of species: Insights from the ecological genomics of whitefish. Philosophical Transactions of the Royal Society of London Series B-Biological Sciences **365**:1783–800.

Bernays, E. and M. Graham. 1988. On the evolution of host specificity in phytophagous arthropods. Ecology **69**:886–92.

Bernays, E. A. 1991. Evolution of insect morphology in relation to plants. Philosophical Transactions of the Royal Society of London Series B-Biological Sciences **333**:257–64.

Bernays, E. A. and W. T. Wcislo. 1994. Sensory capabilities, information-processing, and resource specialization. Quarterly Review of Biology **69**:187–204.

Berner, D., A. C. Grandchamp, and A. P. Hendry. 2009. Variable progress toward ecological speciation in parapatry: stickleback across eight lake-stream transitions. Evolution **63**:1740–53.

Bierbaum, T. J. and G. L. Bush. 1990. Genetic differentiation in the viability of sibling species of *Rhagoletis* fruit-flies on host plants, and the influence of reduced hybrid viability on reproductive isolation. Entomologia Experimentalis Et Applicata **55**:105–18.

Blair, W. F. 1950. Ecological factors in speciation of *Peromyscus*. Evolution **4**:253–75.

Blows, M. W., R. Brooks, and P. G. Kraft. 2003. Exploring complex fitness surfaces: Multiple ornamentation and polymorphism in male guppies. Evolution **57**:1622–30.

Boake, C. R. B., M. P. DeAngelis, and D. K. Andreadis. 1997. Is sexual selection and species recognition a continuum? Mating behavior of the stalk-eyed fly *Drosophila heteroneura*. Proceedings of the National Academy of Sciences of the United States of America **94**:12442–5.

Bolnick, D. I. 2001. Intraspecific competition favours niche width expansion in *Drosophila melanogaster*. Nature **410**:463–6.

Bolnick, D. I. 2004. Can intraspecific competition drive disruptive selection? An experimental test in natural populations of sticklebacks. Evolution **58**:608–18.

Bolnick, D. I., E. J. Caldera, and B. Matthews. 2008. Evidence for asymmetric migration load in a pair of ecologically divergent stickleback populations. Biological Journal of the Linnean Society **94**:273–87.

Bolnick, D. I. and M. Doebeli. 2003. Sexual dimorphism and adaptive speciation: Two sides of the same ecological coin. Evolution **57**:2433–49.

Bolnick, D. I. and B. M. Fitzpatrick. 2007. Sympatric speciation: Models and empirical evidence. Annual Review of Ecology Evolution and Systematics **38**:459–87.

Bolnick, D. I. and O. L. Lau. 2008. Predictable patterns of disruptive selection in stickleback in postglacial lakes. American Naturalist **172**:1–11.

Bolnick, D. I. and T. J. Near. 2005. Tempo of hybrid inviability in centrarchid fishes (Teleostei: Centrarchidae). Evolution **59**:1754–67.

Bolnick, D. I., T. J. Near, and P. C. Wainwright. 2006. Body size divergence promotes post-zygotic reproductive isolation in centrarchids. Evolutionary Ecology Research **8**:903–13.

Bolnick, D. I. and P. Nosil. 2007. Natural selection in populations subject to a migration load. Evolution **61**:2229–43.

Bolnick, D. I., L. K. Snowberg, C. Patenia, W. E. Stutz, T. Ingram, and O. L. Lau. 2009. Phenotypic-dependent native habitat preference facilitates divergence between parapatric lake and stream stickleback. Evolution **63**:2004–16.

Boncoraglio, G. and N. Saino. 2007. Habitat structure and the evolution of bird song: a meta-analysis of the evidence for the acoustic adaptation hypothesis. Functional Ecology **21**:134–42.

Bonduriansky, R. and L. Rowe. 2005. Intralocus sexual conflict and the genetic architecture of sexually dimorphic traits in Prochyliza xanthostoma (Diptera: Piophilidae). Evolution **59**:1965–75.

Bonin, A., P. Taberlet, C. Miaud, and F. Pompanon. 2006. Explorative genome scan to detect candidate loci for adaptation along a gradient of altitude in the common frog (Rana temporaria). Molecular Biology and Evolution **23**:773–83.

Boughman, J. W. 2001. Divergent sexual selection enhances reproductive isolation in sticklebacks. Nature **411**:944–8.

Boughman, J. W. 2002. How sensory drive can promote speciation. Trends in Ecology & Evolution **17**:571–7.

Boughman, J. W. 2007. Condition-dependent expression of red colour differs between stickleback species. Journal of Evolutionary Biology **20**:1577–90.

Boughman, J. W., H. D. Rundle, and D. Schluter. 2005. Parallel evolution of sexual isolation in sticklebacks. Evolution **59**:361–73.

Bowcock, A. M., J. R. Kidd, J. L. Mountain, J. M. Hebert, L. Carotenuto, K. K. Kidd, and L. L. Cavallisforza. 1991. Drift, admixture, and selection in human evolution - a study with DNA polymorphisms. Proceedings of the National Academy of Sciences of the United States of America **88**:839–43.

Bradshaw, H. D. and D. W. Schemske. 2003. Allele substitution at a flower colour locus produces a pollinator shift in monkeyflowers. Nature **426**:176–8.

Brooks, R., J. Hunt, M. W. Blows, M. J. Smith, L. F. Bussiere, and M. D. Jennions. 2005. Experimental evidence for multivariate stabilizing sexual selection. Evolution **59**:871–80.

Brown, K. M., L. M. Burk, L. M. Henagan, and M. A. F. Noor. 2004. A test of the chromosomal rearrangement model of speciation in *Drosophila pseudoobscura*. Evolution **58**:1856–60.

Brumfield, R. T., R. W. Jernigan, D. B. McDonald, and M. J. Braun. 2001. Evolutionary implications of divergent clines in an avian (Manacus: Aves) hybrid zone. Evolution **55**:2070–87.

Brunel-Pons, O., S. Alem, and M. D. Greenfield. 2011. The complex auditory scene at leks: balancing antipredator behaviour and competitive signalling in an acoustic moth. Animal Behaviour **81**:231–9.

Buckling, A. and P. B. Rainey. 2002. The role of parasites in sympatric and allopatric host diversification. Nature **420**:496–9.

Bull, V., M. Beltran, C. D. Jiggins, W. O. McMillan, E. Bermingham, and J. Mallet. 2006. Polyphyly and gene flow between non-sibling *Heliconius* species. BMC Biology **4**:11.

Bulmer, M. G. 1972. Multiple niche polymorphism. American Naturalist **106**:254–257.

Burke, M. K., J. P. Dunham, P. Shahrestani, K. R. Thornton, M. R. Rose, and A. D. Long. 2010. Genome-wide analysis of a long-term evolution experiment with *Drosophila*. Nature **467**:587–90.

Burt, A. and R. Trivers. 2006. Genes in conflict. Harvard University Press, Cambridge, MA.

Bush, G. L. 1969a. Mating behavior, host specificity, and ecological significance of sibling species in frugivorous flies of Genus *Rhagoletis* (Diptera-Tephritidae). American Naturalist **103**:669–672.

Bush, G. L. 1969b. Sympatric host race formation and speciation in frugivorous flies of Genus *Rhagoletis* (Diptera, Tephritidae). Evolution **23**:237–51.

Bush, G. L. 1975. Modes of animal speciation. Annual Review of Ecology and Systematics **6**:339–64.

Butler, M. A., S. A. Sawyer, and J. B. Losos. 2007. Sexual dimorphism and adaptive radiation in Anolis lizards. Nature **447**:202–5.

Butlin, R., J. Bridle, and D. Schluter. 2009. Speciation and patterns of diversity. Cambridge University Press, Cambridge.

Butlin, R. K. 1995. Reinforcement - an idea evolving. Trends in Ecology & Evolution **10**:432–4.

Butlin, R. K. 2005. Recombination and speciation. Molecular Ecology **14**:2621–35.

Butlin, R. K. 2010. Population genomics and speciation. Genetica **138**:409–18.

Butlin, R. K., J. Galindo, and J. W. Grahame. 2008. Sympatric, parapatric or allopatric: the most important way to classify speciation? Philosophical Transactions of the Royal Society B-Biological Sciences **363**:2997–3007.

Cain, M. L., V. Andreasen, and D. J. Howard. 1999. Reinforcing selection is effective under a relatively broad set of conditions in a mosaic hybrid zone. Evolution **53**:1343–53.

Calsbeek, R. 2009. Experimental evidence that competition and habitat use shape the individual fitness surface. Journal of Evolutionary Biology **22**:97–108.

Calsbeek, R. and T. B. Smith. 2008. Experimentally replicated disruptive selection on performance traits in a Caribbean lizard. Evolution **62**:478–84.

Campbell, D. and L. Bernatchez. 2004. Generic scan using AFLP markers as a means to assess the role of directional selection in the divergence of sympatric whitefish ecotypes. Molecular Biology and Evolution **21**:945–56.

Campbell, D. R. 2003. Natural selection in Ipomopsis hybrid zones: implications for ecological speciation. New Phytologist **161**:83–90.

Campbell, D. R., R. Alarcon, and C. A. Wu. 2003. Reproductive isolation and hybrid pollen disadvantage in Ipomopsis. Journal of Evolutionary Biology **16**:536–40.

Campbell, D. R. and N. M. Waser. 2001. Genotype-by-environment interaction and the fitness of plant hybrids in the wild. Evolution **55**:669–76.

Cano, J. M., C. Matsuba, H. Makinen, and J. Merila. 2006. The utility of QTL-Linked markers to detect selective sweeps in natural populations - a case study of the EDA gene and a linked marker in threespine stickleback. Molecular Ecology **15**:4613–21.

Carneiro, M., N. Ferrand, and M. W. Nachman. 2009. Recombination and Speciation: Loci Near Centromeres Are More Differentiated Than Loci Near Telomeres Between Subspecies of the European Rabbit (Oryctolagus cuniculus). Genetics **181**:593–606.

Carroll, L. S. and W. K. Potts. 2006. Functional genomics requires ecology. Pages 173–215 Advances in the Study of Behavior, Vol 36. Elsevier Academic Press San Diego.

Chan, Y. F., M. E. Marks, F. C. Jones, G. Villarreal, M. D. Shapiro, S. D. Brady, A. M. Southwick, D. M. Absher, J. Grimwood, J. Schmutz, R. M. Myers, D. Petrov, B. Jonsson, D. Schluter, M. A. Bell, and D. M. Kingsley. 2010. Adaptive evolution of pelvic reduction in sticklebacks by recurrent deletion of a Pitx1 enhancer. Science **327**:302–5.

Chan, Y. F., G. Villarreal, M. Marks, M. Shapiro, F. Jones, D. Petrov, M. Dickson, A. Southwick, D. Absher, J. Grimwood, J. Schmutz, R. Myers, B. Jnsson, D. Schluter, M. Bell, and D. Kingsley. 2009. From trait to base pairs: Parallel evolution of pelvic reduction in three-spined sticklebacks occurs by repeated deletion of a tissue-specific pelvic enhancer at *Pitx1*. Mechanisms of Development **126**:S14–S15.

Chapman, T., G. Arnqvist, J. Bangham, and L. Rowe. 2003. Sexual conflict. Trends in Ecology & Evolution **18**:41–7.

Charlesworth, B., M. Nordborg, and D. Charlesworth. 1997. The effects of local selection, balanced polymorphism and background selection on equilibrium patterns of genetic diversity in subdivided populations. Genetical Research **70**:155–74.

Charlesworth, D. and B. Charlesworth. 1975. Theoretical genetics of Batesian mimicry. 2. Evolution of supergenes. Journal of Theoretical Biology **55**:305–24.

Chase, J. M., P. A. Abrams, J. P. Grover, S. Diehl, P. Chesson, R. D. Holt, S. A. Richards, R. M. Nisbet, and T. J. Case. 2002. The interaction between predation and competition: a review and synthesis. Ecology Letters **5**:302–15.

Chenoweth, S. F., H. D. Rundle, and M. W. Blows. 2010. The contribution of selection and genetic constraints to phenotypic divergence. American Naturalist **175**:186–96.

Christie, P. and M. R. Macnair. 1984. Complementary lethal factors in 2 North-American populations of the yellow monkey flower. Journal of Heredity **75**:510–11.

Church, S. A. and D. R. Taylor. 2002. The evolution of reproductive isolation in spatially structured populations. Evolution **56**:1859–62.

Claridge, M. F. and J. C. Morgan. 1993. Geographical variation in acoustic signals of the planthopper, Nilaparvata bakeri (Muir), in Asia: species recognition and sexual selection. Biological Journal of the Linnean Society **48**:267–81.

Clarke, B. 1966. Evolution of morph-ratio clines. American Naturalist **100**:389-&.

Clausen, J. 1951. Stages in the evolution of plant species. Cornell University Press, Ithaca, NY.

Cogni, R. and D. J. Futuyma. 2009. Local adaptation in a plant herbivore interaction depends on the spatial scale. Biological Journal of the Linnean Society **97**:494–502.

Cokl, A., M. Zorovic, A. Zunic, and M. Virant-Doberlet. 2005. Tuning of host plants with vibratory songs of Nezara viridula L (Heteroptera: Pentatomidae). Journal of Experimental Biology **208**:1481–8.

Coleman, S. W., A. Harlin-Cognato, and A. G. Jones. 2009. Reproductive isolation, reproductive mode, and sexual selection: empirical tests of the viviparity-driven conflict hypothesis. American Naturalist **173**:291–303.

Colosimo, P. F., K. E. Hosemann, S. Balabhadra, G. Villarreal, M. Dickson, J. Grimwood, J. Schmutz, R. M. Myers, D. Schluter, and D. M. Kingsley. 2005. Widespread parallel evolution in sticklebacks by repeated fixation of ectodysplasin alleles. Science **307**:1928–33.

Conde-Padin, P., R. Cruz, J. Hollander, and E. Rolan-Alvarez. 2008. Revealing the mechanisms of sexual isolation in a case of sympatric and parallel ecological divergence. Biological Journal of the Linnean Society **94**:513–26.

Conner, J. K. 2002. Genetic mechanisms of floral trait correlations in a natural population. Nature **420**:407–10.

Counterman, B. A., F. Araujo-Perez, H. M. Hines, S. W. Baxter, C. M. Morrison, D. P. Lindstrom, R. Papa, L. Ferguson, M. Joron, R. ffrench-Constant, C. P. Smith, D. M. Nielsen, R. Chen, C. D. Jiggins, R. D. Reed, G. Halder, J. Mallet, and W. O. McMillan. 2010. Genomic hotspots for adaptation: the population genetics of Müllerian mimicry in *Heliconius* erato. Plos Genetics **6**:e1000796.

Counterman, B. A. and M. A. F. Noor. 2006. Multilocus test for introgression between the cactophilic species *Drosophila mojavensis* and *Drosophila arizonae*. American Naturalist **168**:682–96.

Coyne, J. A. and H. A. Orr. 1989. Patterns of Speciation in *Drosophila*. Evolution **43**:362–81.

Coyne, J. A. and H. A. Orr. 2004. Speciation. Sinauer Associates, Inc., Sunderland, MA.

Craig, T. P., J. D. Horner, and J. K. Itami. 1997. Hybridization studies on the host races of *Eurosta solidaginis*: Implications for sympatric speciation. Evolution **51**:1552–60.

Craig, T. P., J. D. Horner, and J. K. Itami. 2001. Genetics, experience, and host-plant preference in *Eurosta solidaginis*: Implications for host shifts and speciation. Evolution **55**:773–82.

Craig, T. P., J. K. Itami, W. G. Abrahamson, and J. D. Horner. 1993. Behavioral evidence for host-race formation in *Eurosta solidaginis*. Evolution **47**:1696–710.

Crespi, B. J. 2000. The evolution of maladaptation. Heredity **84**:623–9.

Crespi, B. J., D. C. Morris, and L. A. Mound. 2004. Evolution of ecological and behavioural diversity: Australian Acacia thrips as model organisms. Australian Biological Resources Study & CSIRO Entomology, Canberra, Australia.

Crispo, E., P. Bentzen, D. N. Reznick, M. T. Kinnison, and A. P. Hendry. 2006. The relative influence of natural selection and geography on gene flow in guppies. Molecular Ecology **15**:49–62.

Cummings, M. P., M. C. Neel, and K. L. Shaw. 2008. A genealogical approach to quantifying lineage divergence. Evolution **62**:2411–22.

Cutter, A. D. and J. Y. Choi. 2010. Natural selection shapes nucleotide polymorphism across the genome of the nematode Caenorhabditis briggsae. Genome Research **20**:1103–11.

Cutter, A. D. and A. M. Moses. 2011. Polymorphism, divergence, and the role of recombination in Saccharomyces cerevisiae genome evolution. Molecular Biology and Evolution **28**:1745–54.

Dambroski, H. R. and J. L. Feder. 2007. Host plant and latitude-related diapause variation in *Rhagoletis pomonella*: a test for multifaceted life history adaptation on different stages of diapause development. Journal of Evolutionary Biology **20**:2101–12.

Darwin, C. 1859. On the origin of species by means of natural selection, or the preservation of favoured races in the struggle for life. John Murray, London, UK.

de Aguiar, M. A. M., M. Baranger, E. M. Baptestini, L. Kaufman, and Y. Bar-Yam. 2009. Global patterns of speciation and diversity. Nature **460**:384–7.

de Queiroz, K. 2005. Different species problems and their resolution. Bioessays **27**:1263–9.

Degnan, J. H. and L. A. Salter. 2005. Gene tree distributions under the coalescent process. Evolution **59**:24–37.

Denno, R. F., M. S. McClure, and J. R. Ott. 1995. Interspecific interactions in phytophagous insects - competition reexamined and resurrected. Annual Review of Entomology **40**:297–331.

Derome, N. and L. Bernatchez. 2006. The transcriptomics of ecological convergence between 2 limnetic coregonine fishes (Salmonidae). Molecular Biology and Evolution **23**:2370–8.

Derome, N., B. Bougas, S. M. Rogers, A. R. Whiteley, A. Labbe, J. Laroche, and L. Bernatchez. 2008. Pervasive sex-linked effects on transcription regulation as revealed by expression quantitative trait loci mapping in lake whitefish species pairs (*Coregonus* sp., salmonidae). Genetics **179**:1903–17.

Dettman, J. R., C. Sirjusingh, L. M. Kohn, and J. B. Anderson. 2007. Incipient speciation by divergent adaptation and antagonistic epistasis in yeast. Nature **447**:585–8.

Dice, L. R. 1940. Ecologic and genetic variability within species of *Peromyscus*. American Naturalist **74**:212–21.

Dieckmann, U. and M. Doebeli. 1999. On the origin of species by sympatric speciation. Nature **400**:354–7.

Dieckmann, U., M. Doebeli, J. A. J. Metz, and D. Tautz, editors. 2004. Adaptive Speciation. Cambridge University Press, Cambridge, UK.

Diehl, S. D. and G. L. Bush. 1989. The role of habitat preference in adaptation and speciation. Pages 345–65 *in* D. Otte and J. A. Endler, editors. Speciation and Its Consequences. Sinauer, Sunderland, MA.

Diehl, S. R. and G. L. Bush. 1984. An evolutionary and applied perspective of insect biotypes. Annual Review of Entomology **29**:471–504.

Dijkstra, P. D., O. Seehausen, B. L. A. Gricar, M. E. Maan, and T. G. G. Groothuis. 2006. Can male-male competition stabilize speciation? A test in Lake Victoria haplochromine cichlid fish. Behavioral Ecology and Sociobiology **59**:704–13.

Dijkstra, P. D., O. Seehausen, M. E. R. Pierotti, and T. G. G. Groothuis. 2007. Male-male competition and speciation: aggression bias towards differently coloured rivals varies between stages of speciation in a Lake Victoria cichlid species complex. Journal of Evolutionary Biology **20**:496–502.

Dobzhansky, T. 1937. Genetics and the Origin of Species. 1st edition. Columbia University Press, New York, NY.

Dobzhansky, T. 1940. Speciation as a stage in evolutionary divergence. American Naturalist **74**:312–21.

Dobzhansky, T. 1951. Genetics and the Origin of Species. 3rd edition. Columbia University Press, New York, NY.

Dodd, D. M. B. 1989. Reproductive isolation as a consequence of adaptive divergence in *Drosophila-Pseudoobscura*. Evolution **43**:1308–11.

Doebeli, M. 1996. An explicit genetic model for ecological character displacement. Ecology **77**:510–20.

Doebeli, M. and U. Dieckmann. 2003. Speciation along environmental gradients. Nature **421**:259–64.

Dolgin, E. S., M. C. Whitlock, and A. F. Agrawal. 2006. Male *Drosophila melanogaster* have higher mating success when adapted to their thermal environment. Journal of Evolutionary Biology **19**:1894–900.

Dolman, G. and C. Moritz. 2006. A multilocus perspective on refugial isolation and divergence in rainforest skinks (Carlia). Evolution **60**:573–82.

Dolman, G. and D. Stuart-Fox. 2010. Processes driving male breeding colour and ecomorphological diversification in rainbow skinks: a phylogenetic comparative test. Evolutionary Ecology **24**:97–113.

Dopman, E. B., L. Perez, S. M. Bogdanowicz, and R. G. Harrison. 2005. Consequences of reproductive barriers for genealogical discordance in the European corn borer. Proceedings of the National Academy of Sciences of the United States of America **102**:14706–11.

Dopman, E. B., P. S. Robbins, and A. Seaman. 2010. Components of reproductive isolation between North American pheromones strains of the European corn borer. Evolution **64**:881–902.

Drès, M. and J. Mallet. 2002. Host races in plant-feeding insects and their importance in sympatric speciation. Philosophical Transactions of the Royal Society of London Series B-Biological Sciences **357**:471–92.

Ebert, D., C. Haag, M. Kirkpatrick, M. Riek, J. W. Hottinger, and V. I. Pajunen. 2002. A selective advantage to immigrant genes in a Daphnia metapopulation. Science **295**:485–8.

Egan, S. P. and D. J. Funk. 2006. Individual advantages to ecological specialization: insights on cognitive constraints from three conspecific taxa. Proceedings of the Royal Society B-Biological Sciences **273**:843–8.

Egan, S. P. and D. J. Funk. 2009. Ecologically dependent postmating isolation between sympatric host forms of *Neochlamisus bebbianae* leaf beetles. Proceedings of the National Academy of Sciences of the United States of America **106**:19426–31.

Egan, S. P., P. Nosil, and D. J. Funk. 2008. Selection and genomic differentiation during ecological speciation: Isolating the contributions of host association via a comparative genome scan of *Neochlamisus bebbianae* leaf beetles. Evolution **62**:1162–81.

Ehrlich, P. R. and P. H. Raven. 1964. Buttferlies and plants - a study in coevolution. Evolution **18**:586–608.

Eizaguirre, C., T. L. Lenz, A. Traulsen, and M. Milinski. 2009. Speciation accelerated and stabilized by pleiotropic major histocompatibility complex immunogenes. Ecology Letters **12**:5–12.

Ellegren, H. 2008. Sequencing goes 454 and takes large-scale genomics into the wild. Molecular Ecology **17**:1629–31.

Elliot, M. G. and B. J. Crespi. 2006. Placental invasiveness mediates the evolution of hybrid inviability in mammals. American Naturalist **168**:114–20.

Ellison, C. K., C. Wiley, and K. L. Shaw. 2011. The genetics of speciation: genes of small effect underlie sexual isolation in the Hawaiian cricket Laupala. Journal of Evolutionary Biology **24**:1110–19.

Emelianov, I., F. Marec, and J. Mallet. 2004. Genomic evidence for divergence with gene flow in host races of the larch budmoth. Proceedings of the Royal Society of London Series B-Biological Sciences **271**:97–105.

Emerson, B. C. and N. Kolm. 2005. Species diversity can drive speciation. Nature **434**:1015–17.

Emms, S. K. and M. L. Arnold. 1997. The effect of habitat on parental and hybrid fitness: Transplant experiments with Louisiana irises. Evolution **51**:1112–19.

Emms, S. K. and M. L. Arnold. 2000. Site-to-site differences in pollinator visitation patterns in a Louisiana iris hybrid zone. Oikos **91**:568–78.

Endler, J. A. 1973. Gene flow and population differentiation. Science **179**:243–50.

Endler, J. A. 1977. Geographic Variation, Speciation, and Clines. Princeton University Press, Princeton, NJ.

Endler, J. A. 1986. Natural selection in the wild. Princeton University Press, Princeton, NJ.

Endler, J. A. 1992. Signals, signal conditions, and the direction of evolution. American Naturalist **139**:S125–S153.

Endler, J. A. 1993. Some general comments on the evolution and design of animal communication systems. Philosophical Transactions of the Royal Society of London Series B-Biological Sciences **340**:215–25.

Endler, J. A. and M. Thery. 1996. Interacting effects of lek placement, display behavior, ambient light, and color patterns in three neotropical forest-dwelling birds. American Naturalist **148**:421–52.

Eroukhmanoff, F., A. Hargeby, N. N. Arnberg, O. Hellgren, S. Bensch, and E. I. Svensson. 2009a. Parallelism and historical contingency during rapid ecotype divergence in an isopod. Journal of Evolutionary Biology **22**:1098–110.

Eroukhmanoff, F., A. Hargeby, and E. I. Svensson. 2009b. Rapid adaptive divergence between ecotypes of an aquatic isopod inferred from F-ST-Q(ST) analysis. Molecular Ecology **18**:4912–23.

Erwin, D. H. 2005. Seeds of diversity. Science **308**:1752–3.

Etges, W. J. and M. A. Ahrens. 2001. Premating isolation is determined by larval-rearing substrates in cactophilic *Drosophila mojavensis*. V. Deep geographic variation in epicuticular hydrocarbons among isolated populations. American Naturalist **158**:585–98.

Excoffier, L., T. Hofer, and M. Foll. 2009. Detecting loci under selection in a hierarchically structured population. Heredity **103**:285–98.

Faubet, P. and O. E. Gaggiotti. 2008. A new Bayesian method to identify the environmental factors that influence recent migration. Genetics **178**:1491–504.

Feder, J. L. 1995. The effects of parasitoids on sympatric host races of *Rhagoletis pomonella* (Diptera, Tephritidae). Ecology **76**:801–13.

Feder, J. L., S. H. Berlocher, J. B. Roethele, H. Dambroski, J. J. Smith, W. L. Perry, V. Gavrilovic, K. E. Filchak, J. Rull, and M. Aluja. 2003a. Allopatric genetic origins for sympatric host-plant shifts and race formation in *Rhagoletis*. Proceedings of the National Academy of Sciences of the United States of America **100**:10314–19.

Feder, J. L., C. A. Chilcote, and G. L. Bush. 1988. Genetic differentiation between sympatric host races of the Appple Maggot fly *Rhagoletis pomonella*. Nature **336**:61–4.

Feder, J. L., R. Geiji, S. Yeaman, and P. Nosil. 2011a. Establishment of new mutations under divergence and genome hitchhiking. Philosophical Transactions of the Royal Society B-Biological Sciences, in press.

Feder, J. L., R. Geji, T. H. Q. Powell, and P. Nosil. 2011b. Adaptive chromosomal divergence driven by mixed geographic mode of evolution. Evolution **65:**2157–2170.

Feder, J. L. and P. Nosil. 2009. Chromosomal inversions and species differences: when are genes affecting adaptive divergence and reproductive isolation expected to reside within inversions? Evolution **63**:3061–75.

Feder, J. L. and P. Nosil. 2010. The efficacy of divergence hitchhiking in generating genomic islands during ecological speciation. Evolution **64**:1729–47.

Feder, J. L., S. B. Opp, B. Wlazlo, K. Reynolds, W. Go, and S. Spisak. 1994. Host fidelity is an effective premating barrier between sympatric races of the apple maggotfly. Proceedings of the National Academy of Sciences of the United States of America **91**:7990–4.

Feder, J. L., F. B. Roethele, K. Filchak, J. Niedbalski, and J. Romero-Severson. 2003b. Evidence for inversion polymorphism related to sympatric host race formation in the apple maggot fly, *Rhagoletis pomonella*. Genetics **163**:939–53.

Felsenstein, J. 1976. Theoretical population genetics of variable selection and migration. Annual Review of Genetics **10**:253–80.

Felsenstein, J. 1981. Skepticism towards Santa Rosalia, or why are there so few kinds of animals? Evolution **35**:124–38.

Feulner, P. G. D., M. Plath, J. Engelmann, F. Kirschbaum, and R. Tiedemann. 2009. Electrifying love: electric fish use species-specific discharge for mate recognition. Biology Letters **5**:225–8.

Filchak, K. E., J. B. Roethele, and J. L. Feder. 2000. Natural selection and sympatric divergence in the apple maggot *Rhagoletis pomonella*. Nature **407**:739–42.

Fisher, R. A. 1930. The genetical theory of natural selection. Oxford University Press, Oxford.

Fisher, R. A. 1950. Gene frequencies in a cline determined by selection and diffusion. Biometrics **6**:353–61.

Fishman, L., J. Aagaard, and J. C. Tuthill. 2008. Toward the evolutionary genomics of gametophytic divergence: patterns of transmission distortion in monkeyflower (*Mimulus*) hybrids reveal a complex genetic basis for conspecific pollen precedence. Evolution **62**:2958–70.

Fishman, L. and J. H. Willis. 2006. A cytonuclear incompatibility causes anther sterility in Mimulus hybrids. Evolution **60**:1372–81.

Fitzpatrick, B. M. 2002. Molecular correlates of reproductive isolation. Evolution **56**:191–8.

Fitzpatrick, B. M. 2004. Rates of evolution of hybrid inviability in birds and mammals. Evolution **58**:1865–70.

Fitzpatrick, B. M. 2008a. Dobzhansky-Muller model of hybrid dysfunction supported by poor burst-speed performance in hybrid tiger salamanders. Journal of Evolutionary Biology **21**:342–51.

Fitzpatrick, B. M. 2008b. Hybrid dysfunction: Population genetic and quantitative genetic perspectives. American Naturalist **171**:491–8.

Fitzpatrick, B. M., J. A. Fordyce, and S. Gavrilets. 2008. What, if anything, is sympatric speciation? Journal of Evolutionary Biology **21**:1452–9.

Foll, M. and O. Gaggiotti. 2006. Identifying the environmental factors that determine the genetic structure of populations. Genetics **174**:875–91.

Foll, M. and O. Gaggiotti. 2008. A genome-scan method to identify selected loci appropriate for both dominant and codominant Markers: a Bayesian perspective. Genetics **180**:977–93.

Forbes, A. A., J. Fisher, and J. L. Feder. 2005. Habitat avoidance: Overlooking an important aspect of host-specific mating and sympatric speciation? Evolution **59**:1552–9.

Forbes, A. A., T. H. Q. Powell, L. L. Stelinski, J. J. Smith, and J. L. Feder. 2009. Sequential sympatric speciation across trophic levels. Science **323**:776–9.

Fordyce, J. A. 2010. Host shifts and evolutionary radiations of butterflies. Proceedings of the Royal Society B-Biological Sciences **277**:3735–43.

Fordyce, J. A., C. C. Nice, M. L. Forister, and A. M. Shapiro. 2002. The significance of wing pattern diversity in the Lycaenidae: mate discrimination by two recently diverged species. Journal of Evolutionary Biology **15**:871–9.

Forister, M. L. 2004. Oviposition preference and larval performance within a diverging lineage of lycaenid butterflies. Ecological Entomology **29**:264–72.

Forister, M. L. 2005. Independent inheritance of preference and performance in hybrids between host races of Mitoura butterflies (Lepidoptera: Lycaenidae). Evolution **59**:1149–55.

Freeland, J. R., P. Biss, K. F. Conrad, and J. Silvertown. 2010. Selection pressures have caused genome-wide population differentiation of Anthoxanthum odoratum despite the potential for high gene flow. Journal of Evolutionary Biology **23**:776–82.

Friesen, M. L., G. Saxer, M. Travisano, and M. Doebeli. 2004. Experimental evidence for sympatric ecological diversification due to frequency-dependent competition in Escherichia coli. Evolution **58**:245–60.

Fry, J. D. 1996. The evolution of host specialization: Are trade-offs overrated? American Naturalist **148**:S84–S107.

Fry, J. D. 2003. Multilocus models of sympatric speciation: Bush versus Rice versus Felsenstein. Evolution **57**:1735–46.

Fuller, R. C. 2008. Genetic incompatibilities and the role of environment. Evolution **62**:3056–68.

Fuller, R. C., L. J. Fleishman, M. Leal, J. Travis, and E. Loew. 2003. Intraspecific variation in retinal cone distribution in the bluefin killifish, Lucania goodei. Journal of Comparative Physiology a-Neuroethology Sensory Neural and Behavioral Physiology **189**:609–16.

Fuller, R. C. and J. Travis. 2004. Genetics, lighting environment, and heritable responses to lighting environment affect male color morph expression in bluefin killifish, Lucania goodei. Evolution **58**:1086–98.

Funk, D. J. 1998. Isolating a role for natural selection in speciation: Host adaptation and sexual isolation in *Neochlamisus bebbianae* leaf beetles. Evolution **52**:1744–59.

Funk, D. J., S. P. Egan, and P. Nosil. 2011. "Isolation-by-Adaptation" in *Neochlamisus* leaf beetles: host-related selection promotes neutral genomic divergence. Molecular Ecology, in press.

Funk, D. J., K. E. Filchak, and J. L. Feder. 2002. Herbivorous insects: model systems for the comparative study of speciation ecology. Genetica **116**:251–67.

Funk, D. J. and P. Nosil. 2007. Comparative analyses and ecological speciation in herbivorous insects. Pages 117–35 *in* K. Tilmon, editor. Specialization, Speciation, and Radiation: the Evolutionary Biology of Herbivorous Insects. University of California Press, Berkeley, CA, USA.

Funk, D. J., P. Nosil, and W. J. Etges. 2006. Ecological divergence exhibits consistently positive associations with reproductive isolation across disparate taxa. Proceedings of the National Academy of Sciences of the United States of America **103**:3209–13.

Funk, D. J. and K. E. Omland. 2003. Species level paraphyly and polyphyly: Frequency, causes, and consequences,with insights from animal mitochondrial DNA. Annual Review of Ecology, Evolution, and Systematics **34**:397–423.

Futuyma, D. J. and A. A. Agrawal. 2009a. Evolutionary history and species interactions. Proceedings of the National Academy of Sciences of the United States of America **106**:18043–4.

Futuyma, D. J. and A. A. Agrawal. 2009b. Macroevolution and the biological diversity of plants and herbivores. Proceedings of the National Academy of Sciences of the United States of America **106**:18054–61.

Futuyma, D. J., M. C. Keese, and D. J. Funk. 1995. Genetic constraints on macroevolution - the evolution of host affiliation in the leaf beetle genus *Ophraella*. Evolution **49**:797–809.

Futuyma, D. J. and G. C. Mayer. 1980. Non-allopatric speciation in animals. Systematic Zoology **29**:254–71.

Galen, C. 1985. Regulation of seed set in *Polemonium viscosum* - floral scents, pollination, and resources. Ecology **66**:792–7.

Galen, C. and P. G. Kevan. 1980. Scent and color, floral polymorphisms and pollination biology in *Polemonium viscosum* (Nutt). American Midland Naturalist **104**:281–9.

Galen, C., J. S. Shore, and H. Deyoe. 1991. Ecotypic divergence in alpine *Polemonium viscosum* - genetic structure, quantitative variation, and local adaptation. Evolution **45**:1218–28.

Galindo, J., P. Moran, and E. Rolan-Alvarez. 2009. Comparing geographical genetic differentiation between candidate and noncandidate loci for adaptation strengthens support for parallel ecological divergence in the marine snail *Littorina saxatilis*. Molecular Ecology **18**:919–30.

GarciaRamos, G. and M. Kirkpatrick. 1997. Genetic models of adaptation and gene flow in peripheral populations. Evolution **51**:21–8.

Gavrilets, S. 2000. Rapid evolution of reproductive barriers driven by sexual conflict. Nature **403**:886–9.

Gavrilets, S. 2003. Perspective: Models of speciation: What have we learned in 40 years? Evolution **57**:2197–215.

Gavrilets, S. 2004. Fitness landscapes and the origin of species. Princeton University Press, Princeton, NJ.

Gavrilets, S. and M. B. Cruzan. 1998. Neutral gene flow across single locus clines. Evolution **52**:1277–84.

Gavrilets, S. and A. Hastings. 1996. Founder effect speciation: A theoretical reassessment. American Naturalist **147**:466–91.

Gavrilets, S. and T. I. Hayashi. 2005. Speciation and sexual conflict. Evolutionary Ecology **19**:167–98.

Gavrilets, S. and A. Vose. 2005. Dynamic patterns of adaptive radiation. Proceedings of the National Academy of Sciences of the United States of America **102**:18040–5.

Gavrilets, S. and A. Vose. 2007. Case studies and mathematical models of ecological speciation. 2. Palms on an oceanic island. Molecular Ecology **16**:2910–21.

Gavrilets, S., A. Vose, M. Barluenga, W. Salzburger, and A. Meyer. 2007. Case studies and mathematical models of ecological speciation. 1. Cichlids in a crater lake. Molecular Ecology **16**:2893–909.

Geraldes, A., N. Ferrand, and M. W. Nachman. 2006. Contrasting patterns of introgression at X-linked loci across the hybrid zone between subspecies of the European rabbit (Oryctolagus cuniculus). Genetics **173**:919–33.

Gibson, G. and B. Weir. 2005. The quantitative genetics of transcription. Trends in Genetics **21**:616–23.

Gilad, Y., S. A. Rifkin, and J. K. Pritchard. 2008. Revealing the architecture of gene regulation: the promise of eQTL studies. Trends in Genetics **24**:408–15.

Giraud, T. 2006. Selection against migrant pathogens: the immigrant inviability barrier in pathogens. Heredity **97**:316–18.

Giraud, T., P. Gladieux, and S. Gavrilets. 2010. Linking the emergence of fungal plant diseases with ecological speciation. Trends in Ecology & Evolution **25**:387–95.

Giraud, T., G. Refregier, M. Le Gac, D. M. de Vienne, and M. E. Hood. 2008. Speciation in fungi. Fungal Genetics and Biology **45**:791–802.

Goldberg, E. E. and R. Lande. 2006. Ecological and reproductive character displacement on an environmental gradient. Evolution **60**:1344–57.

Gompert, Z. and C. A. Buerkle. 2009. A powerful regression-based method for admixture mapping of isolation across the genome of hybrids. Molecular Ecology **18**:1207–24.

Gompert, Z., M. L. Forister, J. A. Fordyce, C. C. Nice, R. J. Williamson, and C. A. Buerkle. 2010. Bayesian analysis of molecular variance in pyrosequences quantifies population genetic structure across the genome of Lycaeides butterflies. Molecular Ecology **19**:2455–73.

Gourbiere, S. and J. Mallet. 2010. Are species real? The shape of the species boundary with exponential failure, reinforcement, and the "missing snowball." Evolution **64**:1–24.

Gow, J. L., C. L. Peichel, and E. B. Taylor. 2006. Contrasting hybridization rates between sympatric three-spined sticklebacks highlight the fragility of reproductive barriers between evolutionarily young species. Molecular Ecology **15**:739–52.

Gow, J. L., C. L. Peichel, and E. B. Taylor. 2007. Ecological selection against hybrids in natural populations of sympatric threespine sticklebacks. Journal of Evolutionary Biology **20**:2173–80.

Gow, J. L., S. M. Rogers, M. Jackson, and D. Schluter. 2008. Ecological predictions lead to the discovery of a benthic-limnetic sympatric species pair of threespine stickleback in Little Quarry Lake, British Columbia. Canadian Journal of Zoology-Revue Canadienne De Zoologie **86**:564–71.

Grace, T., S. M. Wisely, S. J. Brown, F. E. Dowell, and A. Joern. 2010. Divergent host plant adaptation drives the evolution of sexual isolation in the grasshopper Hesperotettix viridis (Orthoptera: Acrididae) in the absence of reinforcement. Biological Journal of the Linnean Society **100**:866–78.

Grahame, J. W., C. S. Wilding, and R. K. Butlin. 2006. Adaptation to a steep environmental gradient and an associated barrier to gene exchange in *Littorina saxatilis*. Evolution **60**:268–78.

Grant, B. R. and P. R. Grant. 2008a. Fission and fusion of Darwin's finches populations. Philosophical Transactions of the Royal Society B-Biological Sciences **363**:2821–29.

Grant, P. R. 1986. Ecology and Evolution of Darwin's finches. Princeton University Press, Princeton.

Grant, P. R. and B. R. Grant. 1992. Hybridization of bird species. Science **256**:193–7.

Grant, P. R. and B. R. Grant. 2002. Unpredictable evolution in a 30-year study of Darwin's finches. Science **296**:707–11.

Grant, P. R. and B. R. Grant. 2006. Evolution of character displacement in Darwin's finches. Science **313**:224–6.

Grant, P. R. and B. R. Grant. 2008b. Pedigrees, assortative mating and speciation in Darwin's finches. Proceedings of the Royal Society B-Biological Sciences **275**:661–8.

Grant, P. R., B. R. Grant, and K. Petren. 2005. Hybridization in the recent past. American Naturalist **166**:56–67.

Gratton, C. and S. C. Welter. 1998. Oviposition preference and larval performance of *Liriomyza helianthi* (Diptera: Agromyzidae) on normal and novel host plants. Environmental Entomology **27**:926–35.

Gray, S. M., L. M. Dill, F. Y. Tantu, E. R. Loew, F. Herder, and J. S. McKinnon. 2008. Environment-contingent sexual selection in a colour polymorphic fish. Proceedings of the Royal Society B-Biological Sciences **275**:1785–91.

Gray, S. M. and B. W. Robinson. 2002. Experimental evidence that competition between stickleback species favours adaptive character divergence. Ecology Letters **5**:264–72.

Gripenberg, S., P. J. Mayhew, M. Parnell, and T. Roslin. 2010. A meta-analysis of preference-performance relationships in phytophagous insects. Ecology Letters **13**:383–93.

Groot, A. T., M. Marr, D. G. Heckel, and G. Schofl. 2010. The roles and interactions of reproductive isolation mechanisms in fall armyworm (Lepidoptera: Noctuidae) host strains. Ecological Entomology **35**:105–18.

Guerrero, R. F., F. Rousset, and M. Kirkpatrick. 2011. Coalescence Patterns for chromosomal inversions in divergent populations. Philosophical Transactions of the Royal Society B-Biological Sciences, in press.

Gurevitch, J., J. A. Morrison, and L. V. Hedges. 2000. The interaction between competition and predation: A meta-analysis of field experiments. American Naturalist **155**:435–53.

Gurevitch, J., L. L. Morrow, A. Wallace, and J. S. Walsh. 1992. A metaanalysis of competition in field experiments. American Naturalist **140**:539–72.

Haldane, J. B. S. 1930. A mathematical theory of natural and artificial selection. Proceedings of the Cambridge Philosophical Society **26**:220–30.

Haldane, J. B. S. 1932. The Causes of Evolution. Longmans, New York.

Haldane, J. B. S. 1948. The theory of a cline. Journal of Genetics **48**:277–84.

Hamilton, W. D. and M. Zuk. 1982. Heritable true fitness and bright birds - a role for parasites. Science **218**:384–7.

Hanski, I., T. Mononen, and O. Ovaskainen. 2011. Eco-evolutionary metapopulation dynamics and the spatial scale of adaptation. American Naturalist **177**:29–43.

Harr, B. 2006. Genomic islands of differentiation between house mouse subspecies. Genome Research **16**:730–7.

Harrison, R. G. 1991. Molecular changes at speciation. Annual Review of Ecology and Systematics **22**:281–308.

Hatfield, T. and D. Schluter. 1999. Ecological speciation in sticklebacks: Environment-dependent hybrid fitness. Evolution **53**:866–73.

Hawthorne, D. J. and S. Via. 2001. Genetic linkage of ecological specialization and reproductive isolation in pea aphids. Nature **412**:904–7.

Head, M. L., E. A. Price, and J. W. Boughman. 2009. Body size differences do not arise from divergent mate preferences in a species pair of threespine stickleback. Biology Letters **5**:517–20.

Hebets, E. A. and G. W. Uetz. 1999. Female responses to isolated signals from multimodal male courtship displays in the wolf spider genus Schizocosa (Araneae: Lycosidae). Animal Behaviour **57**:865–72.

Hedrick, P. W. 2005. A standardized genetic differentiation measure. Evolution **59**:1633–8.

Hendry, A. P. 2001. Adaptive divergence and the evolution of reproductive isolation in the wild: an empirical demonstration using introduced sockeye salmon. Genetica **112**:515–34.

Hendry, A. P. 2004. Selection against migrants contributes to the rapid evolution of ecologically dependent reproductive isolation. Evolutionary Ecology Research **6**:1219–36.

Hendry, A. P. 2009. Ecological speciation! Or the lack thereof? Canadian Journal of Fisheries and Aquatic Sciences **66**:1383–98.

Hendry, A. P., D. I. Bolnick, D. Berner, and C. L. Peichel. 2009a. Along the speciation continuum in sticklebacks. Journal of Fish Biology **75**:2000–36.

Hendry, A. P. and T. Day. 2005. Population structure attributable to reproductive date: isolation-by-time and adaptation-by-time. Molecular Ecology **14**:901–16.

Hendry, A. P., T. Day, and E. B. Taylor. 2001. Population mixing and the adaptive divergence of quantitative traits in discrete populations: A theoretical framework for empirical tests. Evolution **55**:459–66.

Hendry, A. P., S. K. Huber, L. F. De Leon, A. Herrel, and J. Podos. 2009b. Disruptive selection in a bimodal population of Darwin's finches. Proceedings of the Royal Society B-Biological Sciences **276**:753–9.

Hendry, A. P., P. Nosil, and L. H. Rieseberg. 2007. The speed of ecological speciation. Functional Ecology **21**:455–64.

Hendry, A. P. and E. B. Taylor. 2004. How much of the variation in adaptive divergence can be explained by gene flow? - An evaluation using lake-stream stickleback pairs. Evolution **58**:2319–31.

Hendry, A. P., J. K. Wenburg, P. Bentzen, E. C. Volk, and T. P. Quinn. 2000. Rapid evolution of reproductive isolation in the wild: Evidence from introduced salmon. Science **290**:516–18.

Henry, C. S. and M. L. M. Wells. 2004. Adaptation or random change? The evolutionary response of songs to substrate properties in lacewings (Neuroptera: Chrysopidae: Chrysoperla). Animal Behaviour **68**:879–95.

Hereford, J. 2009. A quantitative survey of local adaptation and fitness trade-offs. American Naturalist **173**:579–88.

Hermisson, J. and P. S. Pennings. 2005. Soft sweeps: Molecular population genetics of adaptation from standing genetic variation. Genetics **169**:2335–52.

Herrel, A., J. Podos, B. Vanhooydonck, and A. P. Hendry. 2009. Force-velocity trade-off in Darwin's finch jaw function: a biomechanical basis for ecological speciation? Functional Ecology **23**:119–25.

Hey, J. 1991. The structure of genealogies and the distribution of fixed differences between DNA-sequence samples from natural populations. Genetics **128**:831–40.

Hey, J. 2006. Recent advances in assessing gene flow between diverging populations and species. Current Opinion in Genetics & Development **16**:592–6.

Hey, J. 2010a. The divergence of chimpanzee species and subspecies as revealed in multipopulation Isolation-with-Migration analyses. Molecular Biology and Evolution **27**:921–33.

Hey, J. 2010b. Isolation with Migration models for more than two populations. Molecular Biology and Evolution **27**:905–20.

Hey, J. and R. Nielsen. 2004. Multilocus methods for estimating population sizes, migration rates and divergence time, with applications to the divergence of *Drosophila pseudoobscura* and D-persimilis. Genetics **167**:747–60.

Hey, J. and R. Nielsen. 2007. Integration within the Felsenstein equation for improved Markov chain Monte Carlo methods in population genetics. Proceedings of the National Academy of Sciences of the United States of America **104**:2785–90.

Hiesey, W. M., M. A. Nobs, and O. Björkman. 1971. Experimental studies on the nature of species. V. Biosystematics, genetics, and physiological ecology of the Erythranthe section of Mimulus. Carnegie Institute of Washington, Washington D.C., USA, publ. no. 628.

Higgie, M. and M. W. Blows. 2007. Are traits that experience reinforcement also under sexual selection? American Naturalist **170**:409–20.

Higgie, M., S. Chenoweth, and M. W. Blows. 2000. Natural selection and the reinforcement of mate recognition. Science **290**:519–21.

Hirai, Y., H. Kobayashi, T. Koizumi, and H. Katakura. 2006. Field-cage experiments on host fidelity in a pair of sympatric phytophagous ladybird beetles. Entomologia Experimentalis Et Applicata **118**:129–35.

Hoekstra, H. E. and J. A. Coyne. 2007. The locus of evolution: Evo devo and the genetics of adaptation. Evolution **61**:995–1016.

Hoekstra, H. E., R. J. Hirschmann, R. A. Bundey, P. A. Insel, and J. P. Crossland. 2006. A single amino acid mutation contributes to adaptive beach mouse color pattern. Science **313**:101–4.

Hoffmann, A. A. and L. H. Rieseberg. 2008. Revisiting the impact of inversions in evolution: from population genetic markers to drivers of adaptive shifts and speciation? Annual Review of Ecology Evolution and Systematics **39**:21–42.

Hohenlohe, P. A., S. Bassham, P. D. Etter, N. Stiffler, E. A. Johnson, and W. A. Cresko. 2010. Population Genomics of Parallel Adaptation in Threespine Stickleback using Sequenced RAD Tags. Plos Genetics **6:**e1000862.

Holt, R. D. 1977. Predation, apparent competition, and structure of prey communities. Theoretical Population Biology **12**:197–229.

Holt, R. D. and J. H. Lawton. 1994. The ecological consequences of shared natural enemies. Annual Review of Ecology and Systematics **25**:495–520.

Hosken, D. J. and P. Stockley. 2004. Sexual selection and genital evolution. Trends in Ecology & Evolution **19**:87–93.

Hoskin, C. J., M. Higgie, K. R. McDonald, and C. Moritz. 2005. Reinforcement drives rapid allopatric speciation. Nature **437**:1353–6.

Hoso, M., T. Asami, and M. Hori. 2007. Right-handed snakes: convergent evolution of asymmetry for functional specialization. Biology Letters **3**:169–U162.

Hoso, M. and M. Hori. 2008. Divergent shell shape as an antipredator adaptation in tropical land snails. American Naturalist **172**:726–32.

Howard, D. J. and S. H. Berlocher, editors. 1998. Endless forms: species and speciation. Oxford University Press, Oxford.

Howard, D. J., P. G. Gregory, J. M. Chu, and M. L. Cain. 1998. Conspecific sperm precedence is an effective barrier to hybridization between closely related species. Evolution **52**:511–16.

Huber, S. K., L. F. De Leon, A. P. Hendry, E. Bermingham, and J. Podos. 2007. Reproductive isolation of sympatric morphs in a population of Darwin's finches. Proceedings of the Royal Society B-Biological Sciences **274**:1709–14.

Hudson, M. E. 2008. Sequencing breakthroughs for genomic ecology and evolutionary biology. Molecular Ecology Resources **8**:3–17.

Hutchinson, G. E. 1957. Population studies - animal ecology and demography - concluding remarks. Cold Spring Harbor Symposia on Quantitative Biology **22**:415–27.

Hutchinson, G. E. 1959. Homage to Santa Rosalia or why are there so many kinds of animals. American Naturalist **93**:145–59.

Ingvarsson, P. K. and M. C. Whitlock. 2000. Heterosis increases the effective migration rate. Proceedings of the Royal Society of London Series B-Biological Sciences **267**:1321–6.

Janson, E. M., J. O. Stireman, M. S. Singer, and P. Abbot. 2008. Phytophagous insect-microbe mutualisms and adaptive evolutionary diversification. Evolution **62**:997–1012.

Jiggins, C. D. 2008. Ecological speciation in mimetic butterflies. Bioscience **58**:541–8.

Jiggins, C. D., C. Estrada, and A. Rodrigues. 2004. Mimicry and the evolution of premating isolation in *Heliconius* melpomene Linnaeus. Journal of Evolutionary Biology **17**:680–91.

Jiggins, C. D. and J. Mallet. 2000. Bimodal hybrid zones and speciation. Trends in Ecology & Evolution **15**:250–5.

Jiggins, C. D., R. E. Naisbit, R. L. Coe, and J. Mallet. 2001. Reproductive isolation caused by colour pattern mimicry. Nature **411**:302–5.

Johannesson, K., B. Johannesson, and E. Rolanalvarez. 1993. Morphological differentiation and genetic cohesiveness over a microenvironmental gradient in the marine snail *Littorina saxatilis*. Evolution **47**:1770–87.

Johnson, N. A. 2002. Sixty years after "Isolating mechanisms, evolution and temperature": Muller's legacy. Genetics **161**:939–44.

Johnson, N. A. and R. M. Kliman. 2002. Hidden evolution: progress and limitations in detecting multifarious natural selection. Genetica **114**:281–91.

Johnson, N. A. and A. H. Porter. 2000. Rapid speciation via parallel, directional selection on regulatory genetic pathways. Journal of Theoretical Biology **205**:527–42.

Johnson, P. A., F. C. Hoppensteadt, J. J. Smith, and G. L. Bush. 1996. Conditions for sympatric speciation: A diploid model incorporating habitat fidelity and non-habitat assortative mating. Evolutionary Ecology **10**:187–205.

Joly, S., P. A. McLenachan, and P. J. Lockhart. 2009. A statistical approach for distinguishing hybridization and incomplete lineage sorting. American Naturalist **174**:E54–E70.

Jones, A. G., G. I. Moore, C. Kvarnemo, D. Walker, and J. C. Avise. 2003. Sympatric speciation as a consequence of male pregnancy in seahorses. Proceedings of the National Academy of Sciences of the United States of America **100**:6598–603.

Joron, M., R. Papa, M. Beltran, N. Chamberlain, J. Mavarez, S. Baxter, M. Abanto, E. Bermingham, S. J. Humphray, J. Rogers, H. Beasley, K. Barlow, R. H. Ffrench-Constant, J. Mallet, W. O. McMillan, and C. D. Jiggins. 2006. A conserved supergene locus controls colour pattern diversity in *Heliconius* butterflies. Plos Biology **4**:1831–40.

Joyce, D. A., D. H. Lunt, M. J. Genner, G. F. Turner, R. Bills, and O. Seehausen. 2011. Repeated colonization and hybridization in Lake Malawi cichlids. Current Biology **21**:R108–R109.

Juan, C., M. T. Guzik, D. Jaume, and S. J. B. Cooper. 2010. Evolution in caves: Darwin's "wrecks of ancient life" in the molecular era. Molecular Ecology **19**:3865–80.

Kane, N. C., M. G. King, M. S. Barker, A. Raduski, S. Karrenberg, Y. Yatabe, S. J. Knapp, and L. H. Rieseberg. 2009. Comparative genomic and population genetic analyses indicate highly porous genomes and high levels of gene flow between divergent *Helianthus* species. Evolution **63**:2061–75.

Karasov, T., P. W. Messer, and D. A. Petrov. 2010. Evidence that adaptation in *Drosophila* is not limited by mutation at single sites. Plos Genetics **6**:e1000924.

Karlsson, K., F. Eroukhmanoff, R. Hardling, and E. I. Svensson. 2010. Parallel divergence in mate guarding behaviour following colonization of a novel habitat. Journal of Evolutionary Biology **23**:2540–9.

Kassen, R. 2009. Toward a general theory of adaptive radiation insights from microbial experimental evolution. Annals of the New York Academy of Sciences **1168**:3–22.

Katakura, H., M. Shioi, and Y. Kira. 1989. Reproductive isolation by host specificity in a pair of phytophagous ladybird beetles. Evolution **43**:1045–53.

Kawecki, T. J. and R. D. Holt. 2002. Evolutionary consequences of asymmetric dispersal rates. American Naturalist **160**:333–47.

Kibota, T. T. and S. P. Courtney. 1991. Jack of one trade, master of none - host choice by *Drosophila magnaquinaria*. Oecologia **86**:251–60.

Kilias, G., S. N. Alahiotis, and M. Pelecanos. 1980. A Multifactorial Genetic Investigation of Speciation Theory Using *Drosophila-Melanogaster*. Evolution **34**:730–7.

Kimura, M. 1956. A model of a genetic system which leads to closer linkage by natural selection. Evolution **10**:278–87.

Kimura, M. 1968. Evolutionary rate at molecular level. Nature **217**:624–6.

King, R. B. and R. Lawson. 1995. Color-pattern Variation in Lake-Erie water snakes - the role of gene flow. Evolution **49**:885–96.

Kingsolver, J. G., H. E. Hoekstra, J. M. Hoekstra, D. Berrigan, S. N. Vignieri, C. E. Hill, A. Hoang, P. Gibert, and P. Beerli. 2001. The strength of phenotypic selection in natural populations. American Naturalist **157**:245–61.

Kingston, T. and S. J. Rossiter. 2004. Harmonic-hopping in Wallacea's bats. Nature **429**:654–7.

Kirkpatrick, M. 1996. Good genes and direct selection in evolution of mating preferences. Evolution **50**:2125–40.

Kirkpatrick, M. 2000. Reinforcement and divergence under assortative mating. Proceedings of the Royal Society of London Series B-Biological Sciences **267**:1649–55.

Kirkpatrick, M. 2001. Reinforcement during ecological speciation. Proceedings of the Royal Society of London Series B-Biological Sciences **268**:1259–63.

Kirkpatrick, M. and N. Barton. 2006. Chromosome inversions, local adaptation and speciation. Genetics **173**:419–34.

Kirkpatrick, M. and N. H. Barton. 1997. The strength of indirect selection on female mating preferences. Proceedings of the National Academy of Sciences of the United States of America **94**:1282–6.

Kirkpatrick, M., T. Johnson, and N. Barton. 2002. General models of multilocus evolution. Genetics **161**:1727–50.

Kirkpatrick, M. and V. Ravigné. 2002. Speciation by natural and sexual selection: Models and experiments. American Naturalist **159**:S22–S35.

Kirkpatrick, M. and M. J. Ryan. 1991. The evolution of mating preferences and the paradox of the lek. Nature **350**:33–8.

Kisel, Y. and T. G. Barraclough. 2010. Speciation has a spatial scale that depends on levels of gene flow. American Naturalist **175**:316–34.

Knowles, L. L. and T. A. Markow. 2001. Sexually antagonistic coevolution of a postmating-prezygotic reproductive character in desert *Drosophila*. Proceedings of the National Academy of Sciences of the United States of America **98**:8692–6.

Kobayashi, Y. and A. Telschow. 2011. The concept of effective recombination rate and its application in speciation theory. Evolution **65**:617–28.

Kondrashov, A. S. 2003. Accumulation of Dobzhansky-Muller incompatibilities within a spatially structured population. Evolution **57**:151–3.

Kondrashov, A. S. and F. A. Kondrashov. 1999. Interactions among quantitative traits in the course of sympatric speciation. Nature **400**:351–4.

Kondrashov, A. S. and M. V. Mina. 1986. Sympatric speciation - when is it possible. Biological Journal of the Linnean Society **27**:201–23.

Kopp, M. and J. Hermisson. 2008. Competitive speciation and costs of choosiness. Journal of Evolutionary Biology **21**:1005–23.

Kotlik, P., S. Markova, L. Choleva, N. G. Bogutskaya, F. G. Ekmekci, and P. P. Ivanova. 2008. Divergence with gene flow between Ponto-Caspian refugia in an anadromous cyprinid Rutilus frisii revealed by multiple gene phylogeography. Molecular Ecology **17**:1076–88.

Kouyos, R. D., S. P. Otto, and S. Bonhoeffer. 2006. Effect of varying epistasis on the evolution of recombination. Genetics **173**:589–97.

Kozak, K. H., A. A. Larson, R. M. Bonett, and L. J. Harmon. 2005. Phylogenetic analysis of ecomorphological divergence, community structure, and diversification rates in dusky salamanders (Plethodontidae: Desmognathus). Evolution **59**:2000–16.

Kronforst, M. R. 2008. Gene flow persists millions of years after speciation in *Heliconius* butterflies. BMC Evolutionary Biology **8**:98.

Kronforst, M. R., L. G. Young, D. D. Kapan, C. McNeely, R. J. O'Neill, and L. E. Gilbert. 2006. Linkage of butterfly mate preference and wing color preference cue at the genomic location of wingless. Proceedings of the National Academy of Sciences of the United States of America **103**:6575–80.

Kruuk, L. E. B., J. S. Gilchrist, and N. H. Barton. 1999. Hybrid dysfunction in fire-bellied toads (*Bombina*). Evolution **53**:1611–16.

Kuwajima, M., N. Kobayashi, T. Katoh, and H. Katakura. 2010. Detection of ecological hybrid inviability in a pair of sympatric phytophagous ladybird beetles (Henosepilachna spp.). Entomologia Experimentalis Et Applicata **134**:280–6.

Lai, Z., N. C. Kane, Y. Zou, and L. H. Rieseberg. 2008. Natural variation in gene expression between wild and weedy populations of *Helianthus annuus*. Genetics **179**:1881–90.

Lamont, B. B., T. He, N. J. Enright, S. L. Krauss, and B. P. Miller. 2003. Anthropogenic disturbance promotes hybridization between Banksia species by altering their biology. Journal of Evolutionary Biology **16**:551–7.

Lande, R. 1979. Quantitative genetic analysis of multivariate evolution, applied to brain-body size allometry. Evolution **33**:402–16.

Lande, R. 1981. Models of speciation by sexual selection on polygenic traits. Proceedings of the National Academy of Sciences of the United States of America-Biological Sciences **78**:3721–5.

Lande, R. 1982. Rapid origin of sexual isolation and character divergence in a cline. Evolution **36**:213–23.

Landry, C. R. and N. Aubin-Horth. 2007. Ecological annotation of genes and genomes through ecological genomics. Molecular Ecology **16**:4419–21.

Landry, C. R., D. L. Hartl, and J. M. Ranz. 2007. Genome clashes in hybrids: insights from gene expression. Heredity **99**:483–93.

Langerhans, R. B., M. E. Gifford, and E. O. Joseph. 2007. Ecological speciation in Gambusia fishes. Evolution **61**:2056–74.

Langerhans, R. B., C. A. Layman, A. M. Shokrollahi, and T. J. DeWitt. 2004. Predator-driven phenotypic diversification in Gambusia affinis. Evolution **58**:2305–18.

Langerhans, R. B. and D. N. Reznick, editors. 2009. Ecology and evolution of swimming performance in fishes: predicting evolution with biomechanics. Science Publishers, Enfield.

Lawniczak, M. K. N., S. J. Emrich, A. K. Holloway, A. P. Regier, M. Olson, B. White, S. Redmond, L. Fulton, E. Appelbaum, J. Godfrey, C. Farmer, A. Chinwalla, S. P. Yang, P. Minx, J. Nelson, K. Kyung, B. P. Walenz, E. Garcia-Hernandez, M. Aguiar, L. D. Viswanathan, Y. H. Rogers, R. L. Strausberg, C. A. Saski, D. Lawson, F. H. Collins, F. C. Kafatos, G. K. Christophides, S. W. Clifton, E. F. Kirkness, and N. J. Besansky. 2010. Widespread divergence between incipient Anopheles gambiae species revealed by whole genome sequences. Science **330**:512–14.

Leal, M. and L. J. Fleishman. 2002. Evidence for habitat partitioning based on adaptation to environmental light in a pair of sympatric lizard species. Proceedings of the Royal Society of London Series B-Biological Sciences **269**:351–9.

Leal, M. and L. J. Fleishman. 2004. Differences in visual signal design and detectability between allopatric populations of Anolis lizards. American Naturalist **163**:26–39.

Lee, H. Y., J. Y. Chou, L. Cheong, N. H. Chang, S. Y. Yang, and J. Y. Leu. 2008. Incompatibility of nuclear and mitochondrial genomes causes hybrid sterility between two yeast species. Cell **135**:1065–73.

Leimu, R. and M. Fischer. 2008. A meta-analysis of local adaptation in plants. Plos One **3**:e4010.

Lemmon, A. R., C. Smadja, and M. Kirkpatrick. 2004. Reproductive character displacement is not the only possible outcome of reinforcement. Journal of Evolutionary Biology **17**:177–83.

Lemmon, E. M. 2009. Divergence of conspecific signals in sympatry: geographic overlap drives multidimensional reproductive character displacement in frogs. Evolution **63**:1155–70.

Lenormand, T. 2002. Gene flow and the limits to natural selection. Trends in Ecology & Evolution **17**:183–9.

Levin, D. A. 2004. Ecological speciation: The role of disturbance. Systematic Botany **29**:225–33.

Lewontin, R. C. and J. Krakauer. 1973. Distribution of gene frequency as a test of theory of selective neutrality of polymorphisms. Genetics **74**:175–95.

Lexer, C., Z. Lai, and L. H. Rieseberg. 2004. Candidate gene polymorphisms associated with salt tolerance in wild sunflower hybrids: implications for the origin of *Helianthus paradoxus*, a diploid hybrid species. New Phytologist **161**:225–33.

Light, J. E., M. A. Toups, and D. L. Reed. 2008. What's in a name: The taxonomic status of human head and body lice. Molecular Phylogenetics and Evolution **47**:1203–16.

Linn, C., J. L. Feder, S. Nojima, H. R. Dambroski, S. H. Berlocher, and W. Roelofs. 2003. Fruit odor discrimination and sympatric host race formation in *Rhagoletis*. Proceedings of the National Academy of Sciences of the United States of America **100**:11490–3.

Linnen, C. R. and B. D. Farrell. 2007. Mitonuclear discordance is caused by rampant mitochondrial introgression in Neodiprion (hymenoptera: diprionidae) sawflies. Evolution **61**:1417–38.

Linnen, C. R., E. P. Kingsley, J. D. Jensen, and H. E. Hoekstra. 2009. On the origin and spread of an adaptive allele in deer mice. Science **325**:1095–8.

Liou, L. W. and T. D. Price. 1994. Speciation by reinforcement of premating isolation. Evolution **48**:1451–9.

Losos, J. B. 1985. An experimental demonstration of the species-recognition role of *Anolis* dewlap color. Copeia **1985**:905–10.

Losos, J. B. and D. Schluter. 2000. Analysis of an evolutionary species-area relationship. Nature **408**:847–50.

Losos, J. B., T. W. Schoener, and D. A. Spiller. 2004. Predator-induced behaviour shifts and natural selection in field-experimental lizard populations. Nature **432**:505–8.

Lowry, D. B., J. L. Modliszewski, K. M. Wright, C. A. Wu, and J. H. Willis. 2008a. The strength and genetic basis of reproductive isolating barriers in flowering plants. Philosophical Transactions of the Royal Society B-Biological Sciences **363**:3009–21.

Lowry, D. B., R. C. Rockwood, and J. H. Willis. 2008b. Ecological reproductive isolation of coast and inland races of Mimulus guttatus. Evolution **62**:2196–214.

Lowry, D. B. and J. H. Willis. 2010. A widespread chromosomal inversion polymorphism contributes to a major life-history transition, local adaptation, and reproductive isolation. Plos Biology **8:**e1000500.

Lu, G. and L. Bernatchez. 1999a. Correlated trophic specialization and genetic divergence in sympatric lake whitefish ecotypes (*Coregonus clupeaformis*): support for the ecological speciation hypothesis. Evolution **53**:1491–505.

Lu, G. Q. and L. Bernatchez. 1999b. Correlated trophic specialization and genetic divergence in sympatric lake whitefish ecotypes (*Coregonus clupeaformis*): Support for the ecological speciation hypothesis. Evolution **53**:1491–505.

Luikart, G., P. R. England, D. Tallmon, S. Jordan, and P. Taberlet. 2003. The power and promise of population genomics: From genotyping to genome typing. Nature Reviews Genetics **4**:981–94.
Lynch, M. and B. Walsh. 1998. Genetics and analysis of quantitative traits. Sinauer Associates Inc., Sunderland, Massachusetts, USA.
Lythgoe, J. N., W. R. A. Muntz, J. C. Partridge, J. Shand, and D. M. Williams. 1994. The ecology of the visual pigments of snappers (Lutjanidae) on the Great Barrier reef. Journal of Comparative Physiology a-Sensory Neural and Behavioral Physiology **174**:461–7.
Maan, M. E. and M. E. Cummings. 2008. Female preferences for aposematic signal components in a polymorphic poison frog. Evolution **62**:2334–45.
Maan, M. E., B. Eshuis, M. P. Haesler, M. V. Schneider, J. J. M. van Alphen, and O. Seehausen. 2008. Color polymorphism and predation in a Lake Victoria cichlid fish. Copeia **2008**:621–9.
Maan, M. E., K. D. Hofker, J. J. M. van Alphen, and O. Seehausen. 2006. Sensory drive in cichlid speciation. American Naturalist **167**:947–54.
Maan, M. E. and O. Seehausen. 2010. Mechanisms of species divergence through visual adaptation and sexual selection: Perspectives from a cichlid model system. Current Zoology **56**:285–99.
Maan, M. E. and O. Seehausen. 2011. Ecology, sexual selection and speciation. Ecology Letters **14**:591–602.
MacArthur, R. H. and E. O. Wilson. 1967. The theory of island biogeography. Princeton University Press, Princeton, NJ.
MacCallum, C. J., B. Nurnberger, and N. H. Barton. 1995. Experimental evidence for habitat dependent selection in a *Bombina* hybrid zone. Proceedings of the Royal Society of London Series B-Biological Sciences **260**:257–64.
MacCallum, C. J., B. Nurnberger, N. H. Barton, and J. M. Szymura. 1998. Habitat preference in the *Bombina* hybrid zone in Croatia. Evolution **52**:227–39.
Machado, C. A., T. S. Haselkorn, and M. A. F. Noor. 2007. Evaluation of the genomic extent of effects of fixed inversion differences on intraspecific variation and interspecific gene flow in *Drosophila pseudoobscura* and D-persimilis. Genetics **175**:1289–306.
Mackenzie, A. 1996. A trade-off for host plant utilization in the black bean aphid, Aphis fabae. Evolution **50**:155–62.
Macnair, M. R. and P. Christie. 1983. Reproductive isolation as a pleiotropic effect of copper tolerance in Mimulus-guttatus. Heredity **50**:295–302.
Makinen, H. S., M. Cano, and J. Merila. 2008a. Identifying footprints of directional and balancing selection in marine and freshwater three-spined stickleback (*Gasterosteus aculeatus*) populations. Molecular Ecology **17**:3565–82.
Makinen, H. S., T. Shikano, J. M. Cano, and J. Merila. 2008b. Hitchhiking mapping reveals a candidate genomic region for natural selection in three-spined stickleback chromosome VIII. Genetics **178**:453–65.
Mallet, J. 1989. The genetics of warning color in Peruvian hybrid zones of *Heliconius erato* and *Heliconius melpomene*. Proceedings of the Royal Society of London Series B-Biological Sciences **236**:163–185.
Mallet, J. 1995. A species definition for the modern synthesis. Trends in Ecology & Evolution **10**:294–9.
Mallet, J. 2006. What does *Drosophila* genetics tell us about speciation? Trends in Ecology & Evolution **21**:386–93.

Mallet, J. 2008a. Hybridization, ecological races and the nature of species: empirical evidence for the ease of speciation. Philosophical Transactions of the Royal Society B-Biological Sciences **363**:2971–86.

Mallet, J. 2008b. Mayr's view of Darwin: was Darwin wrong about speciation? Biological Journal of the Linnean Society **95**:3–16.

Mallet, J., N. Barton, G. Lamas, J. Santisteban, M. Muedas, and H. Eeley. 1990. Estimates of selection and gene flow from measures of cline width and linkage disequilibrium in *Heliconius* hybrid zones. Genetics **124**:921–36.

Mallet, J. and N. H. Barton. 1989. Strong natural selection in a warning color hybrid zone. Evolution **43**:421–31.

Mallet, J., M. Beltran, W. Neukirchen, and M. Linares. 2007. Natural hybridization in heliconiine butterflies: the species boundary as a continuum. BMC Evolutionary Biology **7**:28.

Mallet, J., W. O. McMillan, and C. D. Jiggins. 1998. Mimicry and warning color at the boundary between races and species. Pages 390–403 *in* D. J. Howard and B. S., editors. Endless forms: species and speciation. Oxford University Press, Oxford.

Mallet, J., A. Meyer, P. Nosil, and J. L. Feder. 2009. Space, sympatry and speciation. Journal of Evolutionary Biology **22**:2332–41.

Mani, G. S. and B. C. Clarke. 1990. Mutational order - a major stochastic process in evolution. Proceedings of the Royal Society of London Series B-Biological Sciences **240**:29–37.

Manoukis, N. C., J. R. Powell, M. B. Toure, A. Sacko, F. E. Edillo, M. B. Coulibaly, S. F. Traore, C. E. Taylor, and N. J. Besansky. 2008. A test of the chromosomal theory of ecotypic speciation in Anopheles gambiae. Proceedings of the National Academy of Sciences of the United States of America **105**:2940–5.

Marchetti, K. 1993. Dark habitats and bright birds illustrate the role of environment in species divergence. Nature **362**:149–52.

Marchinko, K. B. 2009. Predation's role in repeated phenotypic and genetic divergence of armor in threespine stickleback. Evolution **63**:127–38.

Margulies, M., M. Egholm, W. E. Altman, S. Attiya, J. S. Bader, L. A. Bemben, J. Berka, M. S. Braverman, Y. J. Chen, Z. T. Chen, S. B. Dewell, L. Du, J. M. Fierro, X. V. Gomes, B. C. Godwin, W. He, S. Helgesen, C. H. Ho, G. P. Irzyk, S. C. Jando, M. L. I. Alenquer, T. P. Jarvie, K. B. Jirage, J. B. Kim, J. R. Knight, J. R. Lanza, J. H. Leamon, S. M. Lefkowitz, M. Lei, J. Li, K. L. Lohman, H. Lu, V. B. Makhijani, K. E. McDade, M. P. McKenna, E. W. Myers, E. Nickerson, J. R. Nobile, R. Plant, B. P. Puc, M. T. Ronan, G. T. Roth, G. J. Sarkis, J. F. Simons, J. W. Simpson, M. Srinivasan, K. R. Tartaro, A. Tomasz, K. A. Vogt, G. A. Volkmer, S. H. Wang, Y. Wang, M. P. Weiner, P. G. Yu, R. F. Begley, and J. M. Rothberg. 2005. Genome sequencing in microfabricated high-density picolitre reactors. Nature **437**:376–80.

Markow, T. A. 1991. Sexual isolation among populations of *Drosophila mojavensis*. Evolution **45**:1525–9.

Maroja, L. S., J. A. Andres, and R. G. Harrison. 2009a. Genealogical discordance and patterns of introgression and selection across a cricket hybrid zone. Evolution **63**:2999–3015.

Maroja, L. S., J. A. Andres, J. R. Walters, and R. G. Harrison. 2009b. Multiple barriers to gene exchange in a field cricket hybrid zone. Biological Journal of the Linnean Society **97**:390–402.

Martin, O. Y. and D. J. Hosken. 2003. The evolution of reproductive isolation through sexual conflict. Nature **423**:979–82.

Martin, R. A. and D. W. Pfennig. 2009. Disruptive selection and the evolution of variation within species. Integrative and Comparative Biology **49**:E109–E109.
Matessi, C., A. Gimelfarb, and G. Gavrilets. 2001. Long term buildup of reproductive isolation promoted by disruptive selection: how far does it go? Selection **2**:41–64.
Matsubayashi, K. W. and H. Katakura. 2009. Contribution of multiple isolating barriers to reproductive isolation between a pair of phytophagous ladybird beetles. Evolution **63**:2563–80.
Matsubayashi, K. W., I. Ohshima, and P. Nosil. 2010. Ecological speciation in phytophagous insects. Entomologia Experimentalis Et Applicata **134**:1–27.
Matsuo, T., S. Sugaya, J. Yasukawa, T. Aigaki, and Y. Fuyama. 2007. Odorant-binding proteins OBP57d and OBP57e affect taste perception and host plant preference in *Drosophila sechellia*. Plos Biology **5**:985–96.
Matute, D. R. 2010. Reinforcement can overcome gene flow during speciation in *Drosophila* Current Biology **20**:2229–33.
Matute, D. R., C. J. Novak, and J. A. Coyne. 2009. Temperature based extrinsic reproductive isolation in two species of *Drosophila*. Evolution **63**:595–612.
May, R. M., J. A. Endler, and R. E. McMurtrie. 1975. Gene frequency clines in presence of selection opposed by gene flow. American Naturalist **109**:659–76.
Maynard-Smith, J. 1966. Sympatric speciation. American Naturalist **100**:637–50.
Mayr, E. 1942. Systematics and the origin of species. Columbia University Press, New York.
Mayr, E. 1947. Ecological Factors in Speciation. Evolution **1**:263–88.
Mayr, E. 1963. Animal species and evolution. Harvard University Press, Harvard, MA.
McAllister, B. F., S. L. Sheeley, P. A. Mena, A. L. Evans, and C. Schlotterer. 2008. Clinal distribution of a chromosomal rearrangement: A precursor to chromosomal speciation? Evolution **62**:1852–65.
McBride, C. S. and M. C. Singer. 2010. Field studies reveal strong postmating isolation between ecologically divergent butterfly populations. Plos Biology **8**:e1000529.
McClintock, W. J. and G. W. Uetz. 1996. Female choice and pre-existing bias: Visual cues during courtship in two Schizocosa wolf spiders (Araneae: Lycosidae). Animal Behaviour **52**:167–81.
McCormack, J. E., A. J. Zellmer, and L. L. Knowles. 2010. Does niche divergence accompany allopatric divergence in *Aphelocoma* jays as predicted under ecological speciation?: insights from tests with niche models. Evolution **64**:1231–44.
McDonald, C. G., T. E. Reimchen, and C. W. Hawryshyn. 1995. Nuptial colour loss and signal masking in *Gasterosteus*: An analysis using video imaging. Behaviour **132**:963–77.
McGaugh, S. E., C. A. Machado, and M. A. F. Noor. 2011. Genomic impacts of chromosomal inversions in parapatric species of *Drosophila*. Philosophical Transactions of the Royal Society B-Biological Sciences, in press.
McKinnon, J. S., S. Mori, B. K. Blackman, L. David, D. M. Kingsley, L. Jamieson, J. Chou, and D. Schluter. 2004. Evidence for ecology's role in speciation. Nature **429**:294–8.
McKinnon, J. S. and M. E. R. Pierotti. 2010. Colour polymorphism and correlated characters: genetic mechanisms and evolution. Molecular Ecology **19**:5101–25.
McMillan, W. O., C. D. Jiggins, and J. Mallet. 1997. What initiates speciation in passion-vine butterflies? Proceedings of the National Academy of Sciences of the United States of America **94**:8628–33.

McNett, G. D. and R. B. Cocroft. 2008. Host shifts favor vibrational signal divergence in *Enchenopa binotata* treehoppers. Behavioral Ecology **19**:650–6.

McPeek, M. A. and G. A. Wellborn. 1998. Genetic variation and reproductive isolation among phenotypically divergent amphipod populations. Limnology and Oceanography **43**:1162–9.

McPheron, B. A., D. C. Smith, and S. H. Berlocher. 1988. Genetic differences between host races of *Rhagoletis pomonella*. Nature **336**:64–6.

Mendelson, T. C. 2003. Sexual isolation evolves faster than hybrid inviability in a diverse and sexually dimorphic genus of fish (Percidae: Etheostoma). Evolution **57**:317–27.

Mendelson, T. C., B. D. Inouye, and M. D. Rausher. 2004. Quantifying patterns in the evolution of reproductive isolation. Evolution **58**:1424–33.

Merrill, R. M., Z. Gompert, L. M. Dembeck, M. R. Kronforst, W. O. McMillan, and C. D. Jiggins. 2011. Mate preference across the speciation continuum in a clade of mimetic butterflies. Evolution **65**:1489–500.

Meyer, J. R. and R. Kassen. 2007. The effects of competition and predation on diversification in a model adaptive radiation. Nature **446**:432–5.

Michel, A. P., S. Sim, T. H. Q. Powell, M. S. Taylor, P. Nosil, and J. L. Feder. 2010. Widespread genomic divergence during sympatric speciation. Proceedings of the National Academy of Sciences **107**:9724–9.

Miklas, N., N. Stritih, A. Cokl, M. Virant-Doberlet, and M. Renou. 2001. The influence of substrate on male responsiveness to the female calling song in Nezara viridula. Journal of Insect Behavior **14**:313–32.

Mims, M. C., C. D. Hulsey, B. M. Fitzpatrick, and J. T. Streelman. 2010. Geography disentangles introgression from ancestral polymorphism in Lake Malawi cichlids. Molecular Ecology **19**:940–51.

Moran, N. A. 1986. Morphological adaptation to host plants in *Uroleucon* (Homoptera, Aphididae). Evolution **40**:1044–50.

Morjan, C. L. and L. H. Rieseberg. 2004. How species evolve collectively: implications of gene flow and selection for the spread of advantageous alleles. Molecular Ecology **13**:1341–56.

Morton, E. S. 1975. Ecological sources of selection on avian songs. American Naturalist **109**:17–34.

Moyle, L. C., M. S. Olson, and P. Tiffin. 2004. Patterns of reproductive isolation in three angiosperm genera. Evolution **58**:1195–208.

Muller, H. J. 1942. Isolating mechanisms, evolution, and temerature. Biology Symposium **6**:71–125.

Muster, C., W. P. Maddison, S. Uhlmann, T. U. Berendonk, and A. P. Vogler. 2009. Arctic-alpine distributions-metapopulations on a continental scale? American Naturalist **173**:313–26.

Nachman, M. W. and B. A. Payseur. 2011. Recombination rate variation and speciation: theoretical predictions and empirical results from rabbits and mice. Philosophical Transactions of the Royal Society B-Biological Sciences, in press.

Nagel, L. and D. Schluter. 1998. Body size, natural selection, and speciation in sticklebacks. Evolution **52**:209–18.

Nagy, E. S. 1997a. Frequency-dependent seed production and hybridization rates: Implications for gene flow between locally adapted plant populations. Evolution **51**:703–14.

Nagy, E. S. 1997b. Selection for native characters in hybrids between two locally adapted plant subspecies. Evolution **51**:1469–80.

Nagy, E. S. and K. J. Rice. 1997. Local adaptation in two subspecies of an annual plant: Implications for migration and gene flow. Evolution **51**:1079–89.

Naisbit, R. E., C. D. Jiggins, and J. Mallet. 2001. Disruptive sexual selection against hybrids contributes to speciation between *Heliconius cydno* and *Heliconius melpomene*. Proceedings of the Royal Society of London Series B-Biological Sciences **268**:1849–54.

Nakazato, T., M. K. Jung, E. A. Housworth, L. H. Rieseberg, and G. J. Gastony. 2007. A genomewide study of reproductive barriers between allopatric Populations of a homosporous fern, Ceratopteris richardii. Genetics **177**:1141–50.

Navarro, A. and N. H. Barton. 2003. Accumulating postzygotic isolation genes in parapatry: A new twist on chromosomal speciation. Evolution **57**:447–59.

Near, T. J., D. I. Bolnick, and P. C. Wainwright. 2005. Fossil calibrations and molecular divergence time estimates in centrarchid fishes (Teleostei: Centrarchidae). Evolution **59**:1768–82.

Nei, M. and W. H. Li. 1973. Linkage disequilibrium in subdivided populations. Genetics **75**:213–19.

Nevins, J. R. and A. Potti. 2007. Mining gene expression profiles: expression signatures as cancer phenotypes. Nature Reviews Genetics **8**:601–9.

Nielsen, R. 2005. Molecular signatures of natural selection. Annual Review of Genetics **39**:197–218.

Nielsen, R. and J. Wakeley. 2001. Distinguishing migration from isolation: A Markov chain Monte Carlo approach. Genetics **158**:885–96.

Niemiller, M. L., B. M. Fitzpatrick, and B. T. Miller. 2008. Recent divergence with gene flow in Tennessee cave salamanders (*Plethodontidae: Gyrinophilus*) inferred from gene genealogies. Molecular Ecology **17**:2258–75.

Niemiller, M. L., P. Nosil, and B. M. Fitzpatrick. 2010. Recent divergence-with-gene-flow in Tennessee cave salamanders (Plethodontidae; Gyrinophilus) inferred from gene genealogies (vol 17, pg 2258, 2008). Molecular Ecology **19**:1513–14.

Nokkala, C. and S. Nokkala. 1998. Species and habitat races in the chrysomelid Galerucella nymphaeae species complex in northern Europe. Entomologia Experimentalis Et Applicata **89**:1–13.

Nolte, A. W., Z. Gompert, and C. A. Buerkle. 2009. Variable patterns of introgression in two sculpin hybrid zones suggest that genomic isolation differs among populations. Molecular Ecology **18**:2615–27.

Noor, M. A. 1995. Speciation driven by natural selection in *Drosophila*. Nature **375**:674–5.

Noor, M. A. F. and S. M. Bennett. 2010. Islands of speciation or mirages in the desert? Examining the role of restricted recombination in maintaining species (vol 103, pg 434, 2009). Heredity **104**:418–418.

Noor, M. A. F. and J. L. Feder. 2006. Speciation genetics: evolving approaches. Nature Reviews Genetics **7**:851–61.

Noor, M. A. F., D. A. Garfield, S. W. Schaeffer, and C. A. Machado. 2007. Divergence between the *Drosophila pseudoobscura* and D-persimilis genome sequences in relation to chromosomal inversions. Genetics **177**:1417–28.

Noor, M. A. F., K. L. Grams, L. A. Bertucci, and J. Reiland. 2001. Chromosomal inversions and the reproductive isolation of species. Proceedings of the National Academy of Sciences of the United States of America **98**:12084–8.

Nosil, P. 2004. Reproductive isolation caused by visual predation on migrants between divergent environments. Proceedings of the Royal Society of London Series B-Biological Sciences **271**:1521–8.

Nosil, P. 2007. Divergent host plant adaptation and reproductive isolation between ecotypes of *Timema cristinae* walking sticks. American Naturalist **169**:151–62.

Nosil, P. 2008a. Ernst Mayr and the integration of geographic and ecological factors in speciation. Biological Journal of the Linnean Society **95**:26–46.

Nosil, P. 2008b. Speciation with gene flow could be common. Molecular Ecology **17**:2103–6.

Nosil, P. 2009. Adaptive population divergence in cryptic color-pattern following a reduction in gene flow. Evolution **63**:1902–12.

Nosil, P. and B. J. Crespi. 2004. Does gene flow constrain adaptive divergence or vice versa? A test using ecomorphology and sexual isolation in *Timema cristinae* walking-sticks. Evolution **58**:102–12.

Nosil, P. and B. J. Crespi. 2006a. Ecological divergence promotes the evolution of cryptic reproductive isolation. Proceedings of the Royal Society B-Biological Sciences **273**:991–7.

Nosil, P. and B. J. Crespi. 2006b. Experimental evidence that predation promotes divergence in adaptive radiation. Proceedings of the National Academy of Sciences of the United States of America **103**:9090–5.

Nosil, P., B. J. Crespi, R. Gries, and G. Gries. 2007. Natural selection and divergence in mate preference during speciation. Genetica **129**:309–27.

Nosil, P., B. J. Crespi, and C. P. Sandoval. 2002. Host-plant adaptation drives the parallel evolution of reproductive isolation. Nature **417**:440–3.

Nosil, P., B. J. Crespi, and C. P. Sandoval. 2003. Reproductive isolation driven by the combined effects of ecological adaptation and reinforcement. Proceedings of the Royal Society of London Series B-Biological Sciences **270**:1911–18.

Nosil, P., B. J. Crespi, C. P. Sandoval, and M. Kirkpatrick. 2006a. Migration and the genetic covariance between habitat preference and performance. American Naturalist **167**:E66–E78.

Nosil, P., S. P. Egan, and D. J. Funk. 2008. Heterogeneous genomic differentiation between walking-stick ecotypes: "Isolation by adaptation" and multiple roles for divergent selection. Evolution **62**:316–36.

Nosil, P. and S. M. Flaxman. 2011. Conditions for mutation-order speciation. Proceedings of the Royal Society B-Biological Sciences **278**:399–407.

Nosil, P., D. J. Funk, and D. Ortíz-Barrientos. 2009a. Divergent selection and heterogeneous genomic divergence. Molecular Ecology **18**:375–402.

Nosil, P. and L. J. Harmon. 2009. Niche dimensionality and ecological speciation. Pages 127–54 *in* R. Butlin, J. Bridle, and D. Schluter, editors. Speciation and Patterns of Diversity. Cambridge University Press, Cambridge, UK.

Nosil, P., L. J. Harmon, and O. Seehausen. 2009b. Ecological explanations for (incomplete) speciation. Trends in Ecology & Evolution **24**:145–56.

Nosil, P. and H. D. Rundle. 2009. Ecological speciation: natural selection and the formation of new species. Pages 134–42. *in* S. Levin, editor. Princeton Guide to Ecology. Princeton University Press, Princeton, N.J.

Nosil, P. and C. P. Sandoval. 2008. Ecological niche dimensionality and the evolutionary diversification of stick insects. Plos One **3**:e1907.

Nosil, P., C. P. Sandoval, and B. J. Crespi. 2006b. The evolution of host preference in allopatric vs. parapatric populations of *Timema* cristinae walking-sticks. Journal of Evolutionary Biology **19**:929–42.

Nosil, P. and D. Schluter. 2011. The genes underlying the process of speciation. Trends in Ecology & Evolution **26**:160–7.

Nosil, P., T. H. Vines, and D. J. Funk. 2005. Perspective: Reproductive isolation caused by natural selection against immigrants from divergent habitats. Evolution **59**:705–19.

Nosil, P. and R. Yukilevich. 2008. Mechanisms of reinforcement in natural and simulated polymorphic populations. Biological Journal of the Linnean Society **95**:305–19.

Nygren, G. H., S. Nylin, and C. Stefanescu. 2006. Genetics of host plant use and life history in the comma butterfly across Europe: varying modes of inheritance as a potential reproductive barrier. Journal of Evolutionary Biology **19**:1882–93.

Nyman, T., V. Vikberg, D. R. Smith, and J.-L. Boeve. 2010. How common is ecological speciation in plant-feeding insects? A "Higher" Nematinae perspective. BMC Evolutionary Biology **10**:266.

Ogden, R. and R. S. Thorpe. 2002. Molecular evidence for ecological speciation in tropical habitats. Proceedings of the National Academy of Sciences of the United States of America **99**:13612–15.

Ohshima, I. 2008. Host race formation in the leaf-mining moth Acrocercops transecta (Lepidoptera: Gracillariidae). Biological Journal of the Linnean Society **93**:135–45.

Ohshima, I. 2010. Host-associated pre-mating reproductive isolation between host races of Acrocercops transecta: mating site preferences and effect of host presence on mating. Ecological Entomology **35**:253–7.

Ohshima, I. and K. Yoshizawa. 2010. Differential introgression causes genealogical discordance in host races of Acrocercops transecta (Insecta: Lepidoptera). Molecular Ecology **19**:2106–19.

Ohta, T. 1992. The nearly neutral theory of molecular evolution. Annual Review of Ecology and Systematics **23**:263–86.

Orr, H. A. 1995. The population-genetics of speciation - the evolution of hybrid incompatibilities. Genetics **139**:1805–13.

Orr, H. A. 2000. Adaptation and the cost of complexity. Evolution **54**:13–20.

Orr, H. A. 2005a. The genetic basis of reproductive isolation: Insights from Drosophila. Proceedings of the National Academy of Sciences of the United States of America **102**:6522–6.

Orr, H. A. 2005b. The genetic theory of adaptation: A brief history. Nature Reviews Genetics **6**:119–27.

Orr, H. A. and A. J. Betancourt. 2001. Haldane's sieve and adaptation from the standing genetic variation. Genetics **157**:875–84.

Orr, H. A., J. P. Masly, and D. C. Presgraves. 2004. Speciation genes. Current Opinion in Genetics & Development **14**:675–9.

Orr, H. A. and D. C. Presgraves. 2000. Speciation by postzygotic isolation: forces, genes and molecules. Bioessays **22**:1085–94.

Orr, H. A. and M. Turelli. 2001. The evolution of postzygotic isolation: Accumulating Dobzhansky-Muller incompatibilities. Evolution **55**:1085–94.

Ortiz-Barrientos, D., B. A. Counterman, and M. A. F. Noor. 2007. Gene expression divergence and the origin of hybrid dysfunctions. Genetica **129**:71–81.

Ortiz-Barrientos, D., A. Grealy, and P. Nosil. 2009. The genetics and ecology of reinforcement implications for the evolution of prezygotic isolation in sympatry and beyond. Annals of the New York Academy of Sciences **1168**:156–182.

Ortíz-Barrientos, D. and M. A. F. Noor. 2005. Evidence for a one-allele assortative mating locus. Science **310**:1467–1467.

Ortiz-Barrientos, D., J. Reiland, J. Hey, and M. A. F. Noor. 2002. Recombination and the divergence of hybridizing species. Genetica **116**:167–78.

Otte, D. and J. A. Endler, editors. 1989. Speciation and its consequences. Sinauer Associates, Inc., Sunderland, MA.

Otto, S. P. 2004. Two steps forward, one step back: the pleiotropic effects of favoured alleles. Proceedings of the Royal Society of London Series B-Biological Sciences **271**:705–14.

Otto, S. P., M. R. Servedio, and S. L. Nuismer. 2008. Frequency-dependent selection and the evolution of assortative mating. Genetics **179**:2091–112.

Panhuis, T. M., R. Butlin, M. Zuk, and T. Tregenza. 2001. Sexual selection and speciation. Trends in Ecology & Evolution **16**:364–71.

Panova, M., J. Hollander, and K. Johannesson. 2006. Site-specific genetic divergence in parallel hybrid zones suggests nonallopatric evolution of reproductive barriers. Molecular Ecology **15**:4021–31.

Pappers, S. M., G. van der Velde, and N. J. Ouborg. 2002a. Host preference and larval performance suggest host race formation in Galerucella nymphaeae. Oecologia **130**:433–40.

Pappers, S. M., G. Van der Velde, N. J. Ouborg, and J. M. Van Groenendael. 2002b. Genetically based polymorphisms in morphology and life history associated with putative host races of the water lily leaf beetle, Galerucella nymphaeae. Evolution **56**:1610–21.

Paterson, S., T. Vogwill, A. Buckling, R. Benmayor, A. J. Spiers, N. R. Thomson, M. Quail, F. Smith, D. Walker, B. Libberton, A. Fenton, N. Hall, and M. A. Brockhurst. 2010. Antagonistic coevolution accelerates molecular evolution. Nature **464**:275–U154.

Patten, M. A., J. T. Rotenberry, and M. Zuk. 2004. Habitat selection, acoustic adaptation, and the evolution of reproductive isolation. Evolution **58**:2144–55.

Pavey, S. A., H. Collin, P. Nosil, and S. Rogers. 2010a. The role of gene expression in ecological speciation. Year in Evolutionary Biology 1206:110–129.

Pavey, S. A., J. L. Nielsen, and T. R. Hamon. 2010b. Recent ecological divergence despite migration in sockeye salmon (*Oncorhynchus nerka*). Evolution **64**:1773–83.

Payseur, B. A., J. G. Krenz, and M. W. Nachman. 2004. Differential patterns of introgression across the X chromosome in a hybrid zone between two species of house mice. Evolution **58**:2064–78.

Peccoud, J., A. Ollivier, M. Plantegenest, and J. C. Simon. 2009. A continuum of genetic divergence from sympatric host races to species in the pea aphid complex. Proceedings of the National Academy of Sciences of the United States of America **106**:7495–500.

Pelletier, F., D. Garant, and A. P. Hendry. 2009. Eco-evolutionary dynamics Introduction. Philosophical Transactions of the Royal Society B-Biological Sciences **364**:1483–9.

Pena-Castillo, L. and T. R. Hughes. 2007. Why are there still over 1000 uncharacterized yeast genes? Genetics **176**:7–14.

Pennings, P. S. and J. Hermisson. 2006a. Soft sweeps II-molecular population genetics of adaptation from recurrent mutation or migration. Molecular Biology and Evolution **23**:1076–84.

Pennings, P. S. and J. Hermisson. 2006b. Soft sweeps III: The signature of positive selection from recurrent mutation. Plos Genetics **2**:1998–2012.

Pennings, P. S., M. Kopp, G. Meszena, U. Dieckmann, and J. Hermisson. 2008. An analytically tractable model for competitive speciation. American Naturalist **171**:E44–E71.

Peterson, M. A., B. M. Honchak, S. E. Locke, T. E. Beeman, J. Mendoza, J. Green, K. J. Buckingham, M. A. White, and K. J. Monsen. 2005. Relative abundance and the species-specific reinforcement of male mating preference in the Chrysochus (Coleoptera: Chrysomelidae) hybrid zone. Evolution **59**:2639–55.

Pfennig, D. W. and A. M. Rice. 2007. An experimental test of character displacement's role in promoting postmating isolation between conspecific populations in contrasting competitive environments. Evolution **61**:2433–43.

Pfennig, D. W., A. M. Rice, and R. A. Martin. 2007. Field and experimental evidence for competition's role in phenotypic divergence. Evolution **61**:257–71.

Pfennig, D. W., M. A. Wund, E. C. Snell-Rood, T. Cruickshank, C. D. Schlichting, and A. P. Moczek. 2010. Phenotypic plasticity's impacts on diversification and speciation. Trends in Ecology & Evolution **25**:459–67.

Pfennig, K. S. 2007. Facultative mate choice drives adaptive hybridization. Science **318**:965–7.

Pfennig, K. S. and D. W. Pfennig. 2005. Character displacement as the "best of a bad situation": Fitness trade-offs resulting from selection to minimize resource and mate competition. Evolution **59**:2200–8.

Pfennig, K. S. and M. J. Ryan. 2006. Reproductive character displacement generates reproductive isolation among conspecific populations: an artificial neural network study. Proceedings of the Royal Society B-Biological Sciences **273**:1361–8.

Phadnis, N. and H. A. Orr. 2009. A single gene causes both male sterility and segregation distortion in Drosophila hybrids. Science **323**:376–9.

Pialek, J. and N. H. Barton. 1997. The spread of an advantageous allele across a barrier: The effects of random drift and selection against heterozygotes. Genetics **145**:493–504.

Pigot, A. L., A. B. Phillimore, I. P. F. Owens, and C. D. L. Orme. 2010. The shape and temporal dynamics of phylogenetic trees arising from geographic speciation. Systematic Biology **59**:660–73.

Pinho, C. and J. Hey. 2010. Divergence with gene flow: models and data. Pages 215–30 Annual Review of Ecology, Evolution, and Systematics, Vol 41.

Podos, J. 2001. Correlated evolution of morphology and vocal signal structure in Darwin's finches. Nature **409**:185–8.

Podos, J. and A. P. Hendry, editors. 2006. The biomechanics of ecological speciation. Taylor and Francis, New York.

Powell, J. R. 1997. Progress and prospects in evolutionary biology: the Drosophila model. Oxford University Press, New York.

Preisser, E. L. and D. I. Bolnick. 2008. When predators don't eat their prey: Nonconsumptive predator effects on prey dynamics(1). Ecology **89**:2414–15.

Presgraves, D. C. 2003. A fine-scale genetic analysis of hybrid Incompatibilities in drosophila. Genetics **163**:955–72.

Presgraves, D. C. 2007a. Does genetic conflict drive rapid molecular evolution of nuclear transport genes in Drosophila? Bioessays **29**:386–91.

Presgraves, D. C. 2007b. Speciation genetics: Epistasis, conflict and the origin of species. Current Biology **17**:R125–R127.

Presgraves, D. C., L. Balagopalan, S. M. Abmayr, and H. A. Orr. 2003. Adaptive evolution drives divergence of a hybrid inviability gene between two species of Drosophila. Nature **423**:715–19.

Presgraves, D. C. and W. Stephan. 2007. Pervasive adaptive evolution among interactors of the Drosophila hybrid inviability gene, Nup96. Molecular Biology and Evolution **24**:306–14.

Price, C. S. C., C. H. Kim, C. J. Gronlund, and J. A. Coyne. 2001. Cryptic reproductive isolation in the Drosophila simulans species complex. Evolution **55**:81–92.

Price, T. D. 2007. Speciation in Birds. Roberts and Company, Woodbury NY.

Price, T. D., A. Qvarnstrom, and D. E. Irwin. 2003. The role of phenotypic plasticity in driving genetic evolution. Proceedings of the Royal Society of London Series B-Biological Sciences **270**:1433–40.

Pritchard, J. K., M. Stephens, and P. Donnelly. 2000. Inference of population structure using multilocus genotype data. Genetics **155**:945–59.

Pritchard, J. R. and D. Schluter. 2001. Declining interspecific competition during character displacement: Summoning the ghost of competition past. Evolutionary Ecology Research **3**:209–20.

Przeworski, M. 2003. Estimating the time since the fixation of a beneficial allele. Genetics **164**:1667–76.

Puebla, O., E. Bermingham, F. Guichard, and E. Whiteman. 2007. Colour pattern as a single trait driving speciation in Hypoplectrus coral reef fishes? Proceedings of the Royal Society B-Biological Sciences **274**:1265–71.

Putnam, A. S., J. M. Scriber, and P. Andolfatto. 2007. Discordant divergence times among Z-chromosome regions between two ecologically distinct swallowtail butterfly species. Evolution **61**:912–27.

Quesada, H., D. Posada, A. Caballero, P. Moran, and E. Rolan-Alvarez. 2007. Phylogenetic evidence for multiple sympatric ecological diversification in a marine snail. Evolution **61**:1600–12.

Raeymaekers, J. A. M., M. Boisjoly, L. Delaire, D. Berner, K. Rasanen, and A. P. Hendry. 2010. Testing for mating isolation between ecotypes: laboratory experiments with lake, stream and hybrid stickleback. Journal of Evolutionary Biology **23**:2694–708.

Ramsey, J., H. D. Bradshaw, and D. W. Schemske. 2003. Components of reproductive isolation between the monkeyflowers Mimulus lewisii and M-cardinalis (Phrymaceae). Evolution **57**:1520–34.

Rand, D. M. and R. G. Harrison. 1989. Ecological genetics of a mosaic hybrid zone - mitochondrial, nuclear, and reproductive differentiation of crickets by soil type. Evolution **43**:432–49.

Ranz, J. M. and C. A. Machado. 2006. Uncovering evolutionary patterns of gene expression using microarrays. Trends in Ecology & Evolution **21**:29–37.

Räsänen, K. and A. P. Hendry. 2008. Disentangling interactions between adaptive divergence and gene flow when ecology drives diversification. Ecology Letters **11**:624–36.

Ratcliffe, L. M. and P. R. Grant. 1983. Species recognition in Darwins finches (*Geospiza*, Gould). 1. Discrimination by morphological cues. Animal Behaviour **31**:1139–53.

Rausher, M. D. 1984. Tradeoffs in performance on difference hosts - evidence from within-site and between-site variation in the beetle *Deloyala guttata*. Evolution **38**:582–95.

Rausher, M. D. 2007. Comment on "Evolutionary paths underlying flower color variation in Antirrhinum". Science **315**:461.

Reimchen, T. E. 1980. Spine deficiency and polymorphism in a population of *Gasterosteus aculeatus* - an adaptation to predators. Canadian Journal of Zoology-Revue Canadienne De Zoologie **58**:1232–44.

Reimchen, T. E. 1983. Structural relationships between spines and lateral plate in threespine stickleback (*Gasterosteus aculeatus*). Evolution **37**:931–46.

Reimchen, T. E. 1989. Loss of nuptial color in threespine stickleback (*Gasterosteus aculeatus*). Evolution **43**:450–60.

Reimchen, T. E. 1995. Predator-induced cyclical changes in lateral plate frequencies of Gasterosteus. Behaviour **132**:1079–94.

Reimchen, T. E. 2000. Predator handling failures of lateral plate morphs in *Gasterosteus aculeatus*: Functional implications for the ancestral plate condition. Behaviour **137**:1081–96.

Reimchen, T. E. and P. Nosil. 2002. Temporal variation in divergent selection on spine number in threespine stickleback. Evolution **56**:2472–83.

Reimchen, T. E. and P. Nosil. 2004. Variable predation regimes predict the evolution of sexual dimorphism in a population of threespine stickleback. Evolution **58**:1274–81.

Reissig, W. H. and D. C. Smith. 1978. Bionomics of *Rhagoletis pomonella* (Diptera, Tephritidae) in *Crataegus*. Annals of the Entomological Society of America **71**:155–9.

Renaut, S., A. W. Nolte, and L. Bernatchez. 2009. Gene expression divergence and hybrid misexpression between lake whitefish species pairs (*Coregonus* spp. Salmonidae). Molecular Biology and Evolution **26**:925–36.

Reynolds, R. G. and B. M. Fitzpatrick. 2007. Assortative mating in poison-dart frogs based on an ecologically important trait. Evolution **61**:2253–9.

Rice, A. M. and D. W. Pfennig. 2008. Analysis of range expansion in two species undergoing character displacement: why might invaders generally "win" during character displacement? Journal of Evolutionary Biology **21**:696–704.

Rice, A. M. and D. W. Pfennig. 2010. Does character displacement initiate speciation? Evidence of reduced gene flow between populations experiencing divergent selection. Journal of Evolutionary Biology **23**:854–65.

Rice, W. R. 1984. Disruptive selection on habitat preference and the evolution of reproductive isolation - a simulation study. Evolution **38**:1251–60.

Rice, W. R. 1998. Intergenomic conflict, interlocus antagonistic coevolution, and the evolution of reproductive isolation. Pages 261–70 *in* D. J. Howard and S. H. Berlocher, editors. Endless Forms: species and speciation. Oxford University Press, New York.

Rice, W. R. and E. E. Hostert. 1993. Laboratory experiments on speciation - what have we learned in 40 years? Evolution **47**:1637–53.

Rice, W. R. and G. W. Salt. 1990. The evolution of reproductive isolation as a correlated character under sympatric conditions - experimental evidence. Evolution **44**:1140–52.

Richmond, J. Q. and T. W. Reeder. 2002. Evidence for parallel ecological speciation in scincid lizards of the Eumeces skiltonianus species group (Squamata: Scincidae). Evolution **56**:1498–513.

Riechert, S. E. 1993. Investigation of potential gene flow limitation of behavioral adaptation in an aridlands spider. Behavioral Ecology and Sociobiology **32**:355–63.

Riechert, S. E. and R. F. Hall. 2000. Local population success in heterogeneous habitats: reciprocal transplant experiments completed on a desert spider. Journal of Evolutionary Biology **13**:541–50.

Riechert, S. E., F. D. Singer, and T. C. Jones. 2001. High gene flow levels lead to gamete wastage in a desert spider system. Genetica **112**:297–319.

Rieseberg, L. H. 2001. Chromosomal rearrangements and speciation. Trends in Ecology & Evolution **16**:351–8.

Rieseberg, L. H., M. A. Archer, and R. K. Wayne. 1999a. Transgressive segregation, adaptation and speciation. Heredity **83**:363–72.

Rieseberg, L. H. and B. K. Blackman. 2010. Speciation genes in plants. Annals of Botany **106**:439–55.

Rieseberg, L. H., A. M. Desrochers, and S. J. Youn. 1995. Interspecific pollen competition as a reproductive barrier between sympatric species of *Helianthus* (Asteraceae). American Journal of Botany **82**:515–19.

Rieseberg, L. H., O. Raymond, D. M. Rosenthal, Z. Lai, K. Livingstone, T. Nakazato, J. L. Durphy, A. E. Schwarzbach, L. A. Donovan, and C. Lexer. 2003. Major ecological transitions in wild sunflowers facilitated by hybridization. Science **301**:1211–16.

Rieseberg, L. H., J. Whitton, and K. Gardner. 1999b. Hybrid zones and the genetic architecture of a barrier to gene flow between two sunflower species. Genetics **152**:713–27.

Rieseberg, L. H. and J. H. Willis. 2007. Plant speciation. Science **317**:910–14.

Rocha, L. A., D. R. Robertson, J. Roman, and B. W. Bowen. 2005. Ecological speciation in tropical reef fishes. Proceedings of the Royal Society B-Biological Sciences **272**:573–9.

Rodriguez, R. L., L. E. Sullivan, and R. B. Cocroft. 2004. Vibrational communication and reproductive isolation in the Enchenopa binotata species complex of treehoppers (Hemiptera: Membracidae). Evolution **58**:571–8.

Rodriguez, R. L., L. M. Sullivan, R. L. Snyder, and R. B. Cocroft. 2008. Host shifts and the beginning of signal divergence. Evolution **62**:12–20.

Roelofs, D., T. K. S. Janssens, M. Timmermans, B. Nota, J. Marien, Z. Bochdanovits, B. Ylstra, and N. M. Van Straalen. 2009. Adaptive differences in gene expression associated with heavy metal tolerance in the soil arthropod *Orchesella cincta*. Molecular Ecology **18**:3227–39.

Roff, D. A. 2007. A centennial celebration for quantitative genetics. Evolution **61**:1017–32.

Rogers, S. M. and L. Bernatchez. 2007a. The genetic architecture of ecological speciation and the association with signatures of selection in natural lake whitefish (Coregonas sp Salmonidae) species pairs. Molecular Biology and Evolution **24**:1423–38.

Rogers, S. M. and L. Bernatchez. 2007b. The genetic architecture of ecological speciation and the association with signatures of selection in natural lake whitefish (*Coregonas* sp. Salmonidae) species pairs. Molecular Biology and Evolution **24**:1423–38.

Rolan-Alvarez, E. 2007. Sympatric speciation as a by-product of ecological adaptation in the galician Littorina saxatilis hybrid zone. Journal of Molluscan Studies **73**:1–10.

Rolan-Alvarez, E., K. Johannesson, and J. Erlandsson. 1997. The maintenance of a cline in the marine snail Littorina saxatilis: The role of home site advantage and hybrid fitness. Evolution **51**:1838–47.

Rosenzweig, M. L. 1978. Competitive speciation. Biological Journal of the Linnean Society **10**:275–89.

Rosenzweig, M. L. 2001. Loss of speciation rate will impoverish future diversity. Proceedings of the National Academy of Sciences of the United States of America **98**:5404–10.

Roughgarden, J. 1972. Evolution of niche width. American Naturalist **106**:683-&.

Rousset, F. 1997. Genetic differentiation and estimation of gene flow from F-statistics under isolation by distance. Genetics **145**:1219–28.

Rowe, L. and D. Houle. 1996. The lek paradox and the capture of genetic variance by condition dependent traits. Proceedings of the Royal Society of London Series B-Biological Sciences **263**:1415–21.

Rueffler, C., T. J. M. Van Dooren, O. Leimar, and P. A. Abrams. 2006. Disruptive selection and then what? Trends in Ecology & Evolution **21**:238–45.

Ruegg, K. 2008. Genetic, morphological, and ecological characterization of a hybrid zone that spans a migratory divide. Evolution **62**:452–66.

Rundell, R. J. and T. D. Price. 2009. Adaptive radiation, nonadaptive radiation, ecological speciation and nonecological speciation. Trends in Ecology & Evolution **24**:394–9.

Rundle, H. D. 2002. A test of ecologically dependent postmating isolation between sympatric sticklebacks. Evolution **56**:322–9.

Rundle, H. D. 2003. Divergent environments and population bottlenecks fail to generate premating isolation in Drosophila pseudoobscura. Evolution **57**:2557–65.

Rundle, H. D., S. F. Chenoweth, P. Doughty, and M. W. Blows. 2005. Divergent selection and the evolution of signal traits and mating preferences. Plos Biology **3**:1988–95.

Rundle, H. D., L. Nagel, J. W. Boughman, and D. Schluter. 2000. Natural selection and parallel speciation in sympatric sticklebacks. Science **287**:306–8.

Rundle, H. D. and P. Nosil. 2005. Ecological speciation. Ecology Letters **8**:336–52.

Rundle, H. D. and D. Schluter. 1998. Reinforcement of stickleback mate preferences: Sympatry breeds contempt. Evolution **52**:200–8.

Rundle, H. D. and D. Schluter. 2004. Natural selection and ecological speciation in sticklebacks. Pages 192–209 *in* U. Dieckmann, M. Doebeli, J. A. J. Metz, and D. Tautz, editors. Adaptive Speciation. Cambridge University Press, Cambridge.

Rundle, H. D., S. M. Vamosi, and D. Schluter. 2003. Experimental test of predation's effect on divergent selection during character displacement in sticklebacks. Proceedings of the National Academy of Sciences of the United States of America **100**:14943–8.

Rundle, H. D. and M. C. Whitlock. 2001. A genetic interpretation of ecologically dependent isolation. Evolution **55**:198–201.

Ryan, M. J. 1991. Sexual selection and communication in frogs. Trends in Ecology & Evolution **6**:351–5.

Ryan, M. J. 2010. The túngara frog: A model for sexual selection and communication. Pages 453–61 *in* M. D. Breed and J. Moore, editors. Encyclopedia of Animal Behavior. Academic Press, Oxford.

Ryan, M. J., R. B. Cocroft, and W. Wilczynski. 1990. The role of environmental selection in intraspecific divergence of mate recognition signals in the cricket frog, *Acris crepitans*. Evolution **44**:1869–72.

Ryan, M. J. and A. S. Rand. 1993. Species recognition and sexual selection as a unitary problem in animal communication. Evolution **47**:647–57.

Ryan, M. J. and W. Wilczynski. 1991. Evolution of intraspecific variation in the advertisement call of a cricket frog (*Acris crepitans*, Hylidae). Biological Journal of the Linnean Society **44**:249–71.

Sadedin, S., J. Hollander, M. Panova, K. Johannesson, and S. Gavrilets. 2009. Case studies and mathematical models of ecological speciation. 3: Ecotype formation in a Swedish snail. Molecular Ecology **18**:4006–23.

Saetre, G. P., T. Moum, S. Bures, M. Kral, M. Adamjan, and J. Moreno. 1997. A sexually selected character displacement in flycatchers reinforces premating isolation. Nature **387**:589–92.

Saint-Laurent, R., M. Legault, and L. Bernatchez. 2003. Divergent selection maintains adaptive differentiation despite high gene flow between sympatric rainbow smelt ecotypes (Osmerus mordax Mitchill). Molecular Ecology **12**:315–30.

Sambatti, J. B. M., D. Ortiz-Barrientos, E. J. Baack, and L. H. Rieseberg. 2008. Ecological selection maintains cytonuclear incompatibilities in hybridizing sunflowers. Ecology Letters **11**:1082–91.

Sanderson, N. 1989. Can gene flow prevent reinforcement. Evolution **43**:1223–35.

Sandoval, C. P. 1993. Geographic, ecological and behavioral factors affecting spatial variation in color or morph frequency in the walking-stick *Timema cristinae*. PhD Thesis. University of California, Santa Barbara.

Sandoval, C. P. 1994a. Differential visual predation on morphs of *Timema-Cristinae* (Phasmatodeae, Timemidae) and its consequences for host-range. Biological Journal of the Linnean Society **52**:341–56.

Sandoval, C. P. 1994b. The effects of relative geographical scales of gene flow and selection on morph frequencies in the walking-stick *Timema cristinae*. Evolution **48**:1866–79.

Sandoval, C. P. and P. Nosil. 2005. Counteracting selective regimes and host preference evolution in ecotypes of two species of walking-sticks. Evolution **59**:2405–13.

Sauer, J. and B. Hausdorf. 2009. Sexual selection is involved in speciation in a land snail radiation on Crete. Evolution **63**:2535–46.

Savolainen, V., M. C. Anstett, C. Lexer, I. Hutton, J. J. Clarkson, M. V. Norup, M. P. Powell, D. Springate, N. Salamin, and W. J. Baker. 2006. Sympatric speciation in palms on an oceanic island. Nature **441**:210–13.

Scascitelli, M., K. D. Whitney, R. A. Randell, M. King, C. A. Buerkle, and L. H. Rieseberg. 2010. Genome scan of hybridizing sunflowers from Texas (Helianthus annuus and H. debilis) reveals asymmetric patterns of introgression and small islands of genomic differentiation. Molecular Ecology **19**:521–41.

Schadt, E. E., S. A. Monks, T. A. Drake, A. J. Lusis, N. Che, V. Colinayo, T. G. Ruff, S. B. Milligan, J. R. Lamb, G. Cavet, P. S. Linsley, M. Mao, R. B. Stoughton, and S. H. Friend. 2003. Genetics of gene expression surveyed in maize, mouse and man. Nature **422**:297–302.

Scheffer, S. J., G. W. Uetz, and G. E. Stratton. 1996. Sexual selection, male morphology, and the efficacy of courtship signalling in two wolf spiders (Araneae: Lycosidae). Behavioral Ecology and Sociobiology **38**:17–23.

Schemske, D. W. and H. D. Bradshaw. 1999. Pollinator preference and the evolution of floral traits in monkeyflowers (Mimulus). Proceedings of the National Academy of Sciences of the United States of America **96**:11910–15.

Schliewen, U. K., T. D. Kocher, K. R. McKaye, O. Seehausen, and D. Tautz. 2006. Evolutionary biology - Evidence for sympatric speciation? Nature **444**:E12–E13.

Schluter, D. 1988. Estimating the form of natural selection on a quantitative trait. Evolution **42**:849–61.

Schluter, D. 1993. Adaptive radiation in sticklebacks - size, shape, and habitat use efficiency. Ecology **74**:699–709.

Schluter, D. 1994. Experimental evidence that competition promotes divergence in adaptive radiation. Science **266**:798–801.

Schluter, D. 1995. Adaptive radiation in sticklebacks - trade-offs in feeding performance and growth. Ecology **76**:82–90.

Schluter, D. 1996a. Adaptive radiation along genetic lines of least resistance. Evolution **50**:1766–74.

Schluter, D. 1996b. Ecological causes of adaptive radiation. American Naturalist **148**: S40–S64.

Schluter, D. 1996c. Ecological speciation in postglacial fishes. Philosophical Transactions of the Royal Society of London Series B-Biological Sciences **351**:807–14.

Schluter, D. 1998. Ecological causes of speciation. Pages 114–29 *in* D. J. Howard and S. H. Berlocher, editors. Endless forms: Species and Speciation. Oxford University Press, Oxford, UK.

Schluter, D. 2000a. Ecological character displacement in adaptive radiation. American Naturalist **156**:S4–S16.

Schluter, D. 2000b. The ecology of adaptive radiation. Oxford University Press, Oxford, UK.

Schluter, D. 2001. Ecology and the origin of species. Trends in Ecology & Evolution **16**:372–80.

Schluter, D. 2003. Frequency dependent natural selection during character displacement in sticklebacks. Evolution **57**:1142–50.

Schluter, D. 2009. Evidence for ecological speciation and its alternative. Science **323**:737–41.

Schluter, D., E. A. Clifford, M. Nemethy, and J. S. McKinnon. 2004. Parallel evolution and inheritance of quantitative traits. American Naturalist **163**:809–22.

Schluter, D. and G. L. Conte. 2009. Genetics and ecological speciation. Proceedings of the National Academy of Sciences of the United States of America **106**:9955–62.

Schluter, D. and P. R. Grant. 1984a. Determinants of morphological patterns in communities of Darwin finches. American Naturalist **123**:175–96.

Schluter, D. and P. R. Grant. 1984b. Ecological correlates of morphological evolution in a darwins finch, *Geospiza difficilis*. Evolution **38**:856–69.

Schluter, D. and J. D. McPhail. 1992. Ecological character displacement and speciation in sticklebacks. American Naturalist **140**:85–108.

Schluter, D. and L. M. Nagel. 1995. Parallel speciation by natural-selection. American Naturalist **146**:292–301.

Schluter, D. and T. Price. 1993a. Honesty, perception, and population divergence in sexually selected traits. Proceedings of the Royal Society of London Series B-Biological Sciences **253**:117–22.

Schluter, D. and T. Price. 1993b. Honesty, perception, and population divergence in sexually-selected traits. Proceedings of the Royal Society of London Series B-Biological Sciences **253**:117–22.

Schrader, M. and J. Travis. 2008. Testing the viviparity-driven-conflict hypothesis: parent-offspring conflict and the evolution of reproductive isolation in a Poeciliid fish. American Naturalist **172**:806–17.

Schwartz, A. K. and A. P. Hendry. 2006. Sexual selection and the detection of ecological speciation. Evolutionary Ecology Research **8**:399–413.

Schwartz, A. K., D. J. Weese, P. Bentzen, M. T. Kinnison, and A. P. Hendry. 2010. Both geography and ecology contribute to mating isolation in Guppies. Plos One **5**:e15659.

Scotti-Saintagne, C., S. Mariette, I. Porth, P. G. Goicoechea, T. Barreneche, K. Bodenes, K. Burg, and A. Kremer. 2004. Genome scanning for interspecific differentiation between two closely related oak species Quercus robur L. and Q. petraea (Matt.) Liebl. Genetics **168**:1615–26.

Seddon, N. 2005. Ecological adaptation and species recognition drives vocal evolution in neotropical suboscine birds. Evolution **59**:200–15.

Seehausen, O. 2004. Hybridization and adaptive radiation. Trends in Ecology & Evolution **19**:198–207.

Seehausen, O. 2008. Progressive levels of trait divergence along a "speciation transect" in the Lake Victoria cichlid fish Pundamilia. Pages 155–76 *in* R. Butlin, J. Bridle, and D. Schluter, editors. Speciation and Patterns of Diversity. Cambridge University Press, Cambridge, UK.

Seehausen, O. and I. S. Magalhaes. 2010. Geographical mode and evolutionary mechanism of ecological speciation in cichlid fish. Pages 282–308 *in* P. R. Grant and R. Grant, editors. In Search of the Causes of Evolution. Princeton University Press, Princeton.

Seehausen, O. and D. Schluter. 2004. Male-male competition and nuptial-colour displacement as a diversifying force in Lake Victoria cichlid fishes. Proceedings of the Royal Society of London Series B-Biological Sciences **271**:1345–53.

Seehausen, O., G. Takimoto, D. Roy, and J. Jokela. 2008a. Speciation reversal and biodiversity dynamics with hybridization in changing environments. Molecular Ecology **17**:30–44.

Seehausen, O., Y. Terai, I. S. Magalhaes, K. L. Carleton, H. D. J. Mrosso, R. Miyagi, I. van der Sluijs, M. V. Schneider, M. E. Maan, H. Tachida, H. Imai, and N. Okada. 2008b. Speciation through sensory drive in cichlid fish. Nature **455**:620–U623.

Seehausen, O., J. J. M. vanAlphen, and F. Witte. 1997. Cichlid fish diversity threatened by eutrophication that curbs sexual selection. Science **277**:1808–11.

Seehausen, O., F. Witte, J. J. M. van Alphen, and N. Bouton. 1998. Direct mate choice maintains diversity among sympatric cichlids in Lake Victoria. Journal of Fish Biology **53**:37–55.

Selander, R. K. 1966. Sexual dimorphism and differential niche utilization in birds. Condor **68**:113-&.

Servedio, M., S. Van Doorn, M. Kopp, Frame, A., and P. Nosil. 2011. Magic traits: magic but not rare? Trends in Ecology & Evolution **26**:389–387.

Servedio, M. R. 2001. Beyond reinforcement: The evolution of premating isolation by direct selection on preferences and postmating, prezygotic incompatibilities. Evolution **55**:1909–20.

Servedio, M. R. 2004. The evolution of premating isolation: Local adaptation and natural and sexual selection against hybrids. Evolution **58**:913–24.

Servedio, M. R. 2009. The role of linkage disequilibrium in the evolution of premating isolation. Heredity **102**:51–6.

Servedio, M. R. and M. Kirkpatrick. 1997. The effects of gene flow on reinforcement. Evolution **51**:1764–72.

Servedio, M. R. and M. A. F. Noor. 2003. The role of reinforcement in speciation: Theory and data. Annual Review of Ecology Evolution and Systematics **34**:339–64.

Shapiro, A. M. and A. H. Porter. 1989. The lock-and-key hypothesis - evolutionary and biosystematic interpretation of insect genitalia. Annual Review of Entomology **34**:231–45.

Shapiro, M. D., M. E. Marks, C. L. Peichel, B. K. Blackman, K. S. Nereng, B. Jonsson, D. Schluter, and D. M. Kingsley. 2004. Genetic and developmental basis of evolutionary pelvic reduction in threespine sticklebacks. Nature **428**:717–23.

Shaw, K. L. and S. C. Lesnick. 2009. Genomic linkage of male song and female acoustic preference QTL underlying a rapid species radiation. Proceedings of the National Academy of Sciences of the United States of America **106**:9737–42.

Sheppard, P. M. 1953. Polymorphism, linkage, and the blood groups. American Naturalist **87**:283–94.

Siepielski, A. M., J. D. DiBattista, and S. M. Carlson. 2009. It's about time: the temporal dynamics of phenotypic selection in the wild. Ecology Letters **12**:1261–76.

Sinervo, B. and E. Svensson. 2002. Correlational selection and the evolution of genomic architecture. Heredity **89**:329–38.

Singer, M. C. and C. S. McBride. 2010. Multitrait, host-associated divergence among sets of butterfly populations: implications for reproductive isolation and ecological speciation. Evolution **64**:921–33.

Slabbekoorn, H. and T. B. Smith. 2002a. Bird song, ecology and speciation. Philosophical Transactions of the Royal Society of London Series B-Biological Sciences **357**:493–503.

Slabbekoorn, H. and T. B. Smith. 2002b. Habitat-dependent song divergence in the little greenbul: An analysis of environmental selection pressures on acoustic signals. Evolution **56**:1849–58.

Slatkin, M. 1973. Gene flow and selection in a cline. Genetics **75**:733–56.

Slatkin, M. 1979. Frequency-dependent and density-dependent selection on a quantitative character. Genetics **93**:755–71.

Slatkin, M. 1980. Ecological character displacement. Ecology **61**:163–77.

Slatkin, M. 1982. Pleiotropy and parapatric speciation. Evolution **36**:263–70.

Slatkin, M. 1985. Gene flow in natural populations. Annual Review of Ecology and Systematics **16**:393–430.

Slatkin, M. 1993. Isolation by distance in equilibrium and nonequilibrium populations. Evolution **47**:264–79.

Smadja, C., J. Galindo, and R. Butlin. 2008. Hitching a lift on the road to speciation. Molecular Ecology **17**:4177–80.

Smith, H. M. 1965. More evolutionary terms. Systematic Zoology **14**:57–8.

Smith, J. W. and C. W. Benkman. 2007. A coevolutionary arms race causes ecological speciation in crossbills. American Naturalist **169**:455–65.

Smith, R. A. and M. D. Rausher. 2008. Selection for character displacement is constrained by the genetic architecture of floral traits in the ivyleaf morning glory. Evolution **62**:2829–41.

Smith, T. B. 1993. Disruptive selection and the genetic basis of bill size polymorphism in the African finch *Pyrenestes*. Nature **363**:618–20.

Smith, T. B., R. Calsbeek, R. K. Wayne, K. H. Holder, D. Pires, and C. Bardeleben. 2005. Testing alternative mechanisms of evolutionary divergence in an African rain forest passerine bird. Journal of Evolutionary Biology **18**:257–68.

Smith, T. B., R. K. Wayne, D. J. Girman, and M. W. Bruford. 1997. A role for ecotones in generating rainforest biodiversity. Science **276**:1855–7.

Snowberg, L. K. and C. W. Benkman. 2009. Mate choice based on a key ecological performance trait. Journal of Evolutionary Biology **22**:762–9.

Snowberg, L. K. and D. I. Bolnick. 2008. Assortative mating by diet in a phenotypically unimodal but ecologically variable population of stickleback. American Naturalist **172**:733–9.

Snyder, R. J. and H. Dingle. 1989. Adaptive, genetically based differences in life-history between estuary and fresh-water threespine sticklebacks (*Gasterosteus aculeatus*, L.). Canadian Journal of Zoology-Revue Canadienne De Zoologie **67**:2448–54.

Sobel, J. M., G. F. Chen, L. R. Watt, and D. W. Schemske. 2010. The biology of speciation. Evolution **64**:295–315.

Sota, T. and K. Kubota. 1998. Genital lock-and-key as a selective agent against hybridization. Evolution **52**:1507–13.

Soto, I. M., V. P. Carreira, E. M. Soto, and E. Hasson. 2008. Wing morphology and fluctuating asymmetry depend on the host plant in cactophilic Drosophila. Journal of Evolutionary Biology **21**:598–609.

St-Cyr, J., N. Derome, and L. Bernatchez. 2008. The transcriptomics of life-history trade-offs in whitefish species pairs (*Coregonus* sp.). Molecular Ecology **17**:1850–70.

Stebbins, G. L. 1959. The role of hybridization in evolution. Proceedings of the American Philosophical Society **103**:231–51.

Steiner, C. C., J. N. Weber, and H. E. Hoekstra. 2007. Adaptive variation in beach mice produced by two interacting pigmentation genes. Plos Biology **5**:1880–9.

Stelkens, R. and O. Seehausen. 2009a. Genetic distance between species predicts novel trait expression in their hybrids. Evolution **63**:884–97.

Stelkens, R. B., M. E. R. Pierotti, D. A. Joyce, A. M. Smith, I. van der Sluijs, and O. Seehausen. 2008. Disruptive sexual selection on male nuptial coloration in an experimental hybrid population of cichlid fish. Philosophical Transactions of the Royal Society B-Biological Sciences **363**:2861–70.

Stelkens, R. B., C. Schmid, O. Selz, and O. Seehausen. 2009. Phenotypic novelty in experimental hybrids is predicted by the genetic distance between species of cichlid fish. BMC Evolutionary Biology **9**:283.

Stelkens, R. B. and O. Seehausen. 2009b. Phenotypic divergence but not genetic distance predicts assortative mating among species of a cichlid fish radiation. Journal of Evolutionary Biology **22**:1679–94.

Stelkens, R. B., K. A. Young, and O. Seehausen. 2010. The accumulation of reproductive incompatibilities in an African cichlid fish. Evolution **64**:617–32.

Stennett, M. D. and W. J. Etges. 1997. Premating isolation is determined by larval rearing substrates in cactophilic Drosophila mojavensis. III. Epicuticular hydrocarbon variation is determined by use of different host plants in Drosophila mojavensis and Drosophila arizonae. Journal of Chemical Ecology **23**:2803–24.

Stern, D. L. and V. Orgogozo. 2008. The loci of evolution: How predictable is genetic evolution? Evolution **62**:2155–77.

Stinchcombe, J. R. and H. E. Hoekstra. 2008. Combining population genomics and quantitative genetics: finding the genes underlying ecologically important traits. Heredity **100**:158–70.

Stoks, R., J. L. Nystrom, M. L. May, and M. A. McPeek. 2005. Parallel evolution in ecological and reproductive traits to produce cryptic damselfly species across the holarctic. Evolution **59**:1976–88.

Storz, J. F. 2005. Using genome scans of DNA polymorphism to infer adaptive population divergence. Molecular Ecology **14**:671–88.

Storz, J. F. and J. K. Kelly. 2008. Effects of spatially varying selection on nucleotide diversity and linkage disequilibrium: Insights from deer mouse globin genes. Genetics **180**:367–79.

Strasburg, J. L. and L. H. Rieseberg. 2008. Molecular demographic history of the annual sunflowers Helianthus annuus and H-petiolaris - Large effective population sizes and rates of long-term gene flow. Evolution **62**:1936–50.

Strasburg, J. L. and L. H. Rieseberg. 2010. How robust are "Isolation with Migration" analyses to violations of the IM Model? A Simulation Study. Molecular Biology and Evolution **27**:297–310.

Strasburg, J. L. and L. H. Rieseberg. 2011. Interpreting the estimated timing of migration events between hybridizing species. Molecular Ecology **20**:2353–66.

Strasburg, J. L., C. Scotti-Saintagne, I. Scotti, Z. Lai, and L. H. Rieseberg. 2009. Genomic patterns of adaptive divergence between chromosomally differentiated sunflower species. Molecular Biology and Evolution **26**:1341–55.

Strasburg, J. L., N. A. Sherman, K. M. Wright, L. C. Moyle, J. H. Willis, and L. H. Rieseberg. 2011. What can patterns of differentiation across plant genomes tell us about adaptation and speciation? Philosophical Transactions of the Royal Society B-Biological Sciences, in press.

Stuessy, T. F. 2006. Evolutionary biology - Sympatric plant speciation in islands? Nature **443**:E12.

Summers, K., R. Symula, M. Clough, and T. Cronin. 1999. Visual mate choice in poison frogs. Proceedings of the Royal Society of London Series B-Biological Sciences **266**:2141–5.

Svanback, R., M. Pineda-Krch, and M. Doebeli. 2009. Fluctuating population dynamics promotes the evolution of phenotypic plasticity. American Naturalist **174**:176–89.

Svensson, E. I., J. K. Abbott, T. P. Gosden, and A. Coreau. 2009. Female polymorphisms, sexual conflict and limits to speciation processes in animals. Evolutionary Ecology **23**:93–108.

Takahashi, M. K., Y. Y. Takahashi, and M. J. Parris. 2010. On the role of sexual selection in ecological divergence: a test of body-size assortative mating in the eastern newt Notophthalmus viridescens. Biological Journal of the Linnean Society **101**:884–97.

Tang, S. W. and D. C. Presgraves. 2009. Evolution of the Drosophila nuclear pore complex results in multiple hybrid incompatibilities. Science **323**:779–82.

Taper, M. L. and T. J. Case. 1992. Models of character displacement and the theoretical robustness of taxon cycles. Evolution **46**:317–33.

Tavormina, S. J. 1982. Sympatric genetic divergence in the leaf mining insect *Liriomyza brassicae* (Diptera, Agromyzidae). Evolution **36**:523–34.

Taylor, E. B., J. W. Boughman, M. Groenenboom, M. Sniatynski, D. Schluter, and J. L. Gow. 2006. Speciation in reverse: morphological and genetic evidence of the collapse of a three-spined stickleback (Gasterosteus aculeatus) species pair. Molecular Ecology **15**:343–55.

Taylor, E. B. and J. D. McPhail. 2000. Historical contingency and ecological determinism interact to prime speciation in sticklebacks, Gasterosteus. Proceedings of the Royal Society of London Series B-Biological Sciences **267**:2375–84.

Teeter, K. C., L. M. Thibodeau, Z. Gompert, C. A. Buerkle, M. W. Nachman, and P. K. Tucker. 2010. The variable genomic architecture of isolation between hybridizing species of house mice. Evolution **64**:472–85.

Templeton, A. R. 2008. The reality and importance of founder speciation in evolution. Bioessays **30**:470–9.

Terai, Y., O. Seehausen, T. Sasaki, K. Takahashi, S. Mizoiri, T. Sugawara, T. Sato, M. Watanabe, N. Konijnendijk, H. D. J. Mrosso, H. Tachida, H. Imai, Y. Shichida, and N. Okada. 2006. Divergent selection on opsins drives incipient speciation in Lake Victoria cichlids. Plos Biology **4**:2244–51.

Thibert-Plante, X. and A. P. Hendry. 2009. Five questions on ecological speciation addressed with individual-based simulations. Journal of Evolutionary Biology **22**:109–23.

Thibert-Plante, X. and A. P. Hendry. 2010. When can ecological speciation be detected with neutral loci? Molecular Ecology **19**:2301–14.

Thibert-Plante, X. and A. P. Hendry. 2011. The consequences of phenotypic plasticity for ecological speciation. Journal of Evolutionary Biology **24**:326–42.

Thoday, J. M. and J. B. Gibson. 1962. Isolation by disruptive selection. Nature **193**:1164-&.

Thoday, J. M. and J. B. Gibson. 1970. Probability of isolation by disruptive selection. American Naturalist **104**:219-&.

Thorpe, R. S. and R. P. Brown. 1989. Microgeographic variation in the color pattern of the lizard *Gallotia galloti* within the island of Tenerife - distribution, pattern, and hypothesis testing. Biological Journal of the Linnean Society **38**:303–22.

Thorpe, R. S. and M. Richard. 2001. Evidence that ultraviolet markings are associated with patterns of molecular gene flow. Proceedings of the National Academy of Sciences of the United States of America **98**:3929–34.

Ting, C. T., S. C. Tsaur, and C. I. Wu. 2000. The phylogeny of closely related species as revealed by the genealogy of a speciation gene, Odysseus. Proceedings of the National Academy of Sciences of the United States of America **97**:5313–16.

Tobler, M. 2009. Does a predatory insect contribute to the divergence between cave- and surface-adapted fish populations? Biology Letters **5**:506–9.

Tobler, M., T. J. DeWitt, I. Schlupp, F. J. G. de Leon, R. Herrmann, P. G. D. Feulner, R. Tiedemann, and M. Plath. 2008. Toxic hyrdogen sulfide and dark caves: phenotypic and genetic divergence across two abiotic environmental gradients in *Poecilia mexicana*. Evolution **62**:2643–59.

Tobler, M., R. Riesch, C. M. Tobler, T. Schulz-Mirbach, and M. Plath. 2009. Natural and sexual selection against immigrants maintains differentiation among micro-allopatric populations. Journal of Evolutionary Biology **22**:2298–304.

Turelli, M., N. H. Barton, and J. A. Coyne. 2001. Theory and speciation. Trends in Ecology & Evolution **16**:330–43.

Turner, J. R. G. 1967a. On supergenes. I. Evolution of supergenes. American Naturalist **101**:195-&.

Turner, J. R. G. 1967b. Why does genotype not congeal. Evolution **21**:645-&.

Turner, T. L., E. C. Bourne, E. J. Von Wettberg, T. T. Hu, and S. V. Nuzhdin. 2010. Population resequencing reveals local adaptation of Arabidopsis lyrata to serpentine soils. Nature Genetics **42**:260–U242.

Turner, T. L. and M. W. Hahn. 2007. Locus- and population-specific selection and differentiation between incipient species of Anopheles gambiae. Molecular Biology and Evolution **24**:2132–8.

Turner, T. L. and M. W. Hahn. 2010. Genomic islands of speciation or genomic islands and speciation? Molecular Ecology **19**:848–50.

Turner, T. L., M. W. Hahn, and S. V. Nuzhdin. 2005. Genomic islands of speciation in Anopheles gambiae. Plos Biology **3**:1572–8.

Turner, T. L., M. T. Levine, M. L. Eckert, and D. J. Begun. 2008. Genomic analysis of adaptive differentiation in Drosophila melanogaster. Genetics **179**:455–73.

Tyerman, J. G., M. Bertrand, C. C. Spencer, and M. Doebeli. 2008. Experimental demonstration of ecological character displacement. BMC Evolutionary Biology **8**:34.

Uy, J. A. C. and G. Borgia. 2000. Sexual selection drives rapid divergence in bowerbird display traits. Evolution **54**:273–8.

Uy, J. A. C. and A. C. Stein. 2007. Variable visual habitats may influence the spread of colourful plumage across an avian hybrid zone. Journal of Evolutionary Biology **20**.1847–58.

Vacquier, V. D., W. J. Swanson, and Y. H. Lee. 1997. Positive Darwinian selection on two homologous fertilization proteins: What is the selective pressure driving their divergence? Journal of Molecular Evolution **44**:S15–S22.

Vamosi, S. M. 2002. Predation sharpens the adaptive peaks: survival trade-offs in sympatric sticklebacks. Annales Zoologici Fennici **39**:237–48.

Vamosi, S. M. and D. Schluter. 2004. Character shifts in the defensive armor of sympatric sticklebacks. Evolution **58**:376–85.

van der Sluijs, I., T. J. M. Van Dooren, K. D. Hofker, J. J. M. van Alphen, R. B. Stelkens, and O. Seehausen. 2008. Female mating preference functions predict sexual selection against hybrids between sibling species of cichlid fish. Philosophical Transactions of the Royal Society B-Biological Sciences **363**:2871–7.

van Doorn, G. S., P. Edelaar, and F. J. Weissing. 2009. On the origin of species by natural and sexual selection. Science **326**:1704–7.

van Doorn, G. S. and F. J. Weissing. 2001. Ecological versus sexual models of sympatric speciation: a synthesis. Selection **2**:17–40.

Van Valen, L. 1965. Morphological variation and width of the ecological niche. American Naturalist **99**:377–90.

van't Hof, A. E., N. Edmonds, M. Dalikova, F. Marec, and I. J. Saccheri. 2011. Industrial melanism in British peppered moths has a singular and recent mutational origin. Science **332**:961–3.

Vasemagi, A. and C. R. Primmer. 2005. Challenges for identifying functionally important genetic variation: the promise of combining complementary research strategies. Molecular Ecology **14**:3623–42.

Vera, J. C., C. W. Wheat, H. W. Fescemyer, M. J. Frilander, D. L. Crawford, I. Hanski, and J. H. Marden. 2008. Rapid transcriptome characterization for a nonmodel organism using 454 pyrosequencing. Molecular Ecology **17**:1636–47.

Via, S. 1984a. The quantitative genetics of polyphagy in an insect herbivore. 2. Genetic correlations in larval performance within and among host plants. Evolution **38**:896–905.

Via, S. 1984b. The quantitative genetics of polyphagy in an insect herbivore. 1. Genotype-environement interaction in larval performance on different host-plant species. Evolution **38**:881–95.

Via, S. 1991. The genetic structure of host plant adaptation in a spatial patchwork - demographic variability among reciprocally transplanted clones. Evolution **45**:827–52.

Via, S. 1999. Reproductive isolation between sympatric races of pea aphids. I. Gene flow restriction and habitat choice. Evolution **53**:1446–57.

Via, S. 2001. Sympatric speciation in animals: the ugly duckling grows up. Trends in Ecology & Evolution **16**:381–90.

Via, S. 2009. Natural selection in action during speciation. Proceedings of the National Academy of Sciences of the United States of America **106**:9939–46.

Via, S. 2011. Hiding in plain sight: divergence hitchhiking and the dynamics of genomic isolation between incipient ecological species. Philosophical Transactions of the Royal Society B-Biological Sciences, in press.

Via, S., A. C. Bouck, and S. Skillman. 2000. Reproductive isolation between divergent races of pea aphids on two hosts. II. Selection against migrants and hybrids in the parental environments. Evolution **54**:1626–37.

Via, S. and D. J. Hawthorne. 2002. The genetic architecture of ecological specialization: Correlated gene effects on host use and habitat choice in pea aphids. American Naturalist **159**:S76–S88.

Via, S. and J. West. 2008. The genetic mosaic suggests a new role for hitchhiking in ecological speciation. Molecular Ecology **17**:4334–45.

Vines, T. H., S. C. Kohler, A. Thiel, I. Ghira, T. R. Sands, C. J. MacCallum, N. H. Barton, and B. Nurnberger. 2003. The maintenance of reproductive isolation in a mosaic hybrid zone between the fire-bellied toads Bombina bombina and B-variegata. Evolution **57**:1876–88.

Vines, T. H. and D. Schluter. 2006. Strong assortative mating between allopatric sticklebacks as a by-product of adaptation to different environments. Proceedings of the Royal Society B-Biological Sciences **273**:911–16.

Vonlanthen, P., D. Roy, A. G. Hudson, C. R. Largiader, D. Bittner, and O. Seehausen. 2009. Divergence along a steep ecological gradient in lake whitefish (Coregonus sp.). Journal of Evolutionary Biology **22**:498–514.

Waage, J. K. 1979. Dual function of the damselfly penis - sperm removal and transfer. Science **203**:916–18.

Wakeley, J. 2008a. Coalescent Theory: An Introduction. Roberts and Company Publishers, Greenwood Village, CO.

Wakeley, J. 2008b. Complex speciation of humans and chimpanzees. Nature **452**:E3–E4.

Wakeley, J. and J. Hey. 1997. Estimating ancestral population parameters. Genetics **145**:847–55.

Walsh, B. J. 1864. On phytophagic varieties and phytophagic species. Proceedings of the entomological society of Philadelphia **3**:403–30.

Walsh, B. J. 1867. The apple-worm and the apple maggot. Journal of Horticulture **2**:338–43.

Wang, H., E. D. McArthur, S. C. Sanderson, J. H. Graham, and D. C. Freeman. 1997. Narrow hybrid zone between two subspecies of big sagebrush (Artemisia tridentata: Asteraceae). 4. Reciprocal transplant experiments. Evolution **51**:95–102.

Wentzell, A. M., H. C. Rowe, B. G. Hansen, C. Ticconi, B. A. Halkier, and D. J. Kliebenstein. 2007. Linking metabolic QTLs with network and *cis*-eQTLs controlling biosynthetic pathways. Plos Genetics **3**:1687–701.

West-Eberhard, M. J. 2005. Developmental plasticity and the origin of species differences. Proceedings of the National Academy of Sciences of the United States of America **102**:6543–9.

Wheat, C. W. 2010. Rapidly developing functional genomics in ecological model systems via 454 transcriptome sequencing. Genetica **138**:433–51.

Whibley, A. C., N. B. Langlade, C. Andalo, A. I. Hanna, A. Bangham, C. Thebaud, and E. Coen. 2006. Evolutionary paths underlying flower color variation in Antirrhinum. Science **313**:963–6.

White, B. J., C. D. Cheng, D. Sangare, N. F. Lobo, F. H. Collins, and N. J. Besansky. 2009. The population genomics of trans-specific inversion polymorphisms in Anopheles gambiae. Genetics **183**:275–88.

White, B. J., C. D. Cheng, F. Simard, C. Costantini, and N. J. Besansky. 2010. Genetic association of physically unlinked islands of genomic divergence in incipient species of Anopheles gambiae. Molecular Ecology **19**:925–39.

White, B. J., M. W. Hahn, M. Pombi, B. J. Cassone, N. F. Lobo, F. Simard, and N. J. Besansky. 2007. Localization of candidate regions maintaining a common polymorphic inversion (2La) in Anopheles gambiae. Plos Genetics **3**:2404–14.

Whitehead, A. and D. L. Crawford. 2006. Variation within and among species in gene expression: raw material for evolution. Molecular Ecology **15**:1197–211.

Whiteley, A. R., N. Derome, S. M. Rogers, J. St-Cyr, J. Laroche, A. Labbe, A. Nolte, S. Renaut, J. Jeukens, and L. Bernatchez. 2008. The phenomics and expression quantitative trait locus mapping of brain transcriptomes regulating adaptive divergence in lake whitefish species pairs (*Coregonus* sp.). Genetics **180**:147–64.

Wiens, J. J. 2004. Speciation and ecology revisited: Phylogenetic niche conservatism and the origin of species. Evolution **58**:193–7.

Wiklund, C. 1975. Evolutionary relationship between adult oviposition preferences and larval host plant range in *Papilio machaon* L. Oecologia **18**:185–97.

Wilding, C. S., R. K. Butlin, and J. Grahame. 2001. Differential gene exchange between parapatric morphs of Littorina saxatilis detected using AFLP markers. Journal of Evolutionary Biology **14**:611–19.

Williams, L. M. and M. F. Oleksiak. 2011. Ecologically and evolutionarily important SNPs identified in natural populations. Molecular Biology and Evolution **28**:1817–26.

Wilson, D. S. and M. Turelli. 1986. Stable underdominance and the evolutionary invasion of empty niches. American Naturalist **127**:835–50.

Winkler, I. S. and C. Mitter. 2008. The phylogenetic dimension of insect-plant interactions: a review of recent evidence. Pages 240–63 *in* K. Tilmon, editor. The evolutionary biology of herbivorous insects: specialization, speciation, and radiation. University of California Press, Berkeley.

Won, Y. J. and J. Hey. 2005. Divergence population genetics of chimpanzees. Molecular Biology and Evolution **22**:297–307.

Wood, H. M., J. W. Grahame, S. Humphray, J. Rogers, and R. K. Butlin. 2008. Sequence differentiation in regions identified by a genome scan for local adaptation. Molecular Ecology **17**:3123–35.

Wood, T. K. and S. I. Guttman. 1982. Ecological and behavioral basis for reproductive isolation in the sympatric *Enchenopa binotata* complex (Homoptera, Membracidae). Evolution **36**:233–42.

Wood, T. K. and M. C. Keese. 1990. Host-plant induced assortative mating in *Enchenopa* treehoppers. Evolution **44**:619–28.

Woods, P. J., R. Muller, and O. Seehausen. 2009. Intergenomic epistasis causes asynchronous hatch times in whitefish hybrids, but only when parental ecotypes differ. Journal of Evolutionary Biology **22**:2305–19.

Wright, S. 1931. Evolution in Mendelian populations. Genetics **16**:0097–159.

Wright, S. 1940. Breeding structure of populations in relation to speciation. American Naturalist **74**:232–48.

Wright, S. 1943. Isolation by distance. Genetics **28**:114–38.

Wright, S. 1968. Evolution and the genetics of populations, vol. 1 (Genetics and biometric foundations). University of Chicago Press, Chicago.

Wu, C. 2001a. The genic view of the process of speciation. Journal of Evolutionary Biology **14**:851–65.

Wu, C. I. 2001b. The genic view of the process of speciation. Journal of Evolutionary Biology **14**:851–65.

Wu, C. I. and C. T. Ting. 2004. Genes and speciation. Nature Reviews Genetics **5**:114–22.

Xie, X. F., J. Rull, A. P. Michel, S. Velez, A. A. Forbes, N. F. Lobo, M. Aluja, and J. L. Feder. 2007. Hawthorn-infesting populations of Rhagoletis pomonella in Mexico and speciation mode plurality. Evolution **61**:1091–105.

Yatabe, Y., N. C. Kane, C. Scotti-Saintagne, and L. H. Rieseberg. 2007. Rampant gene exchange across a strong reproductive barrier between the annual sunflowers, Helianthus annuus and H-petiolaris. Genetics **175**:1883–93.

Yeaman, S. and S. P. Otto. 2011. Establishment and maintenance of adaptive genetic divergence under migration, selection, and drift. Evolution **65**: 2123–2129.

Yeaman, S. and M. C. Whitlock. 2011. The genetic architecture of adaptation under migration-selection balance. Evolution **65**:1897–911.

Young, K. A., J. M. Whitman, and G. F. Turner. 2009. Secondary contact during adaptive radiation: a community matrix for Lake Malawi cichlids. Journal of Evolutionary Biology **22**:882–9.

Young, N. D. 1996. An analysis of the causes of genetic isolation in two Pacific Coast iris hybrid zones. Canadian Journal of Botany-Revue Canadienne De Botanique **74**:2006–13.

Zouros, E. and C. J. Dentremont. 1980. Sexual isolation among populations *Drosophila mojavensis* - response to pressure from a related species. Evolution **34**:421–30.

Index

adaptive divergence - Divergence of new forms from a common ancestor form due to adaptation